Treatise on Materials Science and Technology

VOLUME 13

Wear

TREATISE ON MATERIALS SCIENCE AND TECHNOLOGY

VOLUME 13

WEAR

EDITED BY

DOUGLAS SCOTT

National Engineering Laboratory
Creep and Tribology Division
East Kilbride, Glasgow
Scotland

 1979

ACADEMIC PRESS   New York   San Francisco   London
A Subsidiary of Harcourt Brace Jovanovich, Publishers

ACADEMIC PRESS, INC.
111 Fifth Avenue, New York, New York 10003

United Kingdom Edition published by
ACADEMIC PRESS, INC. (LONDON) LTD.
24/28 Oval Road, London NW1 7DX

Library of Congress Cataloging in Publication Data
  Main entry under title:

  Wear.

    (Treatise on materials science and technology ;
  v. 13)
    Includes bibliographies.
    1. Mechanical wear.   2. Strength of materials.
  I. Scott, Douglas. II. Treatise
  TA403.T74 vol. 13 [TA418.4]   620.1'1'08s
  ISBN 0–12–341813–5 [620.1'1292]   78–27077

Contents

Theories of Wear and Their Significance for Engineering Practice

F. T. Barwell

The Wear of Polymers

D. C. Evans and J. K. Lancaster

The Wear of Carbons and Graphites

J. K. Lancaster

Scuffing

A. Dyson

Abrasive Wear

Martin A. Moore

Fretting

R. B. Waterhouse

Erosion Caused by Impact of Solid Particles

G. P. Tilly

Rolling Contact Fatigue

D. Scott

Wear Resistance of Metals

T. S. Eyre

Wear of Metal-Cutting Tools

E. M. Trent

List of Contributors

Numbers in parentheses indicate the pages on which the authors' contributions begin.

F. T. BARWELL (1), Department of Mechanical Engineering, University College of Swansea, Singleton Park, Swansea SA2 8PP, United Kingdom

A. DYSON (175), Shell Research Limited, Thornton Research Centre, P.O. Box 1, Chester CH1 3SH, England

D. C. EVANS* (85), Ministry of Defence, Procurement Executive, Materials Department, Royal Aircraft Establishment, Farnborough, Hants, England

T. S. EYRE (363), Department of Metallurgy, Brunel University, Uxbridge, Middlesex UB8 3PH, England

J. K. LANCASTER (85, 141), Ministry of Defence, Procurement Executive, Materials Department, Royal Aircraft Establishment, Farnborough, Hants, England

MARTIN A. MOORE† (217), National Institute of Agricultural Engineering, Wrest Park, Silsoe, Bedford, England

D. SCOTT‡ (321), National Engineering Laboratory, East Kilbride, Glasglow, Scotland

G. P. TILLY (287), Transport and Road Research Laboratory, Crowthorne, Berkshire RG11 6AU, Great Britain

E. M. TRENT (443), Department of Industrial Metallurgy, The University of Birmingham, P.O. Box 363, Birmingham B15 2TT, England

R. B. WATERHOUSE (259), Department of Metallurgy and Materials Science, University of Nottingham, England

* *Present address:* The Glacier Metal Company, Limited, Alperton, Wembley, Middlesex, England
† *Present address:* Fulmer Research Institute Ltd., Stoke Poges, Slough, Bucks, England
‡ *Present address:* Paisley College of Technology, Department of Mechanical and Production Engineering, High Street, Paisley PA1 2BE, Scotland

Foreword

Materials limitations are often the major deterrents to the achievement of new technological advances. In modern engineering systems, materials scientists and engineers must continually strive to develop materials that can withstand extreme environmental conditions and maintain their required properties. In the recent past we have seen the emergence of new types of materials, literally designed and processed with specific uses in mind. Many of these materials and the advanced techniques that were developed to produce them came directly or indirectly from basic scientific research.

Clearly, the relationship between utility and fundamental materials science no longer needs justification. This is exemplified in such areas as composite materials, high-strength alloys, electronic materials, and advanced fabricating and processing techniques. It is this association between the science and technology of materials on which we intend to focus in this *Treatise*.

This 13th *Treatise* volume is dedicated to wear of materials. Douglas Scott has brought together authors who have contributed articles on topics that encompass a range of major scientific and engineering disciplines. Workers have long sought the underlying physical principles that govern wear processes. An understanding of the basic properties, an optimistic technologist might assert, will lead to improved design. That this is not an easily achieved victory is exemplified by the complex world of tribological phenomena and the commonness of wear-related materials failure. That materials researchers are finally making headway in an understanding of wear can be seen from the papers contained in this volume. The diversity of the papers indicates both the complexity and the universal nature of wear of engineering materials.

It is hoped that this volume will assist in the development of solutions to these important materials-based industrial problems.

H. HERMAN

The economic implications of wear cause considerable concern in industry, where a reasonable return is expected from the vast expenditure on machinery and its maintenance. Also, in the present and foreseeable world economic situation, material conservation is becoming increasingly important. Wear is a major cause of material wastage, so any reduction in wear can effect considerable savings. Wear may take many forms, depending upon the geometry and nature of interacting surfaces and the environment in which they interact. The various forms of wear may occur singly or in combination; thus, wear is a complex phenomenon.

Progress in wear prevention can only be made when a better understanding of the mechanisms of wear and the controlling factors has been achieved.

This volume is intended to be a state of the art review of wear, which is to form a basis for all future work on the subject and to be a standard work for all in the field. It should be useful to research workers, academic personnel, and students, as well as to tribologists, designers, practicing engineers, material scientists, physicists, chemists, and petroleum technologists.

The volume begins with Barwell's review of the theories of wear and their significance for engineering practice, which acts as a general background for the nine specific topics. Wear resistance and compatibility of metals, including the role of wear resistant surface coatings, are discussed by Eyre. Evans and Lancaster describe the different types of wear behavior exhibited by polymers and polymeric-based composites, which are finding increasing use in a wide range of tribological applications. Lancaster reviews the more fundamental aspects of the wear behavior of carbons in their specific applications.

The various mechanisms of wear are dealt with. Abrasive wear, caused by the scoring or gouging of a surface by a relatively harder mating surface or particle, is probably the most serious type of wear experienced in industry and is covered by Moore. Dyson describes the physical manifestations of scuffing, assesses its importance in practice, and reviews the various experimental criteria and theoretical interpretations advanced to describe and explain its occurrence. Fretting, a specific form of wear caused by oscillatory tangential movement, is described by Waterhouse, and erosion caused by impact of solid particles is described by Tilly. Rolling contact fatigue, with its practical implications for rolling bearings, gears, and cams, is discussed by Scott. Trent emphasizes the great importance to the economics of engineering of the wear of metal-cutting tools and the continuous trend for materials to increase the rate of metal removal with reduced tool wear.

Although wear control and prevention are vitally important in many industrial applications, the achievement of rapid wear in grinding and machining also has economic significance. Advances in wear prevention through a better understanding of wear phenomena can also aid more effective material forming by wear processes.

Contents of Previous Volumes

Theories of Wear and Their Significance for Engineering Practice

F. T. BARWELL

Department of Mechanical Engineering
University College of Swansea
Singleton Park, Swansea
United Kingdom

I. Introduction

A. *Terotechnology and Tribology*

Mechanical wear is caused by disintegration of interacting machine components as the result of overstressing of the material in the immediate vicinity of the surface. Environmental conditions may result in a combination of chemical attack with mechanical overstressing, and the subject is intimately related with that of corrosion. The combined effects of wear and corrosion account for a very large proportion of the degradation of mechanical equipment generally and are the main agencies for limiting the life of capital goods. The manner in which force is applied to interacting surfaces, the environment, and the nature of the relative movement varies widely between different machines, as does the material of construction. The subject of wear is therefore a complex one embracing many disciplines. Unambiguous relationships between the magnitude of wear and applied factors can only be obtained as the result of careful analyses of the system because, as will be shown, wear may take many forms, depending on the applied conditions and the nature of the materials involved. The relationship between the design of the interacting surfaces (bearing systems) and their degradation in use is discussed in the author's recent book (Barwell, 1978), and this chapter represents an expansion of that treatment insofar as it relates to materials and their properties.

The subject is of major concern to those whose duty it is to maintain machinery and therefore to those who design and construct it. The relationship between these two groups is covered by the newly identified subject of *terotechnology*. The name was defined by Maddock (1972) as "the technology of installation, commissioning, maintenance, replacement and the removal of machinery and equipment, of feed-back to design and operation thereof, and of related subjects and practices," and is based on the Greek word *terein*. The definition had been evolved by a working party set up in 1968 by the then Ministry of Technology of the government of Great Britain to investigate the subject of maintenance engineering. They found that very little work had been done on either the cost of maintenance or on the returns to be expected from investment in means for reducing the necessity for maintenance. The working party published its report and the discussion that followed soon made it clear that the problem was much broader than was first thought. Improvement could only come about if a whole range of disciplines that had never been grouped together were brought to bear on the problem.

To a great extent the findings of the Committee on Terotechnology had been anticipated in 1966 by the Committee on Lubrication, which had been chaired by Jost (1966) and which found it necessary to introduce the term *tribology,* derived from the Greek word *tribos* which means *rubbing,* and defined as "the science and technology of interacting surfaces in relative motion." A similar conclusion was reached by the Mechanical Engineering Research Board of the Department of Scientific and Industrial Research, which set up in 1949 the Lubrication and Wear Division of the Mechanical Engineering Research Laboratory (now National Engineering Laboratories), which has done a great deal in the intervening years to put the subject of *wear* onto a sound scientific basis.

B. Interdisciplinary Nature of Wear Studies

Many features enter into a tribological system as is represented in Fig. 1. First, it is necessary to define the general shape of the contacting bodies. In the diagram these are represented by the radii r_1 and r_2, the relationship between them determines the degree of conformity and thus the stress system. Second, it is necessary to define the applied load and the associated friction forces. The rate of movement of the interacting bodies varies between wide limits, and in the general case both surfaces must be considered as moving. These are indicated here as U_1 and U_2 and must be measured relative to some fixed axis in space which could conveniently be the line of loading. In the absence of lubrication the intensity of heat generation will be proportional to the relative velocity $U_1 - U_2$. However, the severity of thermal distress of an elementary portion of a surface as it passed through the region of interaction would depend on the rate of energy dissipated in the contact multiplied by the duration of interaction within that contact. Thus if U_2 were zero, a particular area of the lower component would be heated continuously. If, however, U_2 were finite, fresh material would be continually introduced to the zone of contact and damage would be less severe.

In a lubricated system the interaction of speed and load with lubricant can produce a film of definite thickness, as indicated by h_m. This may vary from as little as 10_m^{-7} in a heavily loaded gear to as much as 10_m^{-4} in a large generator bearing (4 microinches to 4/1000 of an inch). The smallness of this gap emphasizes that the presence of foreign particles in a lubricant may have serious consequences by causing damage to one or

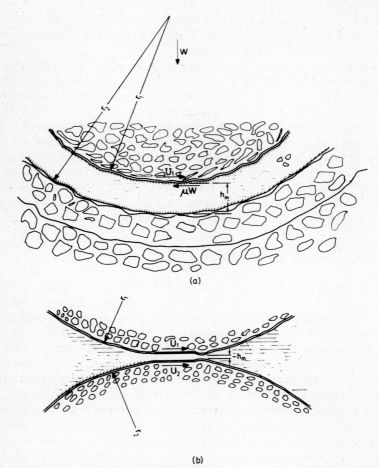

Fig. 1. Factors entering into a wearing situation: (a) conformal and (b) counterformal conditions.

both of the interacting surfaces. It is seldom possible to filter out all particles, and the bearing system must be able to withstand the effects of many small particles, some of them perhaps generated within the system itself by the process of wear. Some relief can result from particles embedding themselves within the soft white-metal bearing liners, but this may result in a hardened journal wearing more rapidly than a nominal soft bearing lining (Baker and Brailey, 1957; Vainshtein and Prondzinski, 1962).

The metals used in engineering practice are covered with oxide films and present a wide variety of properties, e.g., hardness, porosity, chemical reactivity. In addition, chemically reactive ingredients in a lubricant may result in the surfaces being covered with reaction products that may be even thicker than the oxide film. Immediately below the oxide film and often mingled with it in an intimate way is the true surface layer of the metal. Depending on the metal and method of manufacture, this layer may depart considerably from the basic structure of the parent metal, as established by Beilby (1921), whose name is often used to describe a microcrystalline or amorphous "surface region." Most finishing processes will lead to a highly distorted metal, which will often be enriched by oxide removed from the surface layer. Then again, depending on the method of manufacture or on the intensity of loading relative to the material properties of the parent metal, there will be a layer of strained and distorted structure between the superficial oxide layer and the parent metal. Such mixes of strained material may be electronically active with possible catalytic effects on the oxidation of the lubricant itself (Grunberg and Wright, 1955; Ferrante, 1976). The complete study of wearing systems therefore involves a number of academic disciplines.

C. Scales of Measurement

In addition to the dimensions covering the macroscopic configuration of the contact area, the study of wear involves factors such as the thickness of the oxide films, the grain size of the substrate material, the scale of the surface texture, the dimensions of the molecules of the lubricant, and the atomic spacing of the basic material. The SI system of units is used throughout this chapter, but other systems have been used in the past and are frequently encountered in the literature. The microinch used in surface measurement is, as its name implies, one millionth of an inch. The micron, or millionth part of a meter, is frequently used to define such features as particle size and is represented by the symbol μm and is described as the micrometer in modern texts. The angstrom unit, represented by the symbol Å, is used to represent atomic dimensions, and its value is 10^{-10} m.

Figure 2 is intended to assist the reader to assess the scales of magnitude expressed by the different systems of units and to derive some perspective regarding the factors entering into a wear situation.

Considering two interacting components, the dimensions will usually be of the order of magnitude of centimeters or larger. Thus the largest grain size as represented by white metal can be regarded as two orders of magnitude below any typical dimension, so that in calculating of stress

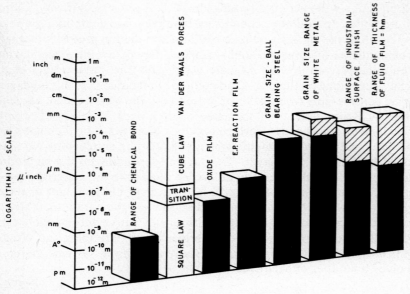

Fig. 2. Scales of measurement.

it is permissible to regard crystalline bodies as being isotropic and uniform. This is particularly the case with hardened ball-bearing steels, where the grain size is one order of magnitude below this. The surface roughness of industrially finished components is usually measured as the root mean square (rms) of the distance of a profile from a central plane. This may be significantly large compared with the thickness of the lubricant films that may be induced between the surfaces. Chemical films, either provided by oxygen or other reagents, tend to be thin relative to the elevation of surface texture. An order of magnitude lower still represents some of the longer molecules constituting lubricants which are again large compared to the spacing of the planes in metal crystals. The study of wear therefore involves consideration of quantities over a range of nine orders of magnitude, and it is important to determine the scale of action when speculating about the mechanism underlying the various forms of wear.

D. Higher and Lower Pairs

A machine has been defined as "an apparatus for applying mechanical power consisting of a number of interrelated parts, each having a definite function" (Morrison and Crossland, 1964). The parts of a machine that are in contact and between which there is relative motion are said to form a "pair," usually consisting of two solid bodies in contact. However complex the machine the pairs can be divided into two main classes, known as "higher" and "lower" pairs, as illustrated in Fig. 3. The three cases of lower pairs can be identified as sliding pairs, turning pairs, and screw pairs. They form the only pairs within which surfaces may touch over a wide area while sliding one relative to the other. All other pairs are known as higher pairs and are characterized by relative motion that is partly sliding and partly turning. An important distinction to be made between the higher and the lower pairs is that the former allow only point or line contact, while the latter permit contact to be dispersed over a surface. From the tribological viewpoint this distinction determines the system of stress applied to the surfaces and therefore the type of wear that will occur.

In considering any wear situation it is important to analyze the system from the point of view of the stress distribution and relative motion because these exert a very powerful influence on the mode of wear that will take place. Thus in a lower pair the two surfaces encountered in practice are never perfectly smooth and the true contact will be confined to a limited number of asperities, usually representing about one-thousandth of the nominal contact area. In the higher pairs the contact is concentrated within a narrow band or an ellipse in the vicinity of a point or line. Because this concentration of stress is high, elastic deformation takes place and the true area of contact is determined by the load, the geometric con-

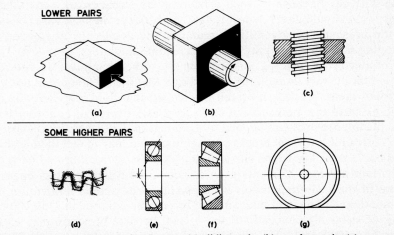

Fig. 3. Higher and lower pairs. Lower: (a) sliding pair; (b) revolute pair; (c) screw pair. Higher: (d) gear tooth contact; (e) ball bearing; (f) taper roller bearing; (g) wheel on rail.

figurations of the contacting bodies, and the elastic constants of the two materials. Hertz (1886) first developed the mathematical method of determination of contact area and these contacts are known as *Hertzian*.

The profound difference between concentrated and dispersed contact is reflected by the properties required of the materials of construction. For Hertzian contact the higher values of stress necessitate the use of very hard materials for all contacting surfaces, usually either through-hardened or case-hardened steel. For dispersed contacts, on the other hand, one of the surfaces is usually made of or lined by a soft material which will conform readily to the other surface. It is desirable to select materials having a considerable difference in hardness and to dispose them so that the permanent shape of the bearing is determined by the harder surface.

The difference between the higher and lower pairs becomes most profound when lubrication is considered. In the case of the lower pairs, oil, water, or gas may be forced between the surfaces at sufficient pressure to balance the applied load. This is known as *hydrostatic lubrication* and requires the interacting parts to be accurately finished. Although an external pressure supply is required, the system has advantages over other types because a pressure film can always be present irrespective of the motion of the surfaces so that wear is virtually eliminated.

When closely conforming surfaces of a lower pair are modified slightly to generate a wedge-shaped gap which is filled with fluid, hydrodynamic lubrication as envisaged by Reynolds may occur. This is possibly the most commonly encountered form of bearing. The load-carrying capacity is proportional to the viscosity of the lubricant and the velocity of the running surface. Because in such a bearing the surfaces are separated by fluid, wear should never occur. In fact, wear occurs in most cases because of the presence of adventitious material and conditions at starting and stopping when the film may be of inadequate thickness.

Counterformal bearing surfaces on the other hand are so intensely loaded that it was thought for many years that a fluid film could not exist between them. Modern research has shown that fluid lubrication is possible because the intense local stress causes a degree of conformity and also the intense pressure applied to the lubricant increases the viscosity of the lubricant by several orders of magnitude and may in fact induce solidification. This is known as *elastohydrodynamic lubrication* and occurs in important machine elements such as gears and rolling contact bearings. Some of the quantities involved in elastohydrodynamic lubrication may be mentioned. The load may amount to several gigapascals (150,000 psi), rates of shear may exceed 10^7 sec^{-1}, and the time of action may extend from 1 to 1000 μsec. The thickness of the film may lie between 0.1 and 1 μm, temperatures may reach 400°C, and the breadth of the contact region may be about 0.1 mm.

E. Surface Texture and Interaction

Until Abbot and Firestone (1933) introduced the stylus method of measurement, the estimation of the quality of a surface was a matter of personal judgment. An important manifestation of the stylus method is the *Talysurf* machine wherein the stylus consists of a diamond, which is finished to a radius of either 2 or 10 μm. When this traverses the surface its up and down motion is measured electronically. The electrical output either can be employed to operate a pen recorder to produce profiles of the surface under examination, or it can be processed to provide a wide range of statistics describing various features of the surface textures. Two statistics which are widely used and which can be obtained directly from the Talysurf apparatus itself (without requiring computer facilities) are the *center-line average* (CLA) as defined in Fig. 4 and the rms value. In order to distinguish between surface texture and waviness, British Standard 1134 (1972) specified a number of alternative cutoff lengths as follows: 0.25, 0.8, 2.5, 8.0, and 25.0 mm, respectively. This standard is based on the recommendation of ISO/R468, "Surface Roughness." The center-line average is redesignated as the *arithmetical mean deviation*.

Although the CLA and rms measures are of great value in controlling the quality of manufacturers' products, they do not uniquely describe a surface and therefore provide limited guidance regarding the interaction of two surfaces. Recent research has produced a number of more sophisticated methods of representing surfaces by mathematical models, and a comprehensive treatment is found in "The Mechanics of the Contact between Deformable Bodies," edited by de Pater and Kalker (1975).

For the wear studies important concepts in addition to the CLA and the rms values are the *mean slope of the asperities,* the *bearing area curve* [originally defined by Abbot and Firestone (1933)], the mean tip radius, and the autocorrelation coefficient. It can be shown that the surface profile can usually be described by a Gaussian distribution of ordinates. Thus

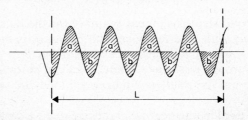

Fig. 4. Definition of center-line average. Center line positioned so that Σ(areas a) = Σ(areas b). The arithmetical mean deviation (formerly CLA) is defined by

$$\frac{\Sigma(\text{areas } a) + \Sigma(\text{areas } b)}{L \times \text{vertical magnification}}$$

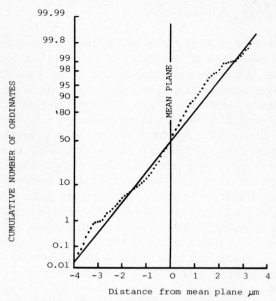

Fig. 5. Gaussian distribution of ordinates.

Fig. 5 shows measurements of an actual surface plotted on probability paper in comparison with a straight line representing a perfectly random distribution. The bearing area curve thus becomes the cumulative probability curve or "ogive," and the rms value assumes an important role. Thus when a hypothetically smooth surface is forced against a randomly distributed profile, a separation of 3σ will produce an interaction of only 1% of the profile.

For real surfaces it has been proposed (Tsukizoe and Hisakado, 1965) that the equivalent roughness is given by the expression

$$\sigma = (\sigma_1^2 + \sigma_1^2)^{1/2} \tag{1}$$

where σ is the equivalent roughness required for the concept of a rough surface interacting with a perfectly smooth plane, σ_1 is the rms value for surface number 1 and σ_2 is that for surface number 2.

The mutual approach of surfaces under load can be estimated from the bearing area curve if the relationship between load and bearing area can be determined.

Most surfaces used in engineering practice exhibit a certain waviness in addition to roughness of texture characterized by short wavelengths. Thus, instead of being distributed uniformly over the entire nominal contact area, the contact spots tend to be grouped into zones or clusters as in-

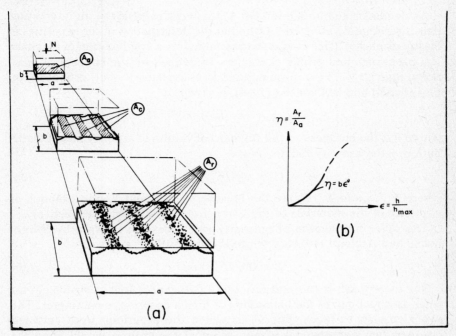

Fig. 6. Contact area: (a) Apparent, contour, and real areas of contact (A_a, A_d, and A_r, respectively); $A_a = a \times b$. Real pressure

$$P_r \leq 0.5\ E^{2\nu/2\nu+1}\Delta^{\nu/2\nu+1}P_c^{1/2\nu+1}$$
$$= 0.5\ E^{4/5}\Delta^{2/5}P_c^{1/5} \text{ if } \nu = 2$$

$P_r = \varphi(P_a)$; $P_r = 0.35\ E\Delta^{2/5}(h_w/R_w)^{2/25}P_a^{1/25}$; after running-in process, $P_r = 0.7(\tau_0 E/d_r)$. Contour pressure $P_c = 0.2E^{4/5}(h_w/R_w)^{2/5}P_a^{1/5}$. Nominal pressure $P_a = N/A_a$. (b) Microgeometrical factor $\Delta = h_{max}/Rb^{1/\nu}$, where h_{max} is the maximal asperity height and R the asperity radius.

dicated in Fig. 6, which is based on the work of Kragelskii and Demkin (1960), who, in addition to the concepts of "real area of contact" and "apparent area of contact," introduced the term "contour area of contact."

For moderate intensities of loading the ratio between real area of contact and apparent area of contact (η) takes the form

$$\eta = b\epsilon^\nu \tag{2}$$

where ϵ is the relative penetration, equal to the actual approach of surfaces divided by the maximum asperity heights, b a constant depending on material and surface characteristics, and ν a constant of value between 2 and 3.

As demonstrated in Section IV,A, the wear behavior of an interacting pair is profoundly affected by whether the deformation of the asperities is elastic or plastic. This is in turn determined by a combination of mechanical properties and profile characters as defined by the concept of *plasticity index* (λ). Two formulations are available; one, proposed by Greenwood and Williamson (1966), is given by

$$\lambda = (E^1/h)(\sigma/r)^{1/2}$$

where h is the hardness, σ the standard deviation of asperity height distribution, r the asperity radius, and

$$1/E^1 = (1 - \gamma_1{}^2)/E_1 + (1 - \gamma_2{}^2)/E_2 \tag{3a}$$

where γ_1, γ_2 and E_1, E_2 are the Poisson's ratios and Young's moduli of elasticity of the materials of the two contacting surfaces, respectively.

The other formulation of plasticity index was put forward by Whitehouse and Archard (1970) in the following form:

$$\lambda = 0.6(E^1/h)(\sigma/L) \tag{3b}$$

The incorporation of r and L in these two expressions is significant insofar as it recognizes the limitations of σ as a descriptive parameter. The two Gaussian surfaces, Fig. 7, in which the asperities were confined between two hypothetical planes of equal separation, would have the same value of σ. The asperities profile on the right-hand side, however, might be expected to be more resistant to crushing under the action of normal forces than those of the left-hand side. Thus a measure of the lateral spread of asperities is required and is provided by r, the hypothetical mean radius—or L, the autocorrelation coefficient. (The autocorrelation coefficient is derived as follows: If a traverse is made of the profile at plane xx and multiplied by the corresponding points of plane yy, then a product would be obtained that would be greater than either. If the distance separating xx and yy is increased, the correspondence of the two profiles will be lessened and the value of the product correspondingly diminished until it reaches a low and constant value. The length at which this happens indicates the mean spacing of profiles that are unrelated and above which the surface may be regarded as being of random nature.)

Direct observation of the contact of conformal surfaces has been reported by a number of workers. For example, Jones (1970) obtained a

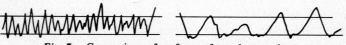

Fig. 7.   Comparison of surfaces of equal rms value.

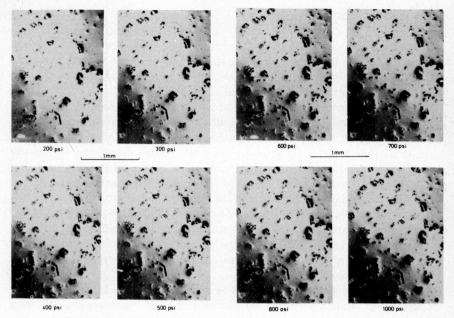

Fig. 8. Distribution of contact area with load. (After Jones, 1976.)

contact pattern under the action of normal applied load, as reproduced in Fig. 8. He adapted a microscope so that metal surfaces could be forced against a glass optical flat, the surface of which had been silvered. The contact spots were observed by Normaski phase-contrast illumination. The wide disperal of contact spots will be noted together with the persistence of the general pattern, in spite of a fivefold increase in load. Sakurai has also studied the interaction between a metal and a quartz surface using phase-contrast illumination. His technique was similar to that used by Jones except that in addition to studying the effect of normal load on dispersion of contact area he introduced relative sliding between the contacting surfaces. A preliminary observation was that damage occurred at the trailing edges of the asperities, the leading edges remaining relatively undisturbed.

F. Modes of Wear

Wear may take many forms, but for convenience four main categories can be identified: adhesive wear, abrasive wear, surface fatigue, and corrosion.

EARLY THEORIES

Rennie (1829) considered that friction arose from the normal motion necessary for the irregularities of the interacting surfaces to pass from one interlocking position to another. Hence wear was attributed to the bending and fracture of these irregularities.

Many investigators have attempted to invoke the atomic or molecular structure of matter as directly determining friction and wear phenomena. Thus Sir Alfred Ewing (1892) considered that when one surface was caused to slide relative to another, the polar forces interacted across the interface to produce a quasi-electric turning of the molecules and that when a certain very limited range of movement had been exceeded, energy was dissipated by breaking of bonds and the establishment of new bonds with oscillation of particles. Tomlinson (1929), taking advantage of recently enunciated ideas of wave mechanics, was able to expand this view. Deryagin (1937, 1957) has also put forward a molecular theory of static friction and has developed an electrostatic theory of adhesion of nonmetallic bodies.

The possibility that friction might be related to the shear strength of the material was first considered by Holm and Guldenpfenning (1937) and Holm *et al.* (1941), and Holm (1946) has introduced the concept of atomic wear in relation to carbon contacts. There is no evidence, however, of atomic wear occurring in machines.

a. Adhesive Wear. Adhesive wear is characterized by the interaction of asperities, causing metal to be transferred from one surface to another, a particularly severe form being known as scuffing. This form of failure is sometimes observed at the "addendum" or "dedendum" of gear flanks and is believed to occur when a transient local temperature exceeds a critical value. An interesting feature of scuffing is a marked tendency for material to be removed from the hotter surface and deposited on the cooler surface. Special lubricants containing additives are employed to combat scuffing. Organic compounds containing oxygen, sulphur, and chlorine are believed to act as antiwelding agents, and compounds containing phosphorus or lead are believed to allow preferential chemical attack on heated protuberances, thereby smoothing the surface.

b. The Adhesion Theory. Although adhesion undoubtedly occurs between metals and can contribute to surface damage, the concept of friction and wear which is widely accepted and which assigns primary importance to adhesion between metals contains certain logical inconsistencies. The arguments for and against the adhesion theory have been well presented by Rabinowicz (1965). It will be recalled that the adhesion theory of friction arose from the early work of Bowden and Tabor (1950), who attributed the force of friction to two principal mechanisms which they denoted by "ploughing" and "junction growth." Basically the adhesion theory holds that the true area of contact is proportional to the load forcing the surfaces together divided by a quantity based on the mechanical strength and known as the mean yield pressure of the asperities p_m. It is assumed that the metals weld together over the area of contact and that the tangential force required to cause relative motion is equal to the shearing strength of the material multiplied by the true area of contact. Thus the coefficient of friction is given by

$$A = W/p_m \tag{4a}$$
$$F = SA \tag{4b}$$
$$\mu = \frac{F}{W} = \frac{S}{p_m} = \frac{\text{shear strength of junctions}}{\text{yield pressure of softer metals}} \tag{4c}$$

Because S and p_m for any given metal may be expected to be related, μ would not differ much among different metals and the coefficient of friction should not vary with load (Amonton's law).

This theory represented a considerable advance over its predecessors because it focused attention on the microscopic as opposed to the molecular scale of interaction. However, flow stress usually assumes a value of about three times the yield stress in compression [due to support of the asperities by surrounding material, (Tabor, 1970)] and shear stress is about one-half of the direct stress so that $S/p_m = \frac{1}{6}$ or 0.17. The experimental value of the friction between clean dry metals lies between 0.6 and 1.2.

The explanation of the discrepancy arises from the assumptions regarding flow stress. The factor of three relating flow stress to compressive yield stress assumes that behavior of the asperities during the frictional process is identical with that which occurs when the interacting bodies are forced together by normal forces and indeed with the action of the indentor of a hardness test. When the force has a tangential component, this no longer applies, and Tabor (1959) describes how the junction grows under the action of tangential force until the area is about three times that predicted by "flow stress," thus reintroducing realistic prediction.

It is the writer's view that it is not necessary to invoke the concept of "adhesion" or "welding" to explain friction and wear processes in general, although there is abundant evidence of its occurrence in particular cases. All that is required is to presuppose some mechanism that will enable tangential force to be transmitted between the interacting bodies. It is well known that failure in a simple compression test is by shear at 45° to the direction of principal stress as shown in Fig. 9a. Figure 9b embodies

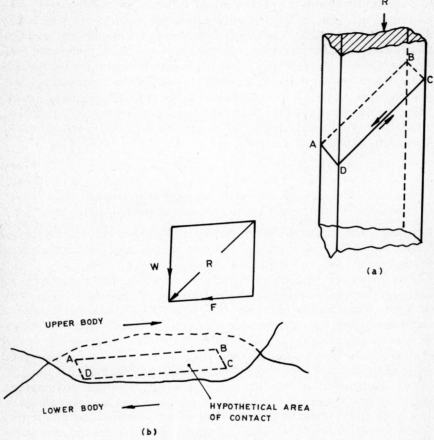

Fig. 9. Simplified contact shear.

a plane *AB* that represents the interface between two asperities, which is coincident with the direction of relative motion. If it is assumed that the strength of interaction between the two bodies is such that shear-type failure occurs, from the stress viewpoint planes *AB* in both parts a and b of Fig. 9 are comparable. For shear to take place along the interfacial plane the plane of the principal stress must be inclined at an angle of 45° thereto. Where *R* is the resultant of the applied normal and tangential forces, $W = R \sin 45°$ and $F = R \cos 45°$, but $\cos 45° = \sin 45°$; therefore,

$$\mu = \frac{F}{W} = \frac{\text{tangential force}}{\text{normal force}} = 1 \tag{5}$$

This simple picture of course requires modification in any practical case, but, because it embodies an absolute minimum of assumptions about the rubbing system and the properties of the interacting materials, produces the most convenient starting point for the evolution of wear theory.

c. Abrasive Wear. Abrasive wear may be defined as damage to a surface by a harder material. This hard material may have been introduced between two rubbing surfaces from outside, it may be formed in situ by oxidation and other chemical processes, or it may be the material forming the second surface.

Resistance to abrasion is related to the hardness of a material but is not directly proportional thereto. Table I is based on tests by Richardson (1967a, b) of the abrasion resistance of a selection of metals. The results of a number of tests in the laboratory and the field have been averaged and shown against the hardness of the material measured by a diamond impression.

It can be seen that wear resistance increases with hardness for any particular steel. Nevertheless, some steels are more wear resistant than others, irrespective of hardness. Thus, austenitic manganese steel at a

TABLE I

RELATIVE WEAR RESISTANCE AGAINST HARDNESS

Material	Hardness (Vickers HV 30)	Relative wear resistance
Carbon and low alloy steels (B.S. 5970)		
EN 24: 0.37%, C, Ni, Cr, Mo	500	1[a]
	350	0.87
EN 8: 0.43% C	600	1.30
	500	1.02
EN 42: 0.74% C	820	1.90
	650	1.30
	500	1.15
Alloy steels		
Austenitic manganese	200	1.3
Hot die (KE 275)	600	1.91
Die: 2% C, 14% Cr	700	1.74
Cast hardfacing alloys		
Delcrome: 3% C, 1.7% Cr, 3% Ni	700	3.47
Stellite: 2.5% C, 33% Cr, 13% W, Co base	700	3.95

[a] Reference material.

measured hardness of 200 is more wear resistant than the reference material at 500.

Where a component that is subject to abrasive wear is of any size, it is often better to use mild steel for the main construction and to apply a coating of wear-resistant alloy such as Delcrome or Stellite.

d.   Surface Fatigue.   Surface fatigue is the predominant mode of failure of rolling contact bearings because of the repeated intense loading of the counterformal contact area. The distribution of Hertzian stress is such that the maximum shearing stress occurs a little way within the surface so that failure commences below the surface. Subsequently the crack reaches the surface so that a piece of metal becomes detached therefrom, leaving a pit; hence the term *pitting failure*.

e.   Hydrogen Embrittlement.   Barwell and Scott (1966) describe the failure of the camshaft of a pump used to provide fluid power for a die casting machine. For many years this pump had operated successfully using a conventional mineral oil. When in the interests of fire prevention a water-based hydraulic fluid was substituted, failure occurred within six weeks. This was simulated in the laboratory using an arrangement wherein one ball was arranged to roll against three others and sections

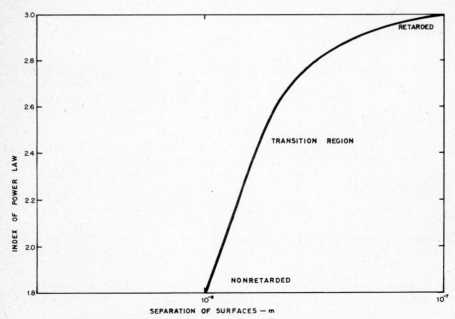

Fig. 10.   Interbody forces. (After Tabor and Winterton, 1969.)

through pitted races revealed numerous subsurface cracks spreading from the pits. Moreover, the nature of the failures was more drastic, rolling elements breaking into two or more parts rather than merely suffering surface distress.

Operational experience indicated that small quantities of water included in petroleum lubricants accelerated the incidence of pitting-type failure of rolling contact bearings. This was confirmed in the laboratory by Grunberg and Scott (1958), who suggested that vacancy-induced diffusion of hydrogen into the highly stressed surface material produced hydrogen embrittlement and accelerated the failure (Grunberg and Scott, 1960). This was confirmed by experiments in which the lubricant contained 3% tritium-labeled water to enable the amount of hydrogen absorbed by the stressed material to be measured (Grunberg *et al.*, 1963). The mechanism suggested is that hydrogen, perhaps in atomic form, is produced by decomposition of water at the surface. Vacancies generated in the metal by plastic deformation under the repeated application of Hertzian stress assist the process of penetration. Eventually hydrogen becomes fixed in traps in the metal, leading to hydrogen embrittlement.

II. Wear in Vacuum or Inert Gases

A. Interbody Forces

When clean metal surfaces are brought into contact a strong chemical bond is formed between the two surfaces. When the surfaces approach closely without actually touching they are attracted by van der Waals forces. The transition has been studied by Ferrante and Smith (1976) and has been shown to occur at a separation of $\frac{1}{2}$ nm.

Two important classes of van der Waals forces are the nonretarded or dispersive (London) forces and the retarded forces. Experimental determination of these is difficult because of the presence on the surface of irregularities which are large relative to the average separation. However, Tabor and Winterton (1969) carried out tests between crossed cylinders of freshly cleaved mica, which indicated that the nonretarded forces obey a law of the form $W = AR/6h^2$ and the retarded forces $W = 2BR/3h^3$, where A and B are constants, h is the separation of the surfaces, and R is the radius of the cylinder [which was 1 cm in the investigations of Tabor and Winterton (1969) and Israelachvili and Tabor (1972)]. The transition from Eq. (4) to Eq. (5) occurs gradually at separation values between 10^{-8} and 10^{-7} m, as shown in Fig. 10. Measured values of A and R were $(1.35 \pm 0.15) \times 10^{19}$ J and $(0.97 \pm 0.06) \times 10^{-28}$ J m, respectively.

The surface energy of the material is determined by the integration of the work necessary to separate two surfaces to infinity. When monolayers of stearic acid were deposited on the mica using a Langmuir trough, the forces for separation above 5×10^{-9} m were virtually unaltered, but when separation was below 3×10^{-9} m, a reduction of about 25% was measured. Thus the force required to separate monolayers of fatty acid is less than that required to separate the two clean mica surfaces. This behavior may therefore be regarded as the basic mode of boundary lubrication.

Direct measurement of the adhesive forces between solid bodies is difficult because the asperities and particles are large in comparison to the value of the separation of the bodies that gives rise to significant forces. This difficulty was overcome by Johnson *et al.* (1971) by the use of rubber and gelatine surfaces which possessed moduli of elasticity that were sufficiently low for the force required to flatten an asperity to be less than the surface attractive forces. An experimental value for the surface energy of gelatine on Perspex was 0.105 J/m^2, showing that under very light loads the magnitude of the surface active forces could be comparable to the forces pressing such forces together. These forces were small, however, in comparison with those which have to be taken into account in practice. Kohno and Hyodo (1974) have brought styli with tips of small radius made of tungsten or fused quartz into contact with optically flat steel surfaces. A microbalance was used to apply loads as light as 10^{-8} N and the forces required to separate the contact were found to be proportional to the tip radius. These forces were sometimes two to three orders of magnitude higher than those applied initially in forcing the surfaces together.

The difficulties in measuring the adhesion forces between solids arise primarily from the fact that significant forces are only detectable when there is a very small separation of the surfaces. Few surfaces are sufficiently smooth for there to be an appreciable area for the forces to act without interruption by contact. In the cases of the Cambridge investigators previously mentioned, Tabor *et al.* were able to overcome this problem by cleaving mica surfaces and Johnson *et al.* by using materials that were so soft that they could conform to the interacting surface with a minimum of force occasioned by elastic deformation. In contrast, the Japanese workers employed hard surfaces and obtained the necessary conditions by using specimens of small radius (less than 15 μm) with correspondingly low forces (10^{-8} N). Notwithstanding the extreme variations and the scale and composition of the materials of the experiments, the basic conclusions of the Japanese workers are fully consistent with those of the British workers, notably that adhesion is to be attributed to the van der Waals dispersion interactions.

Similar experiments by Pollock *et al.* (1977) employed sharply pointed (radius of curvature typically 300 nm) specimens of metal which were forced against flat metallic surfaces, and the electrical resistance of the resulting contact was measured. High values of adhesion were measured for platinum and tungsten surfaces provided that both surfaces were clean. Measurements of adhesion, electrical resistance, and change in resistance as a function of load must be divided into two categories according to whether separation took place at the interface or within the bulk of one or the other of the interacting surfaces. The latter form generally occurs with clean surfaces and involves plastic flow of the weaker of the two materials. When surfaces become contaminated, the work of adhesion is reduced to about 1 J/m² (Skinner and Pollock, 1977) and parting occurs at the interface.

B. Friction in Vacuum

Some understanding of the mechanism whereby material is transferred from one surface to another may be obtained by considering the force of friction. The simplest hypothesis would be that when two bodies rubbed together they adhere until the weakest becomes sheared, the bond being stronger than the original material. Thus where dissimilar metals are rubbed together, it might be expected that the coefficient of friction would correspond to that of the weaker of the two materials. Table II, taken from Kragelskii (1973), shows the results of a number of experiments. The

TABLE II

FRICTION IN VACUUM[a,b]

Combination	Coefficient of friction (μ)	Average of coefficient of metals (μ_1)	Difference ($\mu - \mu_1$)	Lowest of two metals
Ag–Fe	0.30	1.7	−1.40	0.8
Cu–Fe	0.69	1.5	−0.80	0.8
Ni–Fe	0.60	0.9	−0.30	0.8
Ag–Fe	1.28	2.0	−0.72	0.8
Al–Cu	1.67	2.7	−1.23	2.3
Ag–Cu	1.72	2.2	−0.48	2.3
Cu–Ni	2.01	1.7	+0.31	1.1
Al–Ag	2.20	2.8	−0.60	2.6
Al–Ni	2.36	2.1	+0.26	1.1

[a] After Kragelskii (1973).
[b] The values of μ for metals running against themselves are Fe–Fe = 0.8, Ni–Ni = 1.1, Cu–Cu = 2.3, Ag–Ag = 2.6, Al–Al = 3.1.

results for a number of combinations of sliding metals are shown in the
second column in ascending values. The coefficients of the materials
when sliding against themselves are shown in a footnote to the table.
Kragelskii first tested the hypothesis that the friction of the combination
would be the mean of the two materials, and the result is shown in the
third column. The fourth column shows that agreement is not good. The
hypothesis that the coefficient should correspond to the weakest of the
two metals is equally unconvincing. The simple picture of adhesion
behavior must be rejected because it is not borne out by the experimental
facts.

Buckley (1975) has shown that the solid-state structural factors at a surface, such as orientation, lattice registry, crystal lattice defects, and structure, have an imporant effect on the interface that is established when two
metals are brought intó contact, and this interfacial structural character
affects the adhesion or bonding forces of one solid to another. The use of
LEED and Auger emission spectroscopy provides evidence of the adhesion of very clean pure metals. Buckley (1976) has measured the force
necessary to rupture junctions made within a vacuum system evaporated
to 10^{-11} torr. Crystals of copper, gold, silver, nickel, platinum, lead, tantalum, aluminum, and cobalt were cleaned by argon–ion bombardment
before being forced against a clean iron (011) surface by a force of 20 dyn.
When iron was pressed against iron, a force greater than the 400 dyn was
required to separate these surfaces. In the case of the other metals this
force varied from 50 to 250 dyn. Thus in every case the strength of the
junction was greater than the amount of force used to promote it. Even in
the case of lead (which is insoluble in iron) the Auger analysis indicated
transfer of lead to the iron surface. Thus the adhesive bonds of lead to
iron were stronger than the cohesive bonds within the lead. In general it
was found that the cohesively weaker metals adhered and transferred to
the cohesively stronger.

In the case of a metal contacting itself the interface can be compared
with the grain boundary of a polycrystalline metal sample. When dissimilar metals generate the interface, they may differ in atomic size, lattice
spacing, binding energy, and other factors so that the bond strength is
likely to be inferior to that between similar metals. However, it is often
stronger than the cohesion of the weaker of the two metals. Grain orientation is important in the frictional process because of its influence on the
orientation of slip bands. Buckley (1976) reports that examination by
scanning electron microscopy of surfaces after sliding revealed severe
disturbances as a result of contact, surface initiated fracture cracks being
prominent. The wear face of these cracks was extremely smooth, indicating crack initiation along slip planes, which was confirmed by sec-

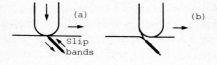

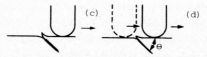

Fig. 11. Mechanism of wear: (a) Slip-band formation; (b) fracture; (c) surface projection; (d) subsequent passes generating wear particle. (After Buckley, 1971.)

tioning the wear track at that angle. Buckley proposes the mechanism shown in Fig. 11. When the copper slide is first brought into touch contact and the load is applied, deformation of surface asperities results in penetration of any surface contaminating films and the metal-to-metal interface formation with strong adhesive bonding. As tangential motion is commenced fracture must occur in the weakest interfacial region. In this circumstance the weakest region is not the interface but rather in the cohesive bonds between the adjacent slip planes in the individual copper grains. Thus when tangential force is applied, atomic bonds along the copper slip line fracture with formation of a surface in the initiated crack, as shown in the figure. Continued application of tangential force will cause the bond to rupture and the slider will move on until adhesion again occurs. This could account for intermittent sliding of a slip character, but the extremely small scale must be noted and the behavior should not be compared with the relaxation oscillations usually referred to as *stick slip*. After the slider-to-grain interfacial bond has fractured, a wake or curl of metal will remain above the plane of the grain surface, as indicated in Fig. 11c; thus subsequent passes will result in the shearing of this protuberance so as to form a wear particle.

It will be noted, therefore, that a measure of adhesion and a measure of crack formation appear to be fundamental to the wear process.

Guidance regarding the relative importance of adhesion and plastic flow is provided by the work of Andarelli *et al.* (1973), who observed the occurrence of dislocations by means of transmission electron microscopy. Glass fibers were slid against aluminum specimens 10^{-5} m thick and normal and tangential forces were determined from the shape assumed by the loaded fiber. Dislocations of the type shown in Fig. 12 were observed.

Further tests by Maugis *et al.* (1976) were carried out also within an electron microscope. Here the stylus was a cold-rolled tungsten wire with a hemispherical tip of radius 2.5×10^{-6} m. The load ranged from 1 to 100 μN and the force required to break the elastic contact was 1–2 μN, as compared with values of 1.2–2.4 μN calculated on the basis of an interfacial energy of 100–200 mJ/m². This indicates that van der Waals forces between metals shielded by adsorbed gases were responsible for the adhesion. In common with the findings of Gane (1970), Maugis *et al.* showed that the load had to exceed the critical value before the stylus suddenly penetrated the surface. This critical value varied from place to place within the surfaces, the variation being attributable to the stress necessary to activate the first dislocation. Measurements of friction were consistent with the observation insofar as nearly zero values were recorded at low loads. These results serve to emphasize the importance of plastic deformation rather than adhesive forces as affecting the generation of frictional forces and wear damage. A nearly zero coefficient of friction was observed as long as deformation remained elastic. When plastic deformation occurred, the coefficient of friction increased with the load and its microscopic value was recorded. The energy expended on friction was

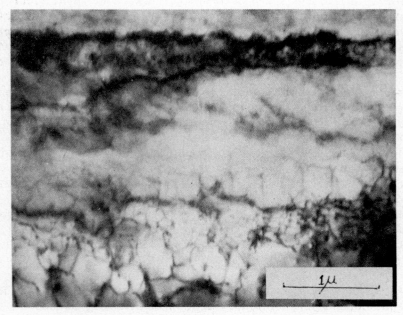

Fig. 12. Dislocations initiated by friction. (After Andarelli *et al.*, 1973.)

evaluated by relating friction behavior to tensile tests producing the same dislocation density, and the proportion of energy stored within the material was shown to be only 1% of that dissipated as frictional heat.

Many metals are polymorphic, that is, they can exist in more than one crystalline form. This ability raises a question as to the effect of the various crystalline forms of a metal on the nature of the interface formed with itself and with other metals. Thus tin can exist with a diamond-type crystal structure, which is then known as *gray tin,* or with a body-centered tetragonal structure known as *white tin*. When an interface was formed on the contact of the two forms of tin with iron, tangential motions indicated differences in frictional behavior (as shown in Fig. 13). The white tin produced a stick-slip or saw-tooth type of friction trace, indicating the formation of strong adhesive junctions at the interface between iron and tin. With continued application of tangential forces the interfacial bonds were broken and slip occurred. The gray tin, however, exhibited a continuous smooth friction trace, and it was shown by Buckley (1975) that with gray tin a continuous uniform interfacial transfer of tin to iron is observed, whereas with the white tin there are random islands of tin on the surface as a result of adhesion and fracture of the bulk tin.

Buckley (1976) has considered the nature of the interface between metals in relation to their classification in the periodic table. Thus when the noble metals silver, copper, and gold are brought into contact with iron (001 surface), interfacial adhesion occurs. The results of subsequent separation, as studied by LEED and Auger emission spectroscopy, have shown a remarkable similarity in the adhesion of these metals.

The group IV elements silicon, germanium, tin, and lead increase in metallic character moving through the group, and experimental results indicate that adhesion and friction are less for the covalent bonded elements than for the metallic-bonded metals.

Small amounts of alloying elements can markedly affect the character of metal surfaces via such a mechanism as differential segregation (Jones,

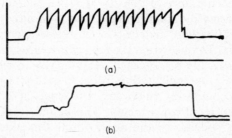

(a)

(b)

Fig. 13. Friction behavior of gray and white tin. (a) White tin, 24°C; (b) gray tin, −46°C.

1976). Thus concentrations of alloy on the surface may far exceed those in the bulk. When two metals are brought into solid-state contact the presence of the segregated species can and does alter the nature of the metal-to-metal interface. Buckley (1976) quotes an example of copper–aluminum alloys in contact with a gold surface. The adhesive bonding of gold to a copper and one atomic aluminum alloy resulted in a measured adhesion force five times that for the elemental copper in identical experiments. The aluminum had segregated out of the iron ore matrix onto the surface of the alloy so that the interface of a contact with the gold was one of gold to aluminum rather than of gold to an alloy of aluminum and copper. For this reason great care must be taken in using bulk metal properties to predict surface behavior because surfaces do not always reflect the characteristics of the bulk material.

Brainard and Buckley (1974) provide insight into the elemental wear processes by carrying them out within a scanning electron miocroscope. Figures 14a and 14b indicate different modes of wear. The existence and nonexistence of stick-slip action were shown to relate to the rigidity with which a specimen was supported. This indicates that perhaps even on this small scale stick-slip motions are in fact relaxation effects rather than representative of basic discontinuities.

Figure 15 shows the formation of cutting-wear particles. Ferrography investigation of real engineering systems (see Section III,H) frequently reveals similar particles, indicating the importance of this mode of wear.

Ruff (1976) reports studies of the initial stages of wear by electron channeling.

III. Wear in Air without Deliberate Lubrication

A. *Effect of Microstructure*

As a first approximation the wear resistance of a material may be identified with its strength insofar as this confirms ability to resist overstressing of the surfaces. A measure of this strength is available in hardness tests, notably the diamond pyramid method, and with any particular material, or closely related family of materials, wear resistance varies in conformity with the hardness. To this extent the microstructures that give rise to hard materials may generally be regarded as giving rise to wear-resistant materials. Nevertheless, as indicated in Table I, hardness provides an inadequate basis for comparing the wear resistance of groups of materials which may differ in other respects. The actual form taken by a wear process is often specific to the application. For example, the case of heavy sliding unlubricated wear in the atmosphere, represented by the

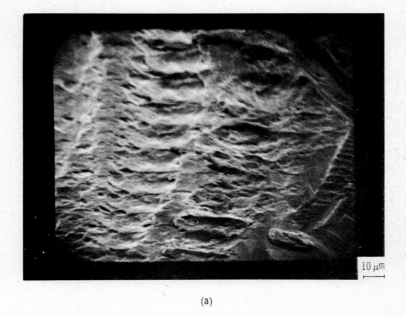

(a)

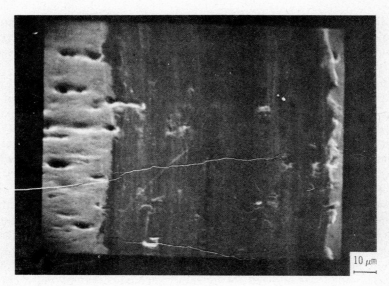

(b)

Fig. 14. Wear pattern after test: (a) Stick-slip; (b) restrained from stick-slip. (After Brainard and Buckley, 1974.)

Fig. 15.   Cutting-type wear. (After Brainard and Buckley, 1974.)

conditions of a flange of a railway wheel when passing around a curve, was investigated by Clayton (1978) and showed that the pearlitic/ferritic steels wore by a plastic flowing action and that the wear resistance was in this particular case related to the work done in a tensile test.

The results of another form of test designed to simulate the action of soil on agricultural implements of the plough type have already been presented in Table I. Field laboratory tests carried out in trials were compared with a manner pioneered by Kruschev and Babichev (1954) and Kruschev (1967). The estimation of the wear resistance of the material, obtained by abrading it with conventional abrasive cloths—carborundum, for example—appears at first sight to provide a simple and satisfactory procedure. However, investigators have reported poor reproducibility of results caused in part by minor variations in the abrasive cloth and partly by change in its abrasive nature with use, either by clogging with debris or by loss of abrasive properties. Kruschev overcame these difficulties by arranging for the test piece to traverse across a rotating disk in a spiral path in the manner of a gramophone needle. Variation in the quality of the abrasive cloth was allowed for by simultaneously carrying out a control experiment in each case. A lead–tin alloy was used as a standard and results were quoted as the relative wear resistance, that is, the ratio of the linear wear rate of the material under test to that of the standard material.

When a series of pure metals was rubbed against electrocarborundum cloth and the grain size averaged 80 μm, the wear rate was shown to be directly proportional to the hardness of the material as determined by a diamond pyramid having an apex angle of 130°. However, with commercial alloys of polycrystalline material, the simple relationship between hardness and relative wear rate no long existed. The behavior of heat-treated steels showed differences in wear rate for materials having identical hardness as measured by the diamond pyramid tests. In the annealed condition, a range of steels gave results that were entirely in accordance with the linear relationship determined between the wear rate and hardness of pure metals. However, when steels were quenched and tempered, a different coefficient of proportionality was obtained. Although for any particular steel the wear rate diminished with increasing hardness, the increase with wear resistance was not proportional to the increase in hardness resulting from quenching and tempering.

Similarly, when the hardness of the test materials had been obtained by a work-hardening process prior to test a simple relationship between hardness (measured prior to the test) and relative wear rate was no longer found. This may be attributed to the possibility that the wear resistance measured in the abrasive wear test depends on the hardness of the material in a fully work-hardened condition, it being immaterial whether this action took place during the abrasive process or prior to the commencement of the test. Richardson (1967a) showed that an important relationship exists between hardness of the abrasive and that of the material under attack. (The hardness of some common abrasives is given in Table III.) Richardson found it necessary to draw a distinction between what he called "hard" abrasive wear and "soft" abrasive wear, the former being relatively unaffected by particle size and the latter dependent thereon.

Transition from hard to soft abrasive wear occurred when the ratio of the metal in a fully work-hardened condition to the hardness of the abrasive was about 0.8. Thus laboratory tests using 180-grit paper known com-

TABLE III

HARDNESS OF ABRASIVES[a]

Material	Hardness (MN/m^2)	Material	Hardness (MN/m^2)
Glass	5790	Corundum	21180
Quartz	10390	Silicon/carbide	29420
Garnet	13370		

[a] After Richardson (1967b).

mercially as "flint paper" gave results of wear resistance of some of the materials containing carbides that were much greater than those recorded in the soils in Table I. However, when larger grit was used on the paper (40–36), the high wear resistance was not reported and good correlation was obtained with the field results.

The explanation of this behavior must relate to the fact that the microstructure of the material embodies some phases which are harder than the abrasive and some which are softer. The large carbide particles appear to offer a powerful obstruction to wear, whereas those which are small relative to the size of the abrasive are relatively ineffective. Thus the abrasive may very well remove the lighter ferrite, leaving cementite grains exposed and liable to be removed by fracture. Evidence of carbide particles in wear debris was provided by Barwell *et al.* (1955) (see Fig. 16).

As indicated in Table I, the steel containing manganese exhibited a resistance to abrasive wear which was disproportionately high in relation to its hardness. Hadfield steel containing 1.0–1.4% carbon and 10–14% manganese is widely used where extreme resistance to impact or abrasive wear is needed or when high contact stresses are applied. Babichev *et al.*

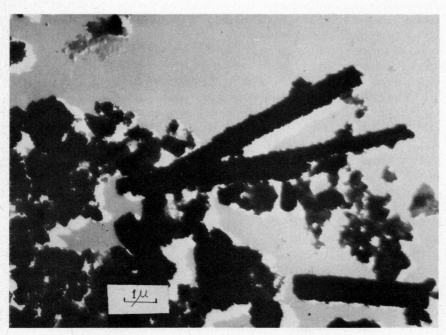

Fig. 16. Cementite particles among wear debris.

(1962) state that the effect of the manganese is to retain the austenitic condition and this is associated with the relatively low hardness. Tests were carried out in which the material was severely cold worked (40–50% deformation), which led to a substantial increase in hardness, i.e., 54% deformation increased hardness from 250 kg/mm² to 669 kg/mm² but made no difference at all to relative wear resistance. The authors conclude that in austenitic (Hadfield) steel there are no essential transformations resulting from work hardening that could explain its hardness. This confirms the previously expressed view that it is the hardness of the steel in the fully work-hardened condition rather than its initially measured hardness that is the important factor determining resistance to abrasive wear.

Larsen-Badse (1966) and Larsen-Badse and Mathew (1969) discuss the effect of the structure of hardened steels on their wear resistance.

A number of investigators, notably Clayton and Jenkins (1951), have noted areas of components that have been subject to severe wearing action which are only slightly attacked by the normal etching agents and exhibit bright white strips or spots. Kislik (1962) showed that the structures of the white layers differ from those of martensite and its decomposition products formed at the same temperature. They have a fine reticular structure with traces of deformation. Barwell *et al.* (1955) showed that the structure could be resolved, using an electron microscope, but was indeed a very fine one.

The occurrence of the white etching layer appears to be associated with systems embracing relative sliding under conditions of concentrated stress (Eyre and Baxter, 1972). A fuller treatment appears in the chapter by Eyre in this volume.

STRUCTURE OF CAST IRON

Cast iron has been used extensively as a wear-resistant material although it is not particularly hard. It is important, however, to draw the distinction from this viewpoint among three major forms: gray, white, and nodular cast iron. The distinction among these types depends on the form in which graphite is present. This can be important for four reasons. First, graphite has a considerable effect on the thermal conductivity, which is greatest in the direction of the basal plane and only one-fifth in the direction of the normal thereto; second it lowers the modulus of elasticity, which in combination with thermal conductivity reduces the susceptibility to thermal shock; and third, it can act as a solid lubricant.

In white cast iron the iron carbide is dissociated during solidification so that the fractured surface appears white rather than gray. The distributed iron carbides produce a brittle nonductile material having, however, a substantial resistance to abrasive wear.

Gray cast iron, on the other hand, contains graphite typically present in the form of flakes, which generally reduce its mechanical properties. Indeed, by analogy with the steel-containing carbides referred to earlier, high-strength irons with fine flakes may have poor resistance to abrasive wear because, as shown by Wright (1957), a disproportional decrease in wear resistance can occur when the distance between graphite laminae is less than the size of the abrasive particle. Angus (1951) reports that when a cast iron of low phosphorus content is employed to engage another metallic surface, the matrix should be fully pearlitic. While ferritic matrices are undesirable, the presence of up to 5–10% ferrite may not be detrimental with high-phosphorus irons (Angus and Lamb, 1957). Under these circumstances carbon should be present in a coarse random graphitic form and should amount to about 3%.

B. Effect of Environment

With the possible exception of gold, metals are never observed directly, but only in their oxide forms. In addition, some metals form nitrates and will of course react with many contaminants present in the atmosphere; for example, in districts close to the sea, traces of chlorine can be found combined within the superficial complex.

Iron can form three stable oxides, namely, wüstite (FeO), magnetite (Fe_3O_4), and hematite (Fe_2O_3), according to the applied conditions. For an equilibrium diagram of the iron/oxygen system reference is made to the literature, notably to a book by Kubaschewski and Hopkins (1967). The oxidation behavior of nonferrous materials has also been treated by these authors. The studies by Vernon *et al.* (1953) of the oxidation of iron in the atmosphere show that an invisible film grows on the surfaces, consisting of a cubic σ-Fe_2O_3, and the subsequent rate of reaction will depend on the rate of diffusion of oxygen inward and the metallic ions outward through the film. After the Fe_2O_3 film has grown to a certain extent, a quantity of the σ variation appears at the top and continued growth depends on the temperature. Above about 200°C the predominant action is for the metallic ion to diffuse through the film to the metallic boundary, where reaction takes place, thus at the higher temperatures the cubic Fe_3O_4 oxide is encountered.

The characteristic behavior of stainless steel chromium nickel varieties, for example, is attributable to the rapid formation of a dense protective film that is rich in chromic oxide. Corrosion resistance is not necessarily coincident with resistance to abrasive wear. Indeed in some mechanisms stainless steel is particularly difficult to lubricate.

Oxidation occurs in most abrasive processes and the resultant oxide becomes embedded in the superficial layer in a fairly complex manner. Moore and Tegart (1952) have shown that under conditions of repeated sliding the detached particles become embedded to a considerable depth within this ruptured zone because of the repeated deformation of the surface layers. Thus bearing surfaces do not generally consist of a simple work-hardened or amorphous juncture but rather of a complicated mixture of strained metallic materials, oxide, and other possible decontamination and reaction products (Earles and Powell, 1968).

An early study of the effect of oxygen on the wear of unlubricated materials was made by Seibel and Kobitzsch (1942), who carried out tests, using different concentrations of oxygen. As the oxygen concentration was increased from zero, the rate of wear increased rapidly to a maximum for an atmosphere of about 10% oxygen in a nitrogen/oxygen mixture. At concentrations above 60% the wear rate fell off rapidly to a negligible value in pure oxygen. Seibel *et al.* (1941) reported that an atmosphere of carbon dioxide results in low friction and negligible wear. Rosenberg and Jordan (1935) showed that the presence of an oxide layer tends to reduce friction between surfaces, and Mailander and Dies (1942) showed that the nature of the oxide formed is determined both by the environment and by the nature of the mechanical interaction. In particular, it has been shown by Kehl and Seibel (1936) and by Welsh (1965) that speed and load are important factors determining wear rate. Sudden transitions occur as indicated in Fig. 17, where certain values of the implied conditions of the wear rate increased or decreased abruptly, the inference being that this is attributable to the effect of the temperatures on the form of oxide that came between the surfaces, either acting as a lubricant or as an abrasive.

Soda and Sasada (1964) studied the wear of pure metals in a controlled atmosphere. The tests were carried out in air under pressures which ranged from 10^{-6} to 760 torr, and the results were divided into two main classes. For combinations of nontransition metals the nature of the wear appeared to be always cohesive, irrespective of running conditions such as loads or speeds, but, on the other hand, the transition metals produced noncohesive wear for moderate speeds and loads, and cohesive wear only occurred under intense conditions. An interesting feature of the tests within the cohesive range as compared with the noncohesive was that the wear particles were always found to be much larger and the rate of the wear much smaller than in the noncohesive range. These authors explained their results by introducing the concept of the mean free time of a small contact point available for the absorption of molecules of the sur-

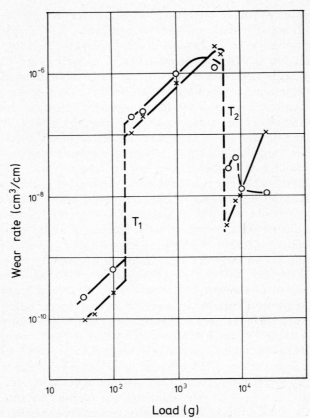

Fig. 17.   Transition between severe and mild wear for 0.52% C steel. Loads: ×, 1kg; ○, 10 kg. (After Welsh, 1965.)

rounding gas. They assumed that contact points would form a bridge that was ultimately sheared off to expose a small surface to the surrounding gas. This exposure would last a time t until the next bridge was formed by sliding. During this time t the gas molecules could attack the clean spot and form a chemisorbed monolayer when the mean surface stress was low or at very low sliding speeds. When either the loading or the velocity was high, the mean free time would be too short to absorb enough molecules for making a monolayer. Thus the time t was deduced to be inversely proportional to pv, where p is the mean load pressure and v the velocity of sliding. Analysis of nickel/nickel combinations verified the fundamental relationship between the mean free time of absorption and pv. The critical point for transition from noncohesive to cohesive wear corresponded to the condition for which the mean free time of absorption was just enough for contact points to form a chemisorbed monolayer from the surrounding gases.

Soda (1975) introduced a concept of a coefficient of friction based on the work of wear, given by

$$\mu = 4\alpha/cp \tag{6}$$

Equation (6) can be transformed by using the relationship that c is proportional to $p^{1/3}$, yielding

$$\mu = \text{const} \cdot c \, \alpha/\sqrt{p} = \text{const} \cdot p^{-1/6} \tag{7}$$

In the same way the amount of wear is proportional to $pl\mu^6$, where p is the load and l the distance of sliding.

The wear of unlubricated metals has been studied by Sasada and his colleagues. Their general approach has been to carry out comparatively simple mechanical tests (usually with a cylindrical pin impinging on the circumference of a cylindrical test piece) and to examine the transfer film and the wear particles with great care. Thus when copper was rubbed against steel at a low speed, the softer copper was transferred initially onto the harder iron. The harder iron was also transferred initially onto the softer copper. Each of the wear particles was a mixture of both iron and copper. As the test proceeded, however, the transfer of copper onto the iron surface became scanty, but iron and iron oxide accumulated on the copper surface. Fine powdery particles were then produced, which were mixtures of copper, iron, and iron oxide. It would normally be supposed that the material from the copper pin would be transferred to the steel disk, but in these experiments it was the iron with its high oxygen affinity that was transferred to the copper surface (Sasada et al. 1972).

Although the basic mechanism of adhesive wear is the transfer of materials from one surface to another, actual wear only occurs when particles are formed. Sasada and Kando (1973) studied the composition of the particles and the distribution of particle size. They found a log-normal distribution ranging over $2\frac{1}{2}$ orders of magnitude. The measurements of particle volume covered the range from 10^{-14} to 10^{-3} mm^3. When an atmosphere of nitrogen was substituted for air under laboratory conditions, the mean size of the particles was increased over a hundredfold, indicating the powerful effect of oxygen in preventing surface adhesion and subsequent particle growth. By means of an optical microscope Sasada et al. (1975c) observed the creation of wear particles. They demonstrated the importance of wedge formation, which is initiated by any irregularity capable of starting a drastic shearing process (asperity interaction or the presence of a foreign particle) and continues until limited strain is attained. Particles so formed tend to be transferred repeatedly from one surface to another (Sasada *et al.*, 1974, 1975) so that their final composition embodies material from both surfaces. Because the successive entrapment of particles on one surface or another must be a matter of chance, the wide distribution of particle size reported is to be expected.

Clark *et al.* (1967) report experiments on the wear of steel in an atmosphere of carbon dioxide.

C. Mild and Severe Wear

Certain wear tests reveal transitions which are so marked as to give rise to the designations of two forms of wear. These are referred to by Hirst (1957) as "severe" and "mild" wear, respectively, and by Kragelskii (1965) as "internal" and "superficial" friction. Lancaster (1963) expressed the view that the transition from mild to severe wear was the result of a competition between two opposing dynamic processes, one being the rate of formation of fresh metal surfaces as the result of the wearing process and the other the rate of formation of surface film by reaction with the surrounding atmosphere. In normal atmospheres the oxide film is rapidly renewed. However, when conditions are such that this is not the case, surfaces in contact will tend to seize with extremely high friction and consequent surface damage. The formation of the oxide prevents or reduces the seizing action as long as the film remains bonded to the surface. Usually, a hard oxide film will provide a basis on which the opposing surface will slide. Alternatively, a soft oxide may be effective in preventing seizure by limiting overall continuous wear to the mild mode. There is some evidence that continuous oxidation may be useful when high temperatures are encountered.

An alternative view has been put forward by Kragelskii (1965). When the interface between the rubbing surfaces is weak, sliding will take place thereon with relatively little damage, but when the bond between the surfaces (whether produced by penetration of asperities or by adhesion) is stronger than the underlying layers, failure will occur within the bulk of the material, causing considerable roughening and superficial damage. This introduces the concept of a gradient of mechanical properties. If the material becomes weaker as the surface is approached, smooth sliding at the interface may be expected. However, where the weakness increases toward the interior, internal rupture will take place. There is therefore a fundamental difference in the way in which running surfaces interact, which can be described as "external" friction, when contact between the surfaces is discrete and the area over which friction occurs is dependent on the applied load and the strength properties of the weaker material, and as "internal" friction, when the surface of action is continuous, independent of load, and laminar displacement of the material occurs in the direction of the relative velocity vector. Conversely, with external friction material may be displaced in a direction perpendicular to the velocity. During external friction the destruction of bounds is localized within a thin surface layer, whereas with internal friction the zone undergoing deformation occupies a considerable volume.

Confirmation of this viewpoint is provided by some experiments carried out by Kragelskii (1965), in which a transition from severe to mild wear occurred when the sliding speed in the experiment was increased to a critical value (0.9 m/sec) and the rate of wear fell off by a factor of between 500 and 600.

Figure 18, taken from the work of Kragelskii (1965), shows experimental results which demonstrated that when a test was so arranged that the interfacial region was cooled by liquid nitrogen, there was a very high rate of wear. When the specimens were heated by passing an electric current across the interface, the wear rate was reduced by a factor of 10^{-3}

These results were explained by the "hardness gradient" concept. When the interface was cooled, the material at the surface was resistant to damage, which therefore occurred within the bulk of the material, causing a considerable volume of material to be strained. When the interfacial temperature was high, the material situated there was weakened, perhaps melted, allowing easy sliding rather than tearing in depth.

D. The Delamination Theory of Wear

Koba and Cook (1974) studied the wear of leaded bronze ($Cu_{83}Sn_7Pb_7Zn_3$) running against steel (SAE 1020), using a slotted annulus impinging on a plane. The intensity of loading varied from 0.067 to 0.938 MN/m^2, and the average velocity of sliding varied from 0.37 to 1.78 m/sec. In their particular experiments the interacting surfaces were lubricated by mineral oil. They demonstrated by scanning electron photomicrographs that under these conditions the metal flowed freely at the surface, smoothing out hills and valleys. Some metal transfer was observed, but this did not appear to be an essential part of the wear process. Suh (1973) and Jahanmir et al. (1974) observed platelike particles of wear debris, which they attributed to the growth of subsurface cracks.

Suh and his coworkers have since investigated a number of wearing systems and have formulated what they describe as the "delamination theory of wear," which they claim provides predictions that are more in accordance with the observed facts than the prediction of the adhesion theory. A quotation from a recent paper by Suh (1977) is as follows:

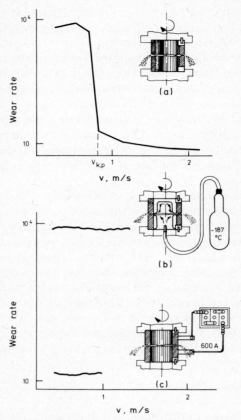

Fig. 18. Wear of Armco iron under various conditions of sliding. (a) normal conditions; (b) specimens cooled with liquid nitrogen; (c) specimens heated by electric current. (From Kragelskii, 1965.)

The delamination theory of wear described the following sequential (or independent, if there are pre-existing subsurface cracks) events which lead to wear particle formation:

(1)   When two sliding surfaces come into contact, normal and tangential loads are transmitted through the contact points by adhesive and ploughing actions. Asperities of the softer surface are easily deformed and some are fractured by the repeated loading action. A relatively smooth surface is generated, either when these asperities are deformed or when they are removed. Once the surface becomes smooth, the contact is not just an asperity-to-asperity contact, but

rather an asperity–plane contact; each point along the softer surface experiences cyclic loading as the asperities of the harder surface plough it.

(2)   The surface traction exerted by the harder asperities on the softer surface induces plastic shear deformation, which accumulates with repeated loading. (The increment of permanent deformation per asperity passage is small compared to the total plastic deformation per passage which is alost completely reversed. This accounts for the dissipation of energy.)

(3)   As the subsurface deformation continued, cracks are nucleated below the surface. Crack nucleation very near the surface is not favored because of a triaxial state of highly compressive stress which exists just below the contact regions.

(4)   Once cracks are present (either due to the crack nucleation process or pre-existing voids), further loading and deformation causes cracks to extend and to propagate, joining neighboring ones. The cracks tend to propagate parallel to the surface at a depth governed by material properties and the coefficient of friction.

(5)   When these cracks finally shear to the surface (at certain weak positions), long and thin wear sheets "delaminate." The thickness of a wear sheet is controlled by the location of subsurface crack growth, which is controlled by the normal and the tangential loads at the surface.

E.   Role of Asperities

The delamination theory is based essentially on the mode of deformation of asperities under applied stress. Thus Jahanmir and Suh (1977) envisage the flow of metal beneath an asperity to take the form shown in Fig. 19a, which shows the slip-line field based on the work of Green (1954, 1955). This figure shows the situation that would pertain if adhesion were to occur across the actual contact area and the material were a rigid perfectly plastic solid.

Figure 19 shows that when $Q = K$ (where K is the yield stress in shear), P falls to zero and the relationship ceases to apply. With this solution when the friction coefficient is greater than 0.38, the slip-line field degenerates into a single slip line along the contact as shown in Fig. 9. The influence of the relationship between normal and shearing stresses (that is, the coefficient of friction) on the distribution of strain in depth must be noted. Shearing stress within the surface beneath asperities would not necessarily cause failure once shakedown has occurred (Johnson and Jef-

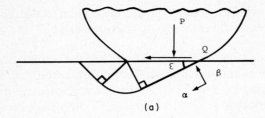

(a)

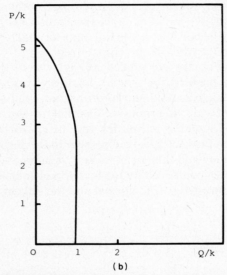

(b)

Fig. 19. Slip-line analysis of adhesive contact. k is the yield strength in shear, P is the normal stress, and Q is the tangential stress. (After Jahanmir and Suh, 1977.)

feries, 1963), and the delamination theory presupposes that planes of weakness are created by a process of nucleation. A very common form of nucleation of voids is attributable to the presence of hard particles which, under the action of repeated shear, either become separated from the parent matrix or themselves become overstressed and fracture. Figure 20, typical of a number of figures published by Jahanmir and Suh, shows the contours of constant values of the interfacial normal stress. This stress is compressive immediately below the surface, as indicated by the minus in front of the 1, and attains its largest value at $y/a = 2$, where a is the characteristic length of the asperity contact. From Fig. 21 two facts emerge. One is that the higher the coefficient of friction the deeper is the distressed region below the surface, and the other that this depth depends on the as-

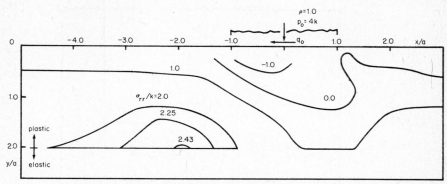

Fig. 20. Contours of constant values of interfacial normal stress around an inclusion for a coefficient of friction of unity. (After Jahanmir and Suh, 1977.)

perities, which can be expected to vary widely in their dimensions. Thus nucleation will be spread to a fair depth although there will be a tendency for the greater concentration of voids to be situated at a depth determined by the average asperity.

Although the presence of the voids must be construed as a weakening of the surface, their main significance is as initiating agencies for cracks. Experimental observation indicates that the number of subsurface voids is far larger than the number of cracks, and it is the rate of crack propagation that determines the mode of failure. Cracks are in fact found to spread most rapidly in a direction parallel to the surface, probably due to the weakening resulting from the presence of voids additional to those actually responsible for the initiation of the crack. The ultimate formation of the particle therefore depends on the operation of two mechanisms—void formation and crack propagation. These will be determined by the nature of the materials. Thus in metals in the medium tensile range with good fracture toughness, nucleation may occur easily so that crack propagation is likely to be the controlling mechanism. However, with other materials in which void nucleation may be extremely difficult and wherein crack propagation can occur readily, void nucleation will become the controlling mechanism.

The rate of wear will be determined not only by the frequency with which particles are removed from the surface but also by their thickness, that is, the depth at which crack formation takes place. Thus with materials of given hardness, high coefficients of friction will produce thicker particles and therefore greater wear. On the other hand, hardness will reduce the characteristic dimension of the asperity, with the result that wear rate will be diminished for given limiting conditions of nucleation or crack propagation.

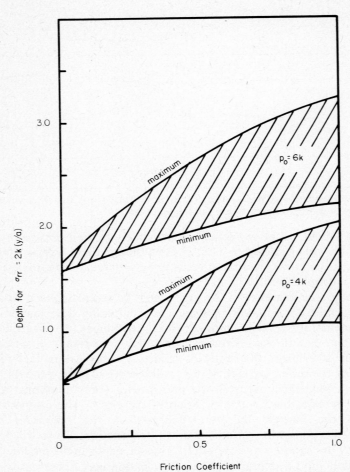

Fig. 21. Depth of void nucleation region as a function of friction coefficient for two applied normal forces. (After Jahanmir and Suh, 1977.)

The preponderance of platelike particles, all under conditions of lubricated mild wear, provides very powerful evidence of a mechanism such as that involved in the delamination theory of wear applying in these cases. The wider application of the theory must await additional experimental evidence.

The effect of speed on delamination is discussed by Saka *et al.* (1977). They carried out tests on two steels and titanium, using a pin and ring geometry at a sliding speed of 0.5–10 m/sec and a normal force of 49 N. Figure 22a shows that the coefficient of friction tended to fall off in each case at a speed between 1 and 2 m/sec. Wear rate did not follow so simple a relationship as is shown in Fig. 22b, but Fig. 22c shows a close correspondence between the coefficient of friction and the estimated surface temperature. The authors attribute the diminishing coefficient of friction to oxide formation. The more complex variation of wear rate with speed is explained in terms of its dependence on the coefficient of friction, hardness, and toughness of the materials concerned. However, microscopic examination of the subsurface (Fig. 23) indicates that there was considerable subsurface deformation, which could lead to crack nucleation and growth processes. The difference between the two materials would appear to depend on their different reactions to the same changes in frictional force and temperature. It is possible that in the case of the stainless steel, reduction in hardness as opposed to sliding speed was sufficient to account for the increasing wear rate in spite of the reduction in the coefficient of friction. However, in the case of steel and titanium, the increase in wear beyond a speed of 5 m/sec may be due to an increase in toughness arising from the dissolution of carbide.

The role of oxide formation is important at higher speeds (Quinn 1963, 1967), but its effect on wear rate may be indirect, as indicated in the previous discussion.

The adhesion theory of wear, being concerned exclusively with asperity contact, predicts that any modification in the metallurgical structure that raises the hardness should increase resistance to wear. Several observed phenomena are not consistent with this view. Saka *et al.* (1977) investigated dispersion-hardened alloys and showed that indeed these produced larger wear rates than the soft copper. Experimental observation showed that wear was primarily due to subsurface crack nucleation, propagation, and delamination to form wear sheets and the greater the degree of dispersion-hardening the greater the amount of wear. This is attributable to the internal oxidation facilitating nucleation of cracks, the wear rate being determined by the crack propagation rate. The influence of crack propagation rate and wear rate has been exemplified by experiments of Fleming and Suh (1977a, b) on two aluminum alloys from which normal fatigue test data was available. In the case of one alloy it had a greater hardness and a faster crack propagation rate in normal fatigue. Wear tests showed the softer material to be more resistant to wear than the harder, indicating the relevance of the crack propagation rate in these circumstances.

The effects of surface texture on dry wear are complex insofar as the initial surface itself is removed by the wearing process in a comparatively short time. Jahanmir and Suh (1977) studied the effect of surface roughness and integrity on sliding wear and found that while the initial

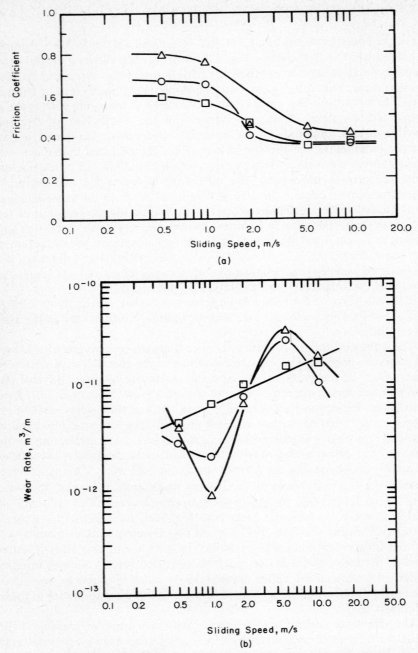

Fig. 22. (a) Friction coefficient versus logarithm of sliding speed, (b) log wear rate versus sliding speed, and (c) average surface temperature versus sliding speed: △ AISI 1020 steel; □ AISI 304 stainless steel; ○ 75A titanium. (After Saka *et al.,* 1977.)

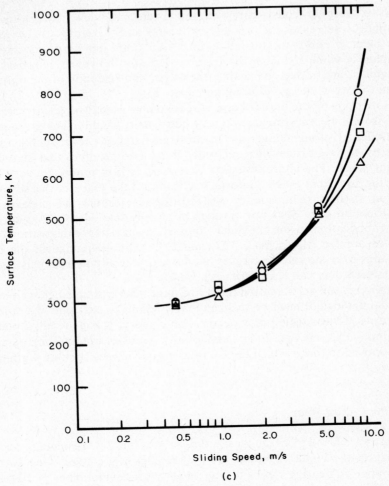

(c)

Fig. 22   Continued.

wear rate was influenced by surface roughness and applied load, the steady-state wear was independent of the initial roughness.

The delamination theory has implications for the formulation of materials for resistance to wear. Other circumstances for which this theory is applicable, that is, for which wear proceeds by subsurface deformation crack nucleation and crack propagation, means can be sought by slowing down or suppressing each of these three processes, especially because lowering the coefficient of friction and raising the hardness of materials reduces the subsurface deformation and hence the crack nucleation rate.

This is entirely consistent with the traditional view. Increasing the toughness decreases the crack growth rates and therefore it is necessary to select a microstructure that increases both the hardness and the toughness while reducing the friction. Usually, however, hardness and toughness are traded one against the other, and selection of the optimum from the wear viewpoint is by no means easy.

Hardening by the introduction of second phases may be counterproductive insofar as crack nucleation may occur more readily. The presence of a thin length of metal layer on the substrate can however reduce the wear rate by several orders of magnitude. In general the thickness should be about $\frac{1}{10}$ μm. The alternative way of hardening is to produce a very hard coating such that plastic deformation cannot take place in the substrate region. In the case of asperity-induced stresses in conformal surfaces, the deformation zone does not occur more deeply than about 200 μm. Possible applications using chemical vapor or physical vapor deposition techniques are for Al_2O_3, WC, TiC, and HfC. The advantage of the hard coatings over the soft coatings is of course that in addition to resisting delamination they can resist abrasive wear.

Although the delamination theory of wear is well established by experimental findings, it must be realized that it relates to only one of a number of forms of wear which may occur in machinery. It is generally found in machines that are operating satisfactorily over reasonably long periods, as opposed to those which fail by catastrophic modes such as scuffing or pitting.

F. Penetrative Wear

Bill and Wisander (1973) have outlined a slip-line field to account for the formation of platelike debris at the trailing edge of a contact zone formed by copper on bearing steel and Fig. 24 shows a development by Bates *et al.* (1974) which accounts for a semicontinuous chip formation in advance of a protuberance on the harder surface. They recognize three regimes of wear. At very low loads (regime I) there is plastic deformation but no visible wear debris. At intermediate loads (regime II) there is plastic deformation permitting sideways flow of material, which builds up to form a ridge at the outside edges of the wear track from which occasionally a wear particle is separated. At high loads (regime III) a semicontinuous chip is evolved in front of the protuberance in a manner analogous to negative-rake machining. This chip appears to be made up of discrete platelets which have adhered to one another. As in regime II, lateral flow occurs in regime III.

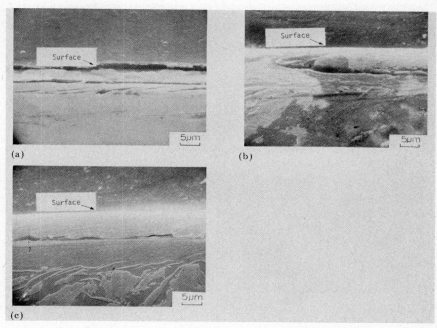

Fig. 23. Scanning electron micrographs of subsurface region. (After Saka *et al.*, 1977.)

Bates *et al.* carried out experiments in which a spherically ended diamond 25 μm in diameter was rubbed against steel and iron and photographed with a scanning electron microscope. They also modeled the contact situation, using a brass cylinder and modeling clay or paraffin wax. In each case they obtained visual evidence of the existence of regimes I–III and of the formation of platelike chips under the conditions of regime III. This may be an alternative explanation for the formation of platelike debris in the experiments of Suh *et al.*

Wedge-shaped particles have also been photographed by Sasada *et al.* (1975c). These particles were generated at the leading edge of the pin of a pin and disk machine and could correspond to the seizure mode suggested by Kragelskii and Aleksev (1976).

G. Seizure

Kragelskii and Aleksev (1976) regard seizure as a two-stage process. The first stage represents the building up of tangential stress until a critical yield value is reached, and the second stage is characterized by the devel-

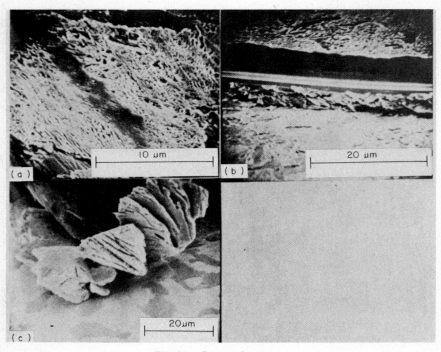

Fig. 24. Penetrative wear.

opment of plastic flow within the macroscopic volume of the surface layer.

The idealized shearing of the junction shown in Fig. 9 represents the first stage in a simple configuration. For a spherical asperity of radius R,

$$Y = (\tfrac{1}{2} - h/R)\,\sigma_s \qquad (8)$$

where Y is the yield shear stress, h the depth of penetration of asperity, and σ_s the compressive stress. For the limiting case where $h = 0$, $Y = \tfrac{1}{2}\sigma_s$, as in the simple theory of Fig. 9. For asperities having the shape of a wedge or cone,

$$Y = \tfrac{1}{2}\cos(\Pi - 2\gamma)\,\sigma_s \qquad (9)$$

where γ is the angle of inclination of asperity when $\gamma = 45°$, $Y = \tfrac{1}{2}\sigma_s$, as before. Kragelskii introduces the quantity ϕ, which is the ratio between the depth of penetration and maximum asperity height of the rougher surface. For low values of ϕ where asperity deformation is either plastic or elastic, the behavior of real surfaces may be characterized by statistical summation of the interaction of a large number of asperity pairs. Assuming a stress/strain law $\bar{\sigma} = \bar{B}e^{-n}$ (where σ denotes generalized stress, e generalized strain, and B and n are material constants), Halling (1976) derived wear equations which suggest that the mean slope of the asperities is the most significant surface characteristic.

The actual value of penetration at which "cutting" wear gives place to seizure depends on the quality of any lubrication present. While in most cases seizure occurs at temperatures that do not exert any significant effect on the properties of materials, the existence of a temperature gradient can lead to seizure, as postulated by Rozeanu (1973). According to this theory, frictional heating gives rise to a gradient of viscosity of the material of the superficial layer directed perpendicularly outward. If this gradient is lower than a certain critical value, macroscopic plastic flow occurs.

H. Fretting Corrosion

This form of wear, which occurs as the result of very small oscillatory displacement between surfaces, consists of interactions among several forms of wear, and can be initiated by adhesion, amplified by corrosion, and have its major effect by abrasion or fatigue (Tomlinson, 1927). The corrosive element is best revealed by studying the effects of the surrounding atmosphere. Thus Wright (1952) found that dry conditions produced very rapid wear, which was reduced as the relative humidity was increased to about 30%, followed by an increase of up to 100%. The presence of oxygen increased the wear rate, but it is not clear whether this was caused by the rapid removal of oxide films, which were continuously regenerated, or by the more rapid oxidation of metallic detritus, forming an abrasive which attacked the base material more vigorously. In one case the presence of oxygen increased the amount of damage threefold.

The nature of the fretting motion is not such as to promote formation of a hydrodynamic film; therefore liquid lubricants were not effective. The use of additives and molybdenum disulfide revealed no benefit. Some advantage revealed by oils and greases without additives is attributable to their action as seals restricting the rate of supply of oxygen to the surfaces. The effect of hardness is not simple. In a range of analogous materials, i.e., cast irons of hardness ranging from 100 to 250 HV 30, fretting wear was reduced as hardness increased, closely following previously reported results for cast iron under unidirectional sliding conditions. Hardened tool steel, some three times as hard as a specimen of cast iron, gave virtually the same rate of surface damage. Similar tests showed that a mild steel suffered only twice as much. It is possible that the explanation may lie in the relationship between the hardness of the metal and the hardness of the oxide formed therefrom.

The effect of environment was stressed by Hurricks (1972, 1974), who showed that at temperatures below 200°C mild steel passed through successive stages of oxidation and abrasive wear but that the main agency of surface degradation was "microfatigue." Waterhouse and Taylor (1974) confirmed this in the case of fretting of some nonferrous materials and cited the "delamination theory of wear" to explain their findings. At temperatures above 200°C the fretting wear of mild steel diminishes with increasing temperature until a second transition temperature is reached of between 500 and 600°C, above which the wear rate increases. Above 380°C the proportion of Fe_3O_4 to Fe_2O_3 increases, with a corresponding reduction in rate of wear. Thus, FeO appears to be the most harmful form of oxide.

There appears to be no way of eliminating fretting corrosion apart from preventing the relative motion by improved fitting, introduction of interference fits, etc. Palliatives are the avoidance of metals that form hard abrasive oxides, phosphating the surfaces, and providing sufficient lubricant to act as an oxygen barrier.

I. Wear of Polymers

While the friction of elastomeric materials has been shown to be dependent on their viscoelastic characteristics (Ludema and Tabor, 1966), other polymeric materials have not behaved in a manner consistent with this explanation.

The influence of surface temperature on wear when metals rub against polymers has been studied by Sasada *et al.* (1976). These authors have proposed a formation mechanism for adhesive wear particles of metal. When a metal–metal junction, formed between interacting surfaces, is sheared off by frictional motion, a small piece of either surface will be removed, which will adhere to the mating surface. These small pieces will be produced from either of the mating surfaces and will pile up to form large wear particles. Consequently, the transfer appears on either rubbing surface and each wear particle is composed of two mating metals. This mutual transfer of migrated wear particles is regarded by these authors as typical of adhesive wear. They demonstrated that similar mechanisms occur when nylon, polyacetal, polythene, and polytetrafluoroethylene (PTFE) are rubbed against metal at a speed of 0.4 m/sec under concentrated conditions of loading. The configuration of the test was such that a flat-ended ferrous pin was rubbed against the side of a polymer disk. Because of the absence of lubricant and the generally poor conductivity of polymers, temperature has figured largely in the explanation of these results. Therefore, Sasada and Kimura (1976) have studied the behavior of hot spots on the sliding surfaces but have substituted sapphire disks for the polymer disks used in the investigations referred to previously. The results revealed that the duration of the hot spots (3–10 μsec) and their temperature (570°C) were inconsistent with the previously accepted reason for particle generation, based on the bulk properties of the rubbing bodies. These authors supposed that the hot spots they observed arose from the wear process. A small piece of material sheared from one of the interacting surfaces would transfer to the opposite surface and further transfer elements would then pile up on it to form a large transfer particle of the type that would eventually break off and become a wear particle. However, before this happened the frictional heat would be generated on this transfer particle because of the transfer to it of the contact load. Such highly heated transfer particles would be revealed as hot spots on the rubbing surface and could account for the observed results.

Polytetrafluoroethylene (PTFE) is recognized as a material having important technological applications, but one that is somewhat anomalous with regard to its tribological properties. Notably, while it sometimes exhibits very low friction, it also can have a very high wear rate. Tanaka et al. (1973a) have studied the effects of heat treatment, speed, and temperature on friction and wear. In common with Makinson and Tabor (1964), they consider that the thin transfer film originates from slippage within amorphous regions. Lumplike wear particles are also frequently visible on the friction surface and are probably due to a pile-up or wrapping of thin films. The sliding speed dependence of frictional wear at various temperatures is attributed to the viscoelastic nature of an amorphous region between the crystalline areas of the PTFE.

In most industrial applications PTFE is reinforced by other materials, usually of a fibrous character. Tanaka et al. (1975) have accordingly carried out measurements of friction and wear of PTFE, incorporating glass and carbon fibers and carbon beads to clarify the wear-reducing action of these fillers. A pin and disk type wear-testing machine was used with a speed range from 0.01 to 0.25 m/sec at a load of 5 kg. The flat ends of 3-mm-diameter pins made of the substance under investigation were rubbed against disks made of glass or mild steel. Scanning electron micrographs were taken of the worn surfaces. The results indicated that the presence of the fibrous reinforcement reduced wear rate by several orders of magnitude. Briscoe et al. (1974), when discussing the action of a filler contained within high-density polythene, attributed its wear reduction to the formation of a strongly adhering film of polymer on the counterface. However, the data of Tanaka et al. indicate that a different mechanism operates in the case of PTFE. First, a fiber-rich layer is formed on the frictional surfaces during rubbing, and the fibers are covered partially with a PTFE film. In all cases the wear rates of the PTFE composites were lower on a steel disk and somewhat higher on a brass disk. The friction and abrasiveness of the glass/fiber system were generally lower than those of the carbon fiber system used. This was considered to be caused by the softening of the exposed glass fibers on the counteracting surface as a result of frictional heating. The carbon-bead system showed very low wear and friction on a glass disk since the friction and wear properties of the composite were largely controlled by those of the carbon itself. On the other hand, the wear rate of the carbon-bead system increased markedly on the steel disk as a result of the abrasive action of small particles produced by the fracture of carbon beads and wear of the steel disk. Broken pieces of carbon fiber were more easily produced during rubbing than were pieces of glass fiber. The breaking of fibers on the frictional surface increased the wear rate of the composite. Microscopic evidence is provided of the generation of fiber-rich layers between the rubbing surfaces during slow sliding.

Other polymers such as polypropylene and polyethylene do not appear to behave in the same way as PTFE. Tanaka (1975) attributes this to their spherulitic structure. The thermal basis for wear behavior was established by the fact that there was a critical dependence of wear rate on sliding speed. Evidence derived from examination of sections, using a polarizing microscope, indicated that the surface of the material had melted and the orientation of the polymer chains had become parallel to the direction of sliding. The depth of melting of polymers rubbed on glass was generally much greater than that of polymers rubbed on steel. The depth of melting of polymers rubbed on steel was only slightly dependent on the speed. It was contended that in these cases the friction must be related to the rheological properties of the molten polymers. As a first approximation the frictional force F may be expressed as

$$F = \eta(V/H)\,A \tag{10}$$

where F is the frictional force, η the viscosity of molten polymer, V the sliding speed, and H the depth of melting over apparent area of contact A. This expression was consistent with measured quantities. The mechanism of wear of spherulitic polymers at low speeds was attributed to the local transfer of molten polymer. However, when polymers were rubbed against glass in the high-speed range, the wear rate increased rapidly with increase in speed, and wear was attributed to the flowing out of molten polymer from the rear edge of the polymer pin as a result of viscous shear stress in the molten layer.

Thus it was necessary to distinguish among the mechanisms of wear typical of the different polymeric materials. For example, in elastomers hysteresis is the predominant factor, in the case of PTFE it is the transfer of material from one surface to the other with retention of the fibrous structure of any reinforcements, whereas with the spherulitic polymers surface melting is the determining factor, at least for the higher range of speeds.

An important use of nonmetallic materials is for the lining of brakes and clutches. These linings are generally composed of fibrous or granular material bonded together with resin. Tanaka *et al.* (1973b) studied resin-based friction materials (for road vehicles) that were mainly composed of phenolic resin and asbestos. Other materials such as cashew dust, barite, and copper powder are also included in some commercial formulations. While at the commencement of a test high friction attributable to the deformation resistance of the resin was measured, as the temperature rose the resin-based composites, containing asbestos and higher melting point additives, exhibited considerably lower friction during prolonged braking. This low friction was attributed to the lubricating action of the product of decomposition of the resin.

IV. Wear of Lubricated Systems

Under conditions of hydrodynamic lubrication, the film thickness should be sufficient to separate the surfaces and therefore to eliminate mechanical wear. However, contact may occur at low values of the duty parameter $\eta V/p$ and "boundary" friction may occur.

A. Interaction between Surfaces and Lubricants—Boundary Lubrication

Whereas boundary lubricants may act by the process of physical absorption or chemical reaction, the possibility also exists that the texture of the surface and the properties of the fluid may interact to provide a mode of lubrication which resembles hydrodynamic lubrication.

As long ago as 1937, Bradford and Vandergriff (1937) suggested that the mysterious quantity called oiliness (which distinguished good from bad lubricants of identical viscosity) might be attributed to the increase of viscosity with pressure. Smith (1962) suggested that an interaction between surface roughness and oil rheology might be a factor in the lubrication of concentrated contacts and postulated that a microelastohydrodynamic process might occur, in which the approach of asperities was accompanied by the expulsion of compressed fluid from the intervening space. Fowles (1969) applied elastohydrodynamic theory to the interaction of idealized surface irregularities and later developed a thermal elastohydrodynamic theory (Fowles, 1971). Fein and Kreutz (1972) applied the Archard–Cowking formula to an idealized single asperity model and demonstrated the feasibility of microelastohydrodynamic film lubrication.

If the heights of the asperities are randomly distributed, a certain number of interactions can be expected that are sufficiently drastic to cause intense but localized surface damage. The nature and intensity of this reaction will depend on whether the asperity deformation is "plastic" or "elastic" as determined by Eqs. (2) and (3).

Hirst and Hollander (1974) arranged for an 18/8 stainless steel (En SA) ball of hardness 500 HV and diameter 12.7 mm to slide over a flat plate made of the same material but of lower hardness (180 HV). The plates were finished, using nine different grades of abrasive paper in worn and unworn condition to produce 60 different surface conditions covering a range of σ from 0.01 to 2 μm and of L from 2 to 30 μm. The lubricant was a 1% solution of stearic acid in white oil. Effective lubrication was quantified by a transition from smooth sliding to irregular and high values of friction, and it was possible to divide the results into "safe" and "unsafe" regions as indicated in Fig. 25.

From Eq. (3) for a given combination of materials and a given value of λ, a plot of σ against L will represent a straight line. A series of such lines, each relating to a particular value of λ is shown in Fig. 25. For a large part of the diagram these lines are parallel to lines representing the transition from safe to unsafe conditions for loads of 2.5, 10, and 50 N, respectively. The variation for a twentyfold increase in load is not great. Thus it can be concluded that the transition from smooth to rough sliding is dependent on the value of λ, that is, on whether asperity or deformation is primarily plastic or elastic.

Another important conclusion to be drawn from Fig. 25 relates to the horizontal line corresponding to a value of σ just below 0.02 μm. All surfaces smoother than this failed readily under test, indicating that a surface may be too smooth to be lubricated effectively. It is also apparent that L must attain a certain value (depending on the applied load) before smooth sliding can occur. Thus lubrication is dependent on surface texture.

As interaction proceeds, a smoothing action may occur by reason of chemical attack on exposed asperities, but it is also possible that these asperities may become extremely ductile under hydrostatic stress, as studied by Bridgeman.

A series of tests (Czichos, 1977; Avient et al., 1960) of the effects of

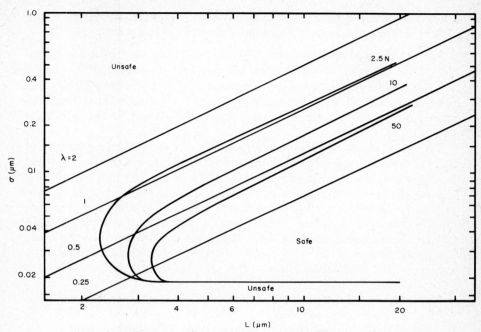

Fig. 25. Map for limits of smooth sliding. (After Hirst and Hollander, 1974.)

surface texture and lubricant on the critical failure data of steel under partial elastohydrodynamic conditions revealed that the ratio of film
thickness to roughness ratio was the determining factor under any given
set of operating variables (speed, load, and temperature). Measurements
of the coefficient of friction and of the duration of contact, measured
under an applied potential of 25 mV, indicated that the film was complete
for values of hm/σ of 2, where σ is the combined rms roughness of the
two surfaces.

When lower values of hm/σ were applied, asperities came into contact
(as detected by the resistance measurements) until the whole load was
carried by asperities. In this region the onset of incipient scuffing was determined by the stability of the asperity surface films as determined by the
chemical reactivity of the lubricant and of any additives contained
therein.

It may be concluded that in addition to their basic action in overcoming
the forces of adhesion between interacting bodies, as outlined in Section
II, boundary lubricants may perform the following three functions:

(1)   act as a hydrostatic medium that separates asperities with a film of
such intense pressure that smoothing may occur because of plastic flow,

(2)   act as a chemical agent to remove protuberances by reaction, and
(3)   provide products of reaction of an antiwelding nature.

Soda (1975) and Mizuno (1967, 1973) consider the wear process to consist of the transfer of metal from one surface to another, followed by transfer in the reverse direction, and finally removal of wear particles, the relative amount of back transfer and particle formation being determined largely by the ambient environment. The concept is similar to that of Kerridge (1955).

Endo and Kotani (1973) made a careful study of the surface texture of their test pieces and recorded high values of residual stress and surface hardening within the superficial layer. They concluded that the surface was roughened initially due to grooves caused by adhesion but that it subsequently became less rough, attaining a steady state after a certain running distance. In the steady state, wear appeared to be attributable to fatigue. There appeared to be a powerful relationship between the topographical features of the worn surface, notably radius and curvature of the asperities and the applied load. The size of wear particles was also dependent on the load.

The effect of the temperature at the surface of sliding contacts has been studied by Makagama and Sakurai (1974), who used a pin and disk type friction machine to investigate chemical wear in a steel–copper sliding system. The experimental results were in good agreement with the treatment of Dayson (1967). For a given sulfur concentration the wear rate versus oil temperature curve for copper gave a minimum which also corresponded to minimum friction and there appeared to be a good correlation between wear rate and chemical reaction rate. Film flaking appeared to be an important contributive factor to the rate of wear.

A particularly comprehensive test program was carried out by Tsuya (1976), using a variety of test arrangements and ambient conditions. Plastic working of the substrate of materials in contact led to what are known by the author as *micronized crystals,* and cracks were found to originate in the vicinity of the boundary between micronized crystals and those near the surface which had been simply distorted. The cracks tend to develop in the direction of material flow until a particle is released. The majority of frictional forces may be regarded as subsurface resistance to plastic flow, although fatigue damage may occur under lubricated wear conditions without extensive plastic flow.

The importance of surface-active substances in determining the wear behavior of metals has been quantified by Hironaka et al. (1975), who have measured the heat of absorption as an indication of the intensity of the surface reaction of different substances. Using a four-ball machine with the surface-active agent dissolved in a hydrofinished mineral oil, they obtained correlation between the heat of absorption of a number of surface-acting agents and wear rate. Heats of absorption were measured on the following powders: α-Fe_2O_3, Fe_3O_4, and FeS, and it was shown that the wear resistance conferred by the various surface-active agents was directly related to the heat of absorption. (This may be due to their repairing abilities after desorption had occurred at the hot spots.) From measurements of the heat of absorption of stearic acid on the iron oxides and sulfides it was shown that the Fe_3O_4 films formed by the solvent type of extreme pressure agent react more readily with the polar additives than does the α-Fe_2O_3 film, which is the usual product when steel oxidizes on exposure to air. Although, for convenience, the delamination theory of wear was introduced in Section III, this mode can also be operative when surfaces are lubricated. It takes the form of "fatigue" and its rate of occurrence may be governed by a general equation of fatigue wear. Such an equation has been proposed by Kragelskii as follows:

$$\text{linear wear rate} = \frac{\Delta h}{L} = (h/r)^{1/2} \frac{Pn}{P} K_1/m \tag{11}$$

where Δh is the depth of wear layer, L the distance of sliding, r the asperity radius, Pn the nominal pressure, P the true asperity pressure, K_1 the parameter governing distribution of asperity heights (usually 0.12–0.15), and m the number of cycles,

$$m = (P_b/P_t)^t$$

where P_b is the strength of the material, $P_t = K_2\mu p$ the surface tangential stress, $t = 5$–7, K_2 the coefficient of tensile stress acting on the surface (value is 3 for elastic material and 5 for brittle material), μ the coefficient of friction, and p the real pressure of contact.

It will be noted that the factors entering into the expression for wear rate are similar to those embodied within the plasticity index [Eq. (3)]. The $(h/r)^{1/2}$ term is obviously related to σ/r and the ratio Pn/p is determined by hardness. The main difference occurs in the term P_b, which relates to strength rather than to modulus of elasticity.

A comprehensive appraisal of world literature on boundary lubrication has been prepared by Ling et al. (1969).

B. Lubrication and Wear of Engine Cylinders

The statement that the wear of cylinders of automobile engines is mainly due to corrosion may be surprising. Williams (1949) fitted thermocouples for measuring cylinder-wall temperature and noted that cylinder wear rate was reduced nearly twentyfold when he increased temperature from 40 to 80°C. Piston ring wear was similarly reduced. Tests during which cylinder-wall temperature was held at 50°C showed the wear rate to be dependent on the proportion of sulfur in the fuel. (The rate of wear of diesel engines is also related to the sulfur in the fuel.)

The conclusion to be drawn from this is that wear is accelerated by the presence of acid-bearing moisture which condenses on the cylinder walls when the engine is operated at low temperatures, particularly during the start-up period when the cylinder walls may be starved of oil. Combustion of hydrocarbons may result in the formation of carbonic, formic, and acetic acids, and sulfur, which is usually present in the fuel, will form sulfur dioxide and sulfur trioxide. The latter forms sulfuric acid in the presence of water. When an engine is operated at low temperature it can be shown (Edgar, 1957) that the wear rate can be reduced by increasing the alkalinity of the oil. Oils containing additives known as "detergent oils" protect the cylinder walls from chemical attack by neutralizing the acids formed by combustion. In addition, by removing or neutralizing the acid radicals formed by oxidation of the lubricant, polymerization (which forms troublesome deposits) is inhibited.

C. Engine Bearings

The bearings of internal combustion engines may also suffer from corrosive attack by acid decomposition products arising from combustion of the fuel. White metals are generally resistant to corrosion, but have insufficient strength to withstand the loads and temperatures imposed on the bearings by some modern engine designs. Copper–lead bearings are therefore employed where the load is very heavy and the lead, often present as a separate phase, could be attacked by the acid present in the crankcase fluid. Bearings are therefore provided with an overlay of less corrodible material such as white metal, silver, and indium, and lubricants can be formulated which, in addition to neutralizing the acid, will lay down a protective lacquer film. Important savings can result from the repair of crankshafts by chromium plating (Byer *et al.*, 1975).

V. Elastohydrodynamic Conditions

Many important machine elements, notably rolling-contact bearings and gears depend on nominal point or line contact with a consequent concentration of stress onto a restricted area. This is known as Hertzian contact. Under these circumstances hard materials must be used so as to withstand the stresses elastically and failure may take the form either of fatigue caused by repeated stressing or scuffing caused by successive generation of heat on the restricted area. Some years ago it was considered that the applied conditions in Hertzian contact were so intense that a fluid film could not be generated to separate the surfaces. However, two effects taken in conjunction have been shown to contribute to the creation of such a film. First, as a result of elastic deformation, the conformity of the interacting elements is greater than would be inferred from their unstrained shape, and second, when mineral lubricants are involved, the viscosity increases exponentially with pressure. Indeed it is now thought that the mineral oils will often solidify into a glasslike condition at the interface.

A. Pitting-Type Failure—Hertzian Stress

Pitting failure, i.e., failure of counterformal surfaces by fatigue, is influenced by the depth within the surface of the zone of maximum shearing stress. Some controversy exists as to whether failures commence at the surface and grow inward to meet the highly stressed plane or whether they originate at inclusions in the vicinity of the plane and grow outward to intercept the surface. The controversy is probably an unreal one insofar as in some circumstances initiation is superficial, while in others it may be subsurface.

The onset of plastic yielding will be determined by the von Mises–Hencky criterion, but the maximum shear stress criterion is more convenient and gives practically the same answer. Thus the elastic limit is reached when $P_0 = 1.7 P_y$, where P_y is the yield stress in compression (Beeching and Nichols, 1948).

The presence of a frictional component of load will affect the distribution of stress within the body, which will no longer be symmetrical about the central ordinate of the zone of contact, resulting in an increase in maximum shear stress. Therefore, pitting occurs at lower loads when friction is present.

In the case of ball bearings, frictional forces are insufficiently high to affect the principal stresses and, as far as gears are concerned, any effect of surface traction is usually ignored (Merritt, 1971).

Under unlubricated conditions, frictional forces may exert a major influence on the magnitude and direction of subsurface stresses. Smith and Lui (1953) showed that if tangential forces are distributed over the zone of contact in exactly the same manner as the normal Hertzian contact pressures, the maximum shearing stress may occur at the surface.

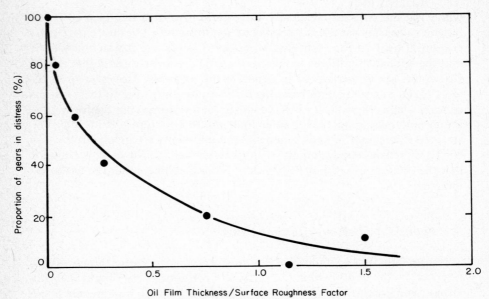

Oil Film Thickness / Surface Roughness Factor

Fig. 26. Gear reliability related to surface roughness. (After Wellauer, 1967.)

They demonstrated that this was the case when $\mu = \frac{1}{3}$. The value of maximum shearing stress was shown to be 43% larger than that attributable to the normal force acting alone. They attribute "shelling" failure of rails to the tangential stresses applied to the rail head because of curvature. Poritsky (1950) reports similar results using a value of 0.3 for μ.

Mindlin (1949) considered the elastic deformation to be associated with the subsurface stresses and demonstrated that the tangential force over part of the zone of contact exceeded the product of the local Hertzian pressure and the coefficient of friction. Thus relative sliding can occur over this region and the magnitude of the tractive stresses will follow the Hertzian ellipse. On the remaining contact area traction will be lower, being determined by elastic deformation of the two surfaces in contact. The effect of the quality of the surface finish can be related to the film thickness by

$$\Lambda = h(\sigma_1{}^2 + \sigma_2{}^2)^{1/2} \tag{12}$$

which defines the film thickness/roughness ratio, where h is the film thickness and σ_1 and σ_2 are the respective rms roughnesses of the interacting surfaces. Figure 26 is based on data provided by Wellauer (1967), which relate gear reliability to Λ. It will be noted that a value of about 2 is desirable. This is difficult to reconcile with the theory that pitting originates from the zone of maximum shear stess situated within the material. It is possible that in any given situation different modes of failure are in competition. In some circumstances of speed, load, and lubrication the failure may originate in the subsurface region, whereas in other circumstances (a low value of Λ, for example) the superficial mechanism may intervene to cause earlier failure.

Because of the difficulty of experimenting with actual gears most published work relates to tests on disk machines. Experiments by Onions and Archard (1974) indicate that the life of actual gears is considerably less than that obtained from tests on disks. The difference may arise from the presence of dynamic loads in the gears which are absent from the disks. Alternatively, they may be attributed to the transient nature of the film-forming process. Again, the method whereby lubricant is injected between two gear flanks is very capricious and simple starvation may be the explanation for the observations of Onions and Archard.

A method of testing metals and lubricants for resistance to pitting-type failure is the rolling four-ball machine (Barwell and Scott, 1956). A central ball is held in a rotating chuck and is forced into a nest of three balls that are free to rotate within a specially designed ball race. Reliable comparisons between different combinations of lubricants and materials can be made that are compatible with industrial experience, provided that sufficient care is taken to use test pieces of correct form, finished in a reproducible manner (Barwell *et al.*, 1976). Figure 27 shows the nest of balls after test. The well-defined running tracks and a pit on one of the balls should be noted.

The incidence of pitting is a random event and, so far as ball bearings are concerned, the "life" of a bearing type is taken to be the period of survival of 90% of a batch. The "scatter" of the life of the bearings is usually quantified by using a Weibull distribution. The Weibull functions take the form

$$P = 1 - \exp[- (x - \mu)_\eta]^{1/\alpha} \tag{13}$$

where α is a shape parameter, η a scale parameter, and μ a location parameter. The average life is usually about five times the rated life based on 90% probability of survival.

B. Scuffing

Scuffing is an important phenomenon, and the success of the extreme pressure lubricants in preventing its occurrence (particularly in automobile axles) provides evidence of the degree of control that is now held over this form of surface degradation. However, the original definition of

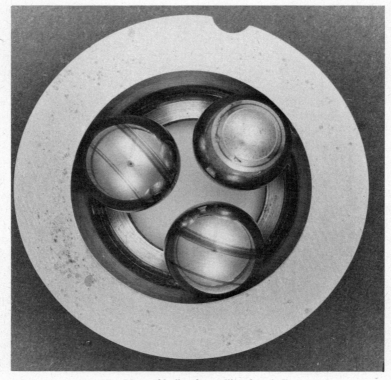

Fig. 27. Nest of balls after rolling four-ball test.

scuffing adopted by the Institution of Mechanical Engineers, namely, "gross damage characterized by the formation of local welds between sliding surfaces," is not entirely adequate as a description of what occurs in practice. Electron-microscopic studies show that initiation of scuffing failure may be primarily plastic and can sometimes be attributed to the ingress of foreign material. Thus, when it is recalled that the film thickness under elastohydrodynamic conditions may be as small as one-tenth of a micrometer, this possibility cannot be ruled out. There appears to be agreement that dynamic processes occur in rubbing surfaces in which mechanical, hydrodynamic, and thermal factors interact and that critical situations occur, below which satisfactory action continues but above which an unstable situation obtains, leading to degradation of a catastrophic nature. When hydrodynamic conditions are improved it can be shown that scuffing load is increased. It appears that scuffing failure is determined by a critical energy condition. At rates of energy input above a certain level, extreme pressure lubricants may cause a selective corrosion of overheated asperities, thus avoiding serious trouble and indeed contributing a beneficial wearing-in action. However, as duty becomes more severe self-accelerating processes occur, aided by a reduction of the viscosity of the lubricant caused by increased temperature. The critical value is affected by such quantities as surface hardness, lubricant activity, and the perfection of surface finish. Experiments in a cross-cylinder machine, wherein the contact area was moved forward onto fresh material to enable the growth of an incipient scuff to be studied, showed intermediate scuffing and recovery at lower loads, recovery becoming less frequent as load was increased until continuous damage occurred. Under the very lightest loads the surfaces were improved by the smoothing out of grinding asperities. As load was increased, the high temperatures at the zone of contact caused burnishing of the surfaces accompanied by local metallurgical changes. With test pieces made of softer steel this burnishing caused hardening of the surface layers, but when hard steel was used (850 HV 30) extensive softening occurred, accompanied by metallurgical transformations of the extreme surface, which restored the initial hardness (Milne *et al.*, 1957). Scuffing damage appeared to grow from the small scores on the surface caused either by adventitious material or by locally hardened material in the mating surface. The notable feature of these score marks was their indeterminateness and their gradual growth in size and frequency with increasing load.

An interesting feature of scuffing is that there is a marked tendency for material to be removed from a slower surface and deposited on the faster surface. This is probably equivalent to saying that material is moved from the hotter surface and deposited onto the cooler surface. This is clearly an example of adhesive wear.

Summarizing, a tentative theory of scuffing failure may be postulated as follows: There will often be points of momentary intimate contact between sliding surfaces, particularly if particles of a size greater than the minimum film thickness exist in the lubricant. At these points of interaction the pressures and coefficients of friction will be high and considerable heat will be generated. At low loads and speeds this heat may be dissipated rapidly within the lubricant and into the surrounding bulk material. At some critical load and speed, depending on the rubbing combinations, the coefficient of friction or some other factor, the rate of heat generation will increase to cause an unstable thermal system with the temperature rising until the melting point of one of the materials is attained.

Dyson (1975) examined the aforementioned hypothesis in the light of a mechanism suggested by Christensen (1973) that invokes the interaction of asperities. He concluded that thermal instability, while possible, was not the cause of scuffing in the particular case examined.

While there remain some difficulties in predicting the incidence of scuffing, there is little doubt that this form of failure is thermal in origin and that rapid changes in surface temperature must occur during the almost instantaneous contact event.

The most significant concept relating to the interaction of counterformal surfaces is the "flash temperature theory" introduced by Blok (1937a,b, 1970). This depends on the estimation of the temperatures generated by frictional heat. The temperature at the surface of a gear flank θ_c is defined as

$$\theta_c = \theta_b + \theta_f \tag{14}$$

where θ_b is the bulk temperature and θ_f stands for the maximum flash temperature. Blok proposed that this should be calculated as

$$\theta_f = \frac{1.11\mu W |U_1 + U_2|}{(K_1\rho_1 C_1 U_1 + K_2\rho_2 C_2 U_2)^{1/2}} (2a)^{-1/2} \tag{15}$$

where $2a$ is the instantaneous width of the band-shaped contact area and K_1 and K_2 are the heat conductivity of the materials of bodies 1 and 2, respectively, ρ is the density, and C the specific heat per unit mass; then ρc represents heat capacity per unit volume.

It can be seen that the grouping $\mu W(U_1 + U_2)$ equals the amount of heat fed into the system and that the other terms represent the ability of the system to receive and store heat.

It is difficult to apply Blok's equation to a practical case because of the need to assume a value for μ. Blok (1937a,b) proposed the value 0.15 and Merrit (1971) an equation of the form

$$\mu \propto [\eta^{0.15}(U_1 + U_2)^{0.15}(U_1 - U_2)^{0.5}]^{-1}[(r_1 + r_2)/(r_1 r_2)]^{1/2} \tag{16}$$

He states that μ reaches a maximum static value of about 0.15 for a steel–bronze combination and 0.2 for steel on steel.

In most counterformal bearing systems the materials of construction are common to both interacting solids so that Eq. (15) is reduced to the forms

$$\theta_f = 0.785\mu W(U_1^{1/2} - U_2^{1/2})/(K\rho Ca)^{1/2} \tag{17}$$

$$\theta_f = 0.62\mu W^{3/4}(U_1^{1/2} - U_2^{1/2})E^1/R^{1/4}K\rho C \tag{18}$$

$$\theta_f = 2.45\mu p_0^{3/2}(U_1^{1/2} - U_2^{1/2})R^{1/2}/(E^1)^{1/2}K\rho C \tag{19}$$

The breakdown of the elastohydrodynamic film does not necessarily result in scuffing. It is possible that a modified surface texture arises from the presence of chemically active (extreme pressure) lubricants, the surface ceasing to consist of clean bright metal but becoming a mixture of oxide, phosphide, polymerized oil, and the occasional metallic flake. The lubricant may cease to be a simple Newtonian fluid and in the immediate proximity to the surface may become a product of reaction such as a viscous soap with non-Newtonian properties. This, acting on the spongy weakened asperities, will allow the surfaces to interact with minimal damage.

VI. General Remarks

A. Analysis of Wearing Systems

As can be gathered from the foregoing, mechanical components are subject to attack by a variety of modes, surfaces being damaged in different ways, depending on the dynamic, thermal, and chemical circumstances surrounding the application and upon the materials of construction and the surface properties of the constituent materials.

Because two or more modes of attack may be combined and because sometimes different modes of attack are in competition, it is impossible to generalize, and attempts to systematize even the nomenclature of wear have not so far received general acceptance.

It is, however, usually possible to provide an adequate description of the mechanisms of wear occurring in any particular machine and to apply quantitative prediction of wear life, albeit of a stochastic nature.

Rather than generalizing about wear, most engineers prefer to study particular systems, e.g., rolling contact bearings and wheels on rails, and to analyze the effect of various factors, e.g., hardness, surface texture, lubrication, on life in the particular circumstances of the application. However, the utility of wear theory is that it provides a basis for the transfer of experience from one application to another. This transfer of experience is particularly valuable when it becomes necessary to diagnose the failure of machinery caused by excessive wear. Such a diagnosis may be carried out on the basis of applied conditions; calculation of Hertzian stress and the number of cycles will enable determination of the probability that failure was caused by surface fatigue to be assessed; calculation of flash temperature will indicate the probability of scuffing; and study of the environment, lubricant in particular, will allow the contribution, if any, of corrosion to be assessed.

B. Systematic Presentation of Data

One proposal for presenting data relating to bearings so as to enable duty to be related to the particular aspects of bearing design that have been shown to be critical in the past is the construction of a "profile,"

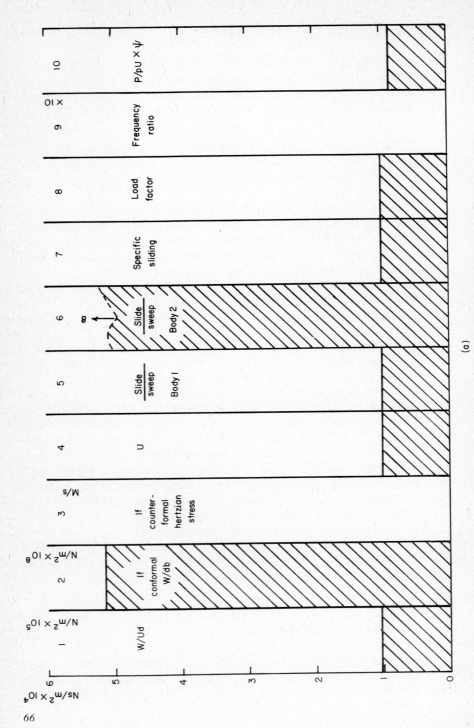

Fig. 28. (a) Profile of failed bearing. (b) Significances of terms used.

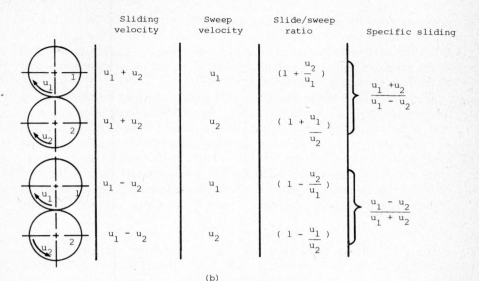

	Sliding velocity	Sweep velocity	Slide/sweep ratio	Specific sliding
	$u_1 + u_2$	u_1	$(1 + \dfrac{u_2}{u_1})$	
	$u_1 + u_2$	u_2	$(1 + \dfrac{u_1}{u_2})$	$\dfrac{u_1 + u_2}{u_1 - u_2}$
	$u_1 - u_2$	u_1	$(1 - \dfrac{u_2}{u_1})$	
	$u_1 - u_2$	u_2	$(1 - \dfrac{u_1}{u_2})$	$\dfrac{u_1 - u_2}{u_1 + u_2}$

(b)

Fig. 28 Continued.

(Barwell and Scott, 1966). Figure 28a shows such a profile and embraces ten characteristics that are displayed in the ten vertical sections of the diagram. Further characteristics may be added in the light of experience. The first section defines the load that must be put on the bearing in relation to diameter and speed, which is a measure of the severity of application. It has the same dimensions as viscosity and measures the difficulty (or ease) of solving the problem hydrodynamically.

Sections 2 and 3 taken together indicate at a glance whether the bearing is conformal or counterformal and then, individually, give stress as a measure of the demands made on one of two classes of material.

Speed, taken alone, is not particularly significant but becomes so if there is a possibility of departure from pure hydrodynamic conditions. Then, taken in conjunction with sections 5 and 6, it indicates the likelihood of one or another of the surfaces becoming overheated. Section 7 is a measure of the ease (or difficulty) of hydrodynamic lubrication when both surfaces are in motion. The significance of the terms used is illustrated in Fig. 28b.

Section 8 gives a measure of the steadiness (or lack of steadiness) of the load. It will be recalled that reversal of the load may be quite important when relative velocities are low. A particular case relates to small-end bearings. Single-acting engines may have a substantially unidirectional load on these bearings that can be relieved only by inertia at top dead center. With two-stroke engines compression forces may exceed this, giving no reversal of load and thus no possibility of creation of a hydrodynamic film. In one case a laboratory investigation led to the conclusion that a series of interconnected axial grooves could be beneficial if situated in the pressure region, in complete contrast to the usual experience that such grooves should be avoided at all cost.

Section 9 relates to the frequency of loading caused by either a rotating load vector or a reciprocating load ω, representing shaft speed, and ω_L that of the vector representing periodicity of load variations. Thus when $\omega_L = \frac{1}{2}\omega$, load capacity vanishes. In most cases, however, there will be some steady load and sections 8 and 9 should be examined together. Negative values in section 9 are equivalent to positive ones. Thus, if the load rotates at the same speed as the shaft, the load capacity is equal to that for unidirectional loading. Perhaps it would be more correct to write

$$(1 - 2\omega_L^2/\omega)^{1/2}.$$

The final column represents the "duty parameter," which embraces hydrodynamic factors and geometric factors. Thus in the case of a journal bearing, ψ is given by the product of the clearance ratio and the breadth/diameter ratio, each raised to the second power.

The quantities used in the diagram are taken from a test to failure described by Barwell (1965).

Following cooperative efforts of the International Research Group on Wear of Engineering Materials (IRG–OECD), Czichos (1976) has proposed a data sheet to facilitate the systematic compilation of parameters relevant to tribological systems. A "checklist" is proposed that divides the main tribological parameters into four groups under the following headings: technical function of the tribosystem, operating variables, structure of the tribosystem, and tribological characteristics.

Figure 29 is a reproduction of the chart published by Czichos. The data have been inserted to correspond with those of Fig. 28. It will be noted that Fig. 29 produces a much more complete description of a system, whereas Fig. 28 emphasizes the quantitative aspects.

Systematic analysis of the total system is necessary if laboratory procedures are to subject matters to exactly the same conditions as those applying in the industrial application. A test of this is that material should fail in exactly the same way in the laboratory as in service. It may, for example, be necessary to carry out laboratory tests at different sliding speeds from those in the full-scale device to account for differences in the thermal environment.

Postfailure examination is difficult because most wearing systems destroy the evidence of their origin. Some forms of damage are readily identifiable from microscopic examinations. Figure 30 shows examples of

Tribological systems data sheet				

General problem statement:

Laboratory test on journal bearing

I Technical function of the tribo-system

Information on load carrying capacity

II Operating variables

Type of motion[1]: Continuous sliding		*Duration of operation t[ ]* Until failure

Load[2] $F_N(t)$ [ ]	*Velocity[3]* $v(t)$ [ ]	*Temperature[4]* $T(t)$ [ ]
311	1.25	20° C

Other op. variables:	*Location:* Ambient

III Structure of the tribo-system

Properties of elements (initial/final)		*Tribo-element (1)*	*Tribo-element (2)*	*Lubricant (3)*	*Atmosphere (4)*
Designation of element and material		Shaft silver steel	Bush brass	Water	Air
Volume properties	*Geometry/Dimensions/ Volume*	0.0238-m diam	0.0254-m wide c/r = 1/1000		
	Chemical composition				
	Phys.-mech. data: Hardness Viscosity $\eta(T,p)$ other[5]	Unhardened		1.01×10^{-3} Ns/m^2	
Surface properties	*Topography descriptors (c.l.a.,etc)*			*Other data:*	
	Surface layer data (if different from volume)				

Contact area A[ ]	6×10^{-4} m^2	*Tribological interactions:*	
Ratio: $\frac{contact\ area}{total\ wear\ track}$ ε [%]	$\varepsilon(1)$ $\varepsilon(2)$	(1) → (2) (3) → (4)	(1) ⟷ (2)
App. lubrication mode: Gravity			

IV Tribological characteristics

Changes in properties of the elements[6]	*Friction data (vs time t or distance s)*	*Wear data (vs time t or distance s)*

Other characteristics (e.g.: contact resistance, vibrations, noise, etc):	*Appearance of worn surfaces:*
Minimum film thickness prior to failure 13 μm	Seized

BAM, Berlin-Dahlem

1) e.g.: sliding, rolling, oscillating, reciprocating, etc
2) the contact pressure p is given by $p = F_N/A$
3) velocity of tribo-element(1) relative to tribo-element(2)
4) temperature at stated location
5) e.g.: density, thermal conductivity, Young's modulus, etc.
6) e.g.: changes in hardness of (1),(2), changes in viscosity of (3), changes in humidity of (4), etc.

Fig. 29. Tribological data sheet.

pitting, Fig. 31 of fretting, and Fig. 32 of scuffing. Microhardness measurements, particularly on taper sections, can sometimes enable the thermal history of the wear event to be reconstructed, and sectioning can reveal informative metallurgical features such as "white etching" constituents and mechanical troostite.

C. *Particle Examination*

Because the material removed in a wear process usually persists in particulate form one of the most convenient means for classification of a wear situation is to examine the wear particles.

The study of particle formation both in the laboratory and in industrial situations has been greatly facilitated by the development of ferrographic methods. One method of particle examination consists of pumping a sample of oil at a slow, steady rate between the poles of a magnet. The fluid runs down an inclined microscope slide which has been treated so that the oil is confined to a centrally situated strip. The net effect of the viscous and magnetic forces acting on the particles is to sort them by size, the larger particles being deposited first, whereas the smaller particles are carried some way along the slide before being deposited. A washing and

Fig. 30.   Pitting (×100).

Fig. 31. Fretting (×100).

Fig. 32. Scuffing (×100).

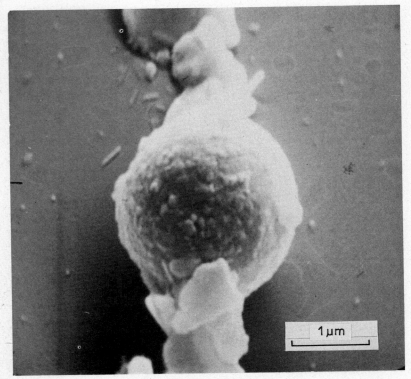

Fig. 33. Spherical particles.

fixing process is used to remove the surplus oil and hold the particles in position. At this stage the slide becomes a permanent record of the condition of the lubricant and is known as a *ferrogram*.

In addition to information about the number and size of particles, a great wealth of information is available from microscopic observation of the nature and shape of the particles. An optical microscope that employs colored transmitted and reflected light simultaneously, is used to distinguish metallic particles from those composed of oxide or polymeric material. The latter are deposited because of transfer of material in which small quantities of metallic substance are dispersed throughout the particle.

Great interest attaches to the formation of spherical particles of the type shown in Fig. 33. These particles appear to be associated with the incidence of pitting-type failure of rolling elements. On this basis the appear-

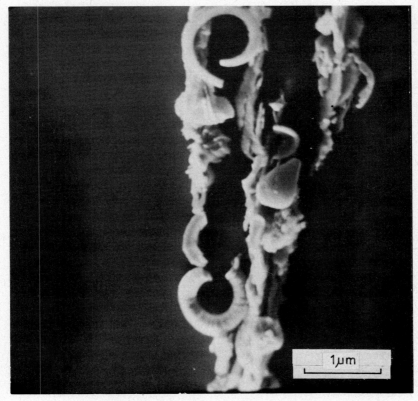

Fig. 34. Coiled (cutting wear) particles.

ance of such particles could form a valuable early warning of the imminence of fatigue failure. The coiled particles of Fig. 34 provide evidence of cutting-type wear. They are often present when the plant is being commissioned and originate from the manufacturing and assembly process.

The particles illustrated in Fig. 35 are representative of those found on a ferrogram of a gear coupling lubricant of a turbine compressor that had been withdrawn from service because of mechanical damage. The particles in Fig. 35a have the platelet appearance normally indicative of a mild rubbing wear situation. The size of the platelets, however, is also indicative of the wear mode and in this instance, although the lubricant was being filtered to 0.5 μm, some very large particles, of the order of 10 μm, were present (see Fig. 35b).

Preliminary tests, wherein the nature of the wear particles has been related to the severity of the rubbing conditions causing the wear, have been

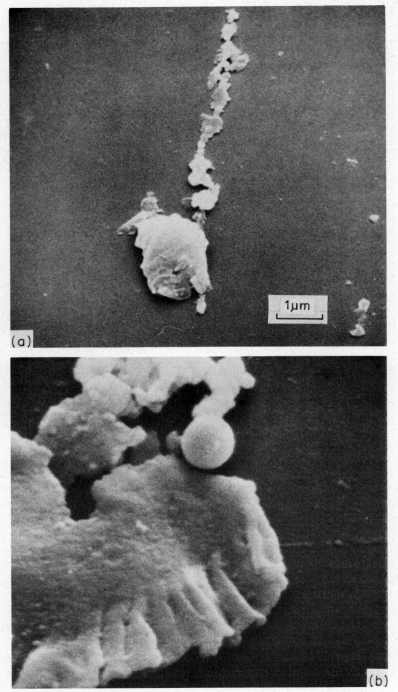

Fig. 35. Platelike particles from failed gear: (a) small particles; (b) large particles.

reported by Reda *et al.* (1977), who identified six regions as indicated in Table IV. The range of conditions tested was necessarily more limited than that encountered in industrial practice so that every mode of wear and particle generation was not represented. The results nevertheless illustrate the feasibility of wear particle examination for diagnostic and monitoring purposes.

Thus by working backward from the particles, it is possible to determine the mode of wear operating in a given situation. For example, regime 2 is characteristic of delamination wear and regime 6 of severe thermally induced scuffing or scoring.

D. Different Mechanisms of Wear and Their Relative Importance

The seriousness (or otherwise) of a manifestation of wear will depend on the detailed functioning of a particular machine element. Generally, a change in dimensions will result in malfunctioning because of excessive clearance, but changes in form (e.g., corrugations of railway track) can

TABLE IV

WEAR REGIMES[a]

Regime	Particle description and major dimension	Surface description	Wear rate
1	Free metal particles usually less than 5 μm	Varies between polished and very rough; one surface can be polished while the opposing surface remains as generated	Approaches zero
2	Free metal particles usually less than 15 μm	Stable, smooth, shear-mixed layer with a few grooves, depending on the number of particles in the oil	Low
3	Free metal particles usually less than 150 μm	Ploughed with evidence of plastic flow and surface cracking	High
4	Red oxide particles as clusters or individually up to 150 μm	Ploughed with areas of oxides on the surface	High
5	Black oxide particles as clusters or individually up to 150 μm	Ploughed with areas of oxides on the surface	High
6	Free metal particles up to 1 mm	Severely ploughed, gross plastic flow and smearing	Catastrophic

[a] Delimited by Reda *et al.* (1975).

lead to objectionable vibrations although the component dimensions may remain within acceptable limits. Some forms of wear, although not serious in themselves, may be the cause of other forms of damage. Thus fretting corrosion can lead to failure by fatigue.

Considering the various forms of wear in turn, atomic wear, as considered by Holme and Tomlinson, is not manifested in any technological application. Regarding adhesive wear, although there is no lack of evidence of the transfer of metal from one surface to another, this does not appear to contribute directly to the damage of components, probably because of the intervention of oxidation or some other chemical reaction determined by the environment (Kerridge, 1955).

Abrasive wear is attributable primarily to local overstress of the surface material, but the main agency of surface deterioration appears to be fatigue, which can operate at different levels, very near to the surface (as determined by the size of asperities) so as to give rise to delamination wear, or in the case of counterformal contact, at the depth of the maximum Hertzian stess so as to give rise to pitting. Thermal effects may lead directly to softening or melting of the surface, but more generally they may transform the mechanical properties of the material by changes in microstructure.

Although the combined attack on a surface by mechanical stress, heat, and chemical action can produce an infinity of wear actions, practical experience shows that these can usually be expressed in terms of a limited number of classes, namely, adhesive wear, abrasive wear, surface fatigue (on Hertzian or delamination scales), scuffing or scoring, fretting, and erosion.

These modes are sometimes combined and sometimes competitive, but an analytical process will usually single out some modification of material, heat treatment, lubricant, or operating and environmental conditions that will enable a satisfactory life to be achieved.

References

Abbot, E. J. and Firestone, F. A. (1933). Specifying surface quality, *Mech. Eng.* **55**, 569–572.

Andarelli, G., Maugis, D., and Courtel, R. (1973). Observation of dislocations created by friction on aluminum thin foils, *Wear* **23**, 21–31.

Angus, H. T. (1951). The wear of cast iron machine tool slides, shears and guideways, *Wear* **1**, 40–51.

Angus, H. T., and Lamb, H. D. (1957). Destruction of cast iron surfaces under conditions of dry sliding wear, *Proc. Conf. Lub. and Wear* pp. 689–693. Inst. Mech. Eng., London.

Avient, B. W. E., Goddard, J., and Wilman, H. (1960). An experimental study of friction and wear during abrasion of metals, *Proc. R. Soc.* **258**, 159–180.

Babichev, M. A., Velikova, A. A., and Kraposlina, L. B. (1962). The influence of management on the abrasive wear of a 1 per cent carbon steel and of ferrous alloys, "Friction and Wear in Machines" (translation), pp. 1–19. American Society of Mechanical Engineers, New York.

Baker, D. W. C., and Brailey, E. D. (1957). Field testing of big end bearings, *Proc. Conf. Lub. and Wear*, p. 770. Inst. Mech. Eng., London.

Barwell, F. T. (1965). In discussion, *Proc. Inst. Mech. Eng.* **179**, Pt. 3J, 304–306.

Barwell, F. T. (1978). "Bearing Systems, Principles and Practice." Oxford Univ. Press (Clarendon), London and New York.

Barwell, F. T., and Scott, D. (1956). The effect of lubricants on the pitting failure of ball bearings, *Engineering* **182**, 9–12.

Barwell, F. T., and Scott, D. (1966). The investigation of unusual bearing failures, *Proc. Inst. Mech. Eng.* **180**, Pt. 3K, 277–297.

Barwell, F. T., Grunberg, L., Milne, A. A., and Scott, D. (1955). Application of some physical methods to lubrication research, *Proc. World Pet. Congr., 4th, Rome* pp. 115–136.

Barwell, F. T., Bunce, J. K., Davies, V. A., and Roylance, B. J. (1976). The rolling four-ball machine—a study of its performance, *Int. Symp. Rolling Contact Fatigue*. Institute of Petroleum, London.

Bates, T. R., Ludema, K. C., and Brainard, W. A. (1974). A rheological mechanism of penetration wear, *Wear* **3**, 365–375.

Beagley, T. M. (1976). Severe wear of rolling/sliding contacts, *Wear* **36**, 317–335.

Beeching, R., and Nichols, W. (1948). A theoretical discussion of pitting in gears, *Proc. Inst. Mech. Eng.* **158(A)**, 317–323.

Beilby, Sir George (1921). "Aggregation and Flow in Solids." Macmillan, New York.

Bill, R. C., and Wisander, D. W. (1973). Role of Plastic Deformation in Copper and Copper 10% Aluminium Alloy in Cryogenic Fuels. NASA Rep. No. TH D-7253, Washington, D.C.

Blok, H. (1937). Measurement of temperature flashes on gear teeth under extreme pressure conditions, *Proc. Inst. Mech. Eng.* General Discussion on Lubricants, Pt. 2, pp. 14–20.

Blok. H. (1937a) Theoretical study of temperature rise on surfaces of actual contact under lubricating conditions, *Proc. Inst. Mech. Eng.* General Discussion on Lubricants, Pt. 2, pp. 222–231.

Blok, H. (1970). The constancy of scoring temperature, *Proc. Interdisciplinary Approach to Lub. Wear* pp. 153–248. NASA Rep. No. SP-237, Washington, D.C.

Bowden, F. P., and Tabor, D. (1950). "The Friction and Lubrication of Solids." Oxford Univ. Press, London and New York.

Bradford, L. J., and Vandergriff, C. G. (1937). Relationship of the pressure–viscosity effect to bearing performances, *Proc. General Disc. Lubr. Lubricants,* **1**, 23–29. Inst. Mech. Eng., London.

Brainard, W. A., and Buckley, D. A. (1974). Dynamic Scanning Electron Microscope Study of Friction and Wear, NASA Tech. Note D-7700, Washington, D.C.

Briscoe, B. J., Pogosian, H. K., and Tabor, D. (1974). The friction and wear of high density polyethylene. The action of lead oxide and copper oxide fillers, *Wear* **27**, 19.

Buckley, D. H. (1975). Friction and Wear of Tin and Tin Alloys from 110° to 150°C, NASA Tech. Note D-8004, Washington, D.C.

Buckley, D. H. (1976). Metal-to-metal interface and its effects on adhesion and friction, *J. Colloid Interface Sci.* **58**, 36–53.

Byer, K., Dallimore, V. J., and Lowe, C. B. (1975). Experiences with chromium plating of medium sized diesel engine crankshafts, *Inst. Marine Eng. Trans.* **87**, 49–65.

Christensen, H. (1973). The reliability of the elasto-hydrodynamic oil film, *Proc. Inst. Eur. Tribol. Congr.* Inst. Mech. Eng. Paper No. C239/73, London.

Clark, W. T., Pritchard, C., and Midgeley, J. W. (1967). Mild wear of unlubricated hard steels in carbon dioxide, *Proc. Inst. Mech. Eng.* **182,** Part 3N 1967–68, 97–106.

Clayton, P. (1978). Lateral Wear of Rails on Curves, *Proc. Inst. Mech. Eng.* (in press).

Clayton, D., and Jenkins, C. H. M. (1951). Physical changes in rubbing surfaces on scuffing, *Brit. J. Appl. Phys. Suppl. No. 1* 69–78.

Cocks, M. (1957). Role of atmospheric oxygen in high speed sliding phenomenon, *J. Appl. Phys.* **28,** 835–843.

Czichos, H. (1977). A systems analysis data sheet for friction and wear tests and an outline for simulative testing, *Wear* **41,** 45–55.

Czichos, H. (1978). "Tribology—a systems approach to the science and technology of friction, lubrication and wear." Elsevier, Amsterdam.

Dayson, C. (1967). Surface temperature of unlubricated sliding contacts, *ASLE Trans.* **10,** 169–174.

de Pater, A. D., and Kalker, J. J. (1975). The mechanics of the contact between deformable bodies, *Proc. Symp. Int. Un. Theoret. Appl. Mech. (IUTAM), Enscheide, Netherlands, 1974.* Delft Univ. Press, Delft, The Netherlands.

Deryagin, B. V. (1937). *Z. Phys.* **88,** 661.

Deryagin, B. V. (1957). In discussion, *Proc. Conf. Frict. Wear, Inst. Mech. Eng.* pp. 767–768.

Dyson, A. (1975). Thermal stability of models of rough elastohydrodynamic systems, *J. Mech. Eng. Sci.* **18,** 11–18.

Earles, S. W. E., and Powell, A. E. (1968). Stability of self-generated oxide films on unlubricated EN1A steel surfaces, *Proc. Inst. Mech. Eng.* **182,** Pt. 31, 167–174.

Edgar, J. A. (1957). Control of wear in piston engines, *Proc. Conf. Lub. Wear* pp. 498–504. Inst. Mech. Eng., London.

Endo, K., and Kotani, S. (1973). Observation of steel surfaces under lubricated wear, *Wear* **26,** 239–281.

Ewing, Sir Alfred (1892). Notices, *Proc. R. Inst. G. B.* **13,** 387.

Eyre, T. S., and Baxter, A. (1972). The formation of white layers at rubbing surfaces, *Met. Mater.,* 435–439.

Farrell, R. H., and Eyre, T. S. (1970). The relationship between load and sliding distance in the initiation of mild wear in steels, *Wear* **15,** 359–372.

Fein, R. S., and Kreutz, K. L. (1972). Contribution to Discussion, Interdisciplinary Approach to Friction and Wear, NASA SP-181, pp. 358–376, Washington, D.C.

Ferrante, J. (1976). Exo-electron emission from a clean annealed, magnesium single crystal during oxygen adsorption, *Trans. ASLE Preprint Ann. Meeting, 31st, Philadelphia.*

Ferrante, J., and Smith, J. R. (1976). Metal interfaces, adhesive energies and electronic barriers, *Solid State Commun.* **20,** 393–396.

Fleming, J. R., and Suh, N. P. (1977a). Mechanics of crack propagation in delamination wear, *Wear* **44,** 39–56.

Fleming, J. R., and Suh, N. P. (1977b). The relationship between crack propagation rates and wear rates, *Wear* **44,** 57–64.

Fowles, P. E. (1969). The application of elastohydrodynamic lubrication theory and asperity–asperity collisions, *Trans. ASME J. Lub. Tech.* **91,** 464–476.

Fowles, P. E. (1971). A thermal elastohydrodynamic theory for individual asperity–asperity collisions, *Wear* **93,** 383–397.

Gane, N. (1970). The direct measurement of the strength of metals on a submicroscopic scale, *Proc. R. Soc. London Ser. A* **317,** 367–391.

Green, A. P. (1954). The plastic yielding of metal junction due to combined shear and pressure, *J. Mech. Phys. Solids* **2,** 197–211.

Green, A. P. (1955). Friction between unlubricated metals, a theoretical model of the junction model, *Proc. R. Soc. London Ser. A* **228**, 191–204.

Greenwood, F. A., and Williamson, J. B. P. (1966). Contact of nominally plane surfaces, *Proc. R. Soc. London Ser. A* **295**, 300–319.

Grunberg, L., and Scott, D. (1958). The acceleration of pitting failure by water in the lubricant, *J. Inst. Pet.* **44**, 406–410.

Grunberg, L., and Scott, D. (1960). The effect of additives on the water-induced pitting of ball bearings, *J. Inst. Pet.* **46**, 259–266.

Grunberg, L., and Wright, K. H. R. (1955) A study of the structure of abraded metal surfaces, *Proc. R. Soc. London Ser. A* **232**, 403–423.

Grunberg, L., Jamieson, D. T., and Scott, D. (1963). Hydrogen penetration in water accelerated fatigue of rolling surfaces, *Philos. Mag.* **8**, 1553–1568.

Halling, J. (1976). A contribution to the theory of friction and wear and the relation between them, *Proc. Inst. Mech. Eng.* **190**, 477–483.

Hertz, H. (1886). *J. Reinl. Angew. Math.* **92**, 156. [See also Timoshenko and Goodier (1934).]

Hironaka, S., Yahagi, Y., and Sakurai, T. (1975). Heats of adsorption and anti-wear properties of some surface active substances, *Bull. Jpn. Pet. Inst.* **17**, 201–205.

Hirst, W. (1957). Wear of unlubricated metals, *Proc. Conf. Lub. Wear* pp. 674–681. Institution of Mechanical Engineers, London.

Hirst, W., and Hollander, A. E. (1974). Surface finish and damage in sliding contact, *Proc. R. Soc. London Ser. A* **337**, 379–394.

Holm, R. (1946). "Electric Contacts." Hugo Gebers Forlag, Stockholm.

Holm, R., and Guldenpfennig, F. (1937). Uber die stoffwaunderun in elektrischen ausschalt-contaken, *Wiss. Veroeff. Siemens-Werk* **16/1**, 81.

Holm, R., Frith, H. P., and Guldenpfenning, F. (1941). Beitrag zur Kenntnis der reibung, *Wiss. Veroeff. Siemens-Werk* **20/1**, 68.

Hurricks, P. L. (1972). The fretting wear of mild steel from room temperature to 200°C, *Wear* **19**, 207–229.

Hurricks, P. L. (1974). The fretting of mild steel from 200°C to 500°C, *Wear* **30**, 189–212.

Israelachvili, J. N., and Tabor, D. (1972). The measurement of Van der Waals dispersion forces in the range 1.5 to 130, *Proc. R. Soc. London Ser. A* **331**, 19–38.

Jahanmir, S. and Suh, N. P. (1977a). Mechanics of subsurface void nucleation in delamination wear, *Wear* **44**, 17–38.

Jahanmir, S., and Suh, N. P. (1977b). Surface topography and integrity effects on sliding wear, *Wear* **44**, 87–99.

Jahanmir, S., Suh, N. P., and Abrahamson, E. P. (1974). Microscopic observation of the wear sheet formation by delamination, *Wear* **28**, 230–249.

Johnson, K. L., and Jefferies, S. A. (1963). Plastic flow and remedial stresses in rolling and sliding contact, *Proc. Symp. Fatigue Rolling Contact,* pp. 56–65. Institution of Mechanical Engineers, London.

Johnson, K. L., Kendall, K., and Roberts, A. D. (1971). Surface energy and the contact of elastic solids, *Proc. R. Soc. London Ser. A* **324**, 301–313.

Jones, M. H. (1970). Private communication.

Jones, M. H. (1976). Element concentration analysis of films generated on a phosphor bronze pin worn against steel under conditions of boundary lubrication, *ASLE Trans.* Preprint No. 76-LC-2B-3.

Jost, H. P. (1966). "Lubrication (Tribology) Education and Research." H. M. Stationery Office, London.

Kehl, E., and Siebel, E. (1936). Investigation of wear behaviours of metals during sliding friction, *Arch. Eisenhutt Wes.* **9**, 563.

Kerridge, M. (1955). Metal transfer and the wear process, *Proc. Phys. Soc.* **B68**, 400–407.

Kislik, V. A. (1962). The nature of white layers formed on friction surfaces, "Friction and Wear in Machinery" (translation) pp. 153–171. American Society of Mechanical Engineers, New York.

Koba, H., and Cook, N. H. (1974). "Wear Particle Formation Mechanics." MIT, Cambridge, Massachusetts.

Kohno, A., and Hyodo, S. (1974). The effect of surface energy on the micro adhesion between hard solids, *J. Phys. Appl. Phys.* **7**, 1243–1246.

Kragelskii, I. V. (1965). "Friction and Wear." Butterworth, London.

Kragelskii, I. V. (1973). "Trenie i Iznos v Vakume" (in Russian). Mashnostroenie, Moscow.

Kragelskii, I. V., and Aleksev, N. M. (1976). On the calculation of seizure considering the plastic flow of the superficial layers, *Trans. ASME J. Lub. Tech.* **98**, 133–138.

Kragelskii, I. V., and Demkin, N. B. (1960). Contact area of rough surfaces, *Wear* **3**, 170–187.

Kruschev, M. H. (1967). Resistance of metal to wear by abrasion as related to hardness, *Proc. Conf. Lub. Wear* pp. 655–659. Institution of Mechanical Engineers, London.

Kruschev, M. M., and Babichev, M. A. (1954). Investigation of resistance of steels to wear by rubbing on abrasive surface, *in* "Collected Papers IX. Trenie i Iznos v Mash." (in Russian). Inst. Mash. Akad. Nauk SSR.

Kubashewski, O., and Hopkins, B. E. (1967). "Oxidation of Metals and Alloys." Butterworth, London.

Lancaster, J. K. (1963). The relationship between the wear of carbon brush materials and their elastic modulus, *Proc. J. Appl. Phys.* **14**, 497.

Larsen-Badse, J. (1966). The abrasive resistance of some hardened and tempered carbon steels, *Trans. Metall. Soc. AIME* **236**, 1461–1466.

Larsen-Badse, J., and Mathew, K. G. (1969). Influence of structure on the abrasion resistance of A1040 Steel, *Wear* **14**, 199–206.

Ling, F. E., Klaus, E. E., and Fein, R. S. (eds.) (1969). "Boundary lubrication—an appraisal of world literature." American Society of Mechanical Engineers, New York.

Ludema, K. C., and Tabor, D. (1966). The friction and visco-elastic properties of polymeric solids, *Wear* **9**, 329–348.

Maddock, I. (1972). The scope and nature of terotechnology, *Symp. Terotechnol.* pp. 1–3. Institution of Mechanical Engineers, London.

Mailander, R., and Dies, K. (1963). Beitrag zur enforschung der vorgänge beim verschleigs, *Arch. Eisenhuttenw.* **16**, 385.

Makagama, I., and Sakurai, T. (1974). The effect of surface temperature on chemical wear, *Wear* **29**, 273–389.

Makinson, K. R., and Tabor, D. (1964). The friction and transfer of polytetrafluoroethylene, *Proc. R. Soc. London Ser. A* **281**, 49–61.

Maugis, D., Desalos-Andarelli, G., Heurtel, A., and Courtel, R. (1976). Adhesion and Friction on A1 Thin foils Related to Observed Dislocation Density. ASLE Paper No. 76-LC-2B-2, Boston, Massachusetts, October.

Merritt, H. E. (1971) "Gear Engineering." Pitmans, London.

Milne, A. A., Scott, D., and Macdonald, D. (1957). Some studies of scuffing with crossed-cylinder machines, *Proc. Inst. Mech. Eng. Conf. Lub. Wear* pp. 735–741.

Mindlin, R. D. (1949). Compliance of Elastic Bodies, *Trans. ASME J. Appl. Mech.* **72**, 259–268.

Mizuno, M. (1967). Macroscopic analysis of wear process, *Bull. Jpn. Soc. Precision Eng.* **2**, 201–213.

Mizuno, M. (1973). Experimented study of wear using an electron probe x-ray microanalyzer, *J. Jpn. Soc. Lub. Eng.* 305–314.

Moore, A. W. J., and Tegart, W. J. McG. (1952). Effects of included oxide films on the structure of the Beilby layers, *Proc. R. Soc. London Ser. A* **212**, 458–459.

Morrison, J. L. M., and Crossland, B. (1964). "An Introduction to the Mechanics of Machines." Longman, London.

Onions, R. A., and Archard, J. F. (1974). Pitting of gears and discs, *Proc. Inst. Mech. Eng.* **22**, 297–322.

Pollock, H. M., Shufflebottom, P., and Skinner, S. (1977). Contact adhesion between solids in vacuum: I Single-asperity experiments, *J. Phys. D. Appl. Phys.* **10**, 127–138.

Poritsky, H. (1950). Stresses and deflection of cylindrical bodies in contact with applications to the contact of gears and locomotive wheels, *Trans. ASME J. Appl. Mech.* **72**, 191–201.

Quinn, T. F. J. (1963). The role of oxidation in the mild wear of steel, *Br. J. Appl. Phys.* **13**, 33–37.

Quinn, T. F. J. (1967). The effect of hot spot temperature on the unlubricated wear of steel, *ASLE Trans.* **10**, 158–168.

Rabinowicz, E. (1965). "Friction and Wear of Materials." Wiley, New York.

Reda, A. A., Bowen, R., and Westcott, V. C. (1975). Characteristics of the particles generated at the interface between sliding surfaces, *Wear* **34**, 261–273.

Rennie, G. (1829). *Philos. Trans. R. Soc.* **134**, 143.

Richardson, R. C. D. (1967a). Laboratory simulation of abrasive wear, such as that imposed by soil, *Proc. Inst. Mech. Eng.* **182**, Part 3A, 29–31.

Richardson, R. C. D. (1967b). The abrasive wear of metals. *Proc. Inst. Mech. Eng.* **182**, Part 3A, 410–414.

Rosenberg, S. L., and Jordan, L. (1935). *Trans. Am. Soc. Met.* **28**, 577.

Rozeanu, L. (1973). A model for seizure, *Trans. ASLE* **16**, 115–118.

Ruff, A. W. (1976). Study of Initial Stages of Wear by Electron Channelling, Nat. Bur. Std. Rep. NBS-1R76-1141, Washington, D.C.

Saka, H., Eleiche, A. M., and Suh, N. P. (1976). Delamination wear of dispersion hardened alloys, *Trans. ASME Product. Eng. Div.* Winter Ann. Meeting, December.

Saka, H., Eleiche, A. M., and Suh, N. P. (1977). Wear of materials at high sliding speeds, *Wear* **44**, 109–125.

Sasada, T., and Kando, H. (1973). Formation of wear particles by the mutual transfer and wear process, *Proc. Jpn. Congr. Mater. Res., 16th* pp. 32–35. Society of Materials Science.

Sasada, T., and Kimura, Y. (1976). Behaviour of hot spots on sliding surfaces, *Proc. Jpn. Congr. Mater. Res., 19th.* Society of Materials Science.

Sasada, T., Ohmura, H., and Norose, S. (1972). The wear and mutual transfer in Cu/Fe rubbing, *Jpn. Congr. Mater. Res., 15th* pp. 1–6. Society of Materials Science.

Sasada, T., Norose, S., and Sugimoto, K. (1974). The mutual metallic transfer at friction surfaces in lubricating oil, *Proc. Jpn. Congr. Mater. Res., 17th* pp. 32–35. Society of Materials Science.

Sasada, T., Norose, S. and Shimura, Y. (1975a). Composition of wear particles produced under rotary friction of different metal combinations, *Proc. Jpn. Congr. Mater. Res., 18th* pp. 77–81. Society of Materials Science.

Sasada, T., Norose, S., and Shimura, Y. (1975b). The relation between "wear-distance" and "wear velocity" curves, *Proc. Jpn. Congr. Mater. Res., 18th* pp. 82–86.

Sasada, T., Norose, S., and Thiam (1975c). Birth and growth of a wear particle observed through the relative transversal movement of rubbing surfaces, *Proc. Jpn. Congr. Mater. Res., 18th* pp. 72–76. Society of Materials Science.

Sasada, T., Norose, S., and Morohasi, H. (1976). Adhesive wear in metal/polymer rubbing, *Proc. Jpn. Congr. Mater. Res., 19th.* Society of Materials Science.

Seibel, E., and Kobitzsch, R. (1942). *Z. Ver. Deutsch. Ing.* **84,** 157.

Skinner, J., and Pollock, H. M. (1977). Contact adhesion between surfaces in vacuum and deformation and surface energy, *J. Phys. D. Appl. Phys.* (in press).

Smith, F. W. (1962). The effect of temperature on concentrated contact lubrication, *Trans. ASLE* **5,** 142–148.

Smith, J. O., and Lui, C. K. S. (1953). Stresses due to tangential and normal loads on an elastic solid with some application to some contact stress problems, *Trans. ASME J. Appl. Mech.* **75,** 157–166.

Soda, N. (1975). On the correspondence of the amount of wear to the frictional work, *J. Jpn. Soc. Lub. Eng.* **20,** 360–370.

Soda, N., and Sasada, T. (1964). Studies in adhesive wear, effect of gas-absorbed films on metallic wear (abstract).

Suh, N. P. (1973). The delamination theory of wear, *Wear* **25,** 17–24.

Suh, N. P. (1977). An overview of the delamination theory of wear, *Wear* **44,** 1–16.

Tabor, D. (1959). Junction growth in metallic friction, the role of combined stresses and surface contaminants, *Proc. R. Soc. London Ser. A* **25,** 378–392.

Tabor, D. (1970). Hardness of Solids, *Rev. Phys. Tech.* **1,** 145–179.

Tabor, D., and Winterton, R. H. S. (1969). The direct measurement of normal and retarded van der Waals forces, *Proc. R. Soc. London Ser. A* **312,** 435–450.

Tanaka, K. (1975). A review of recent studies on polymer friction and wear in Japan, *Proc. Int. Solid Lub. Symp., Tokyo.*

Tanaka, K. Uchiyama, Y., and Toyooka, S. (1973a). The mechanism of wear of polytetra-fluoroethylene, *Wear* **23,** 153–177.

Tanaka, K., Ueda, S., and Noguchi, N. (1973b). Fundamental studies on the break friction of resin-based friction materials *Wear* **23,** 349–365.

Tanaka, K., Uchiyama, Y., Ueda, S., and Shimzizu, T. (1975). Friction and wear of PTFE-based composites, *J.S.L.E–A.S.L.E. Int. Lub. Conf.*

Timoshenko, S. P., and Goodier, J. N. (1934). "Theory of Elasticity." McGraw-Hill, New York.

Tomlinson, G. A. (1927). The rusting of steel surfaces in contact, *Proc. R. Soc. A* **115,** 472–483.

Tomlinson, G. A. (1929). A Molecular Theory of Friction, *Philos. Mag.* **7,** 905–939.

Tsukizoe, T., and Hisakada, T. (1965). On the mechanism of contact between metal surfaces, *Trans. Am. Soc. Mech. Eng.* **87,** 666–674.

Tsuya, Y. (1976). Microstructure of Wear, Friction and Solid Lubrication, Tech. Rep. of Mech. Eng. Lab. No. 81.

Vainshtein, V. E., and Prondzinskii, (1962). Use of radioactive traces for evaluation of the ability of bearing materials to absorb abrasive particles from the lubricant, "Friction and Wear in Machinery" (translation), pp. 35–44. American Society of Mechanical Engineers, New York.

Vernon, W. H. J., *et al.* (1953). *Proc. R. Soc.* **A216,** 375–397.

Waterhouse, R. B., and Taylor, V. E. (1974). Fretting debris and the delamination theory of wear, *Wear* **29,** 337–346.

Wellauer, E. S. (1967). A.G.M.A experience in establishing co-ordinated gear rating standards, *Semi-Int. Symp. Jpn. Soc. Mech. Eng.*

Welsh, N. C. (1965). The dry wear of steel, *Philos. Trans. R. Soc. A* **257**, 31–72.

Whitehouse, D. J., and Archard, J. F. (1970). *Proc. R. Soc.* **A316**, 97–121.

Williams, C. G. (1949). "Collected research on cylinder wear." Institute of Automobile Engineers, London.

Wright, K. H. R. (1952). An investigation of fretting corrosion, *Proc. Inst. Mech. Eng.* **1B**, 556–563.

Wright, K. H. R. (1957). Fretting corrosion of cast iron, *Proc. Conf. Lub. Wear* pp. 623–634, Institution of Mechanical Engineers, London.

TREATISE ON MATERIALS SCIENCE AND TECHNOLOGY, VOL. 13

The Wear of Polymers

D. C. EVANS† and J. K. LANCASTER

Ministry of Defence, Procurement Executive
Materials Department
Royal Aircraft Establishment
Farnborough, Hants, England

† Present address: The Glacier Metal Company Limited, Alperton, Wembley, Middlesex, England.

I. Introduction

During the past decade there has been a continuous increase in the utilization of polymers and polymer-based composites in a wide variety of tribological applications. These include such diverse components as piston rings, seals, brakes, prosthetic joints, gears, tires, and dry bearings, and the materials involved cover a wide spectrum of mechanical properties from elastomers to hard reinforced thermosetting resins. This chapter is intended to describe, in general terms, the different types of wear behavior exhibited by this range of materials in different conditions of sliding, but with particular emphasis on the more rigid thermoplastic and thermosetting materials. Much of the information presented has been derived from laboratory work rather than from evaluation in service conditions because only in the laboratory is it possible to restrict and control the many variables that play an important part in wear.

There have been many attempts to classify wear, based either on the phenomena observed (e.g., Archard and Hirst, 1956) or on the causative agents believed to be responsible (e.g., Burwell, 1957). For present purposes, it is most convenient to adopt the latter course and to divide the wear processes for polymers into four main groups—abrasion, adhesion, fatigue, and thermal/oxidative degradation. Definitions of these processes are given elsewhere (OECD, 1969). There are, however, two difficulties which arise from such a formal classification. The first is that for any one process, the detailed mechanisms of surface failure leading to the removal of debris depend significantly on the mechanical properties of the material. For example, abrasive wear against rough surfaces may occur via tensile tearing for elastomers, localized fatigue for more rigid polymers, or cutting and shear for very high modulus polymer composites

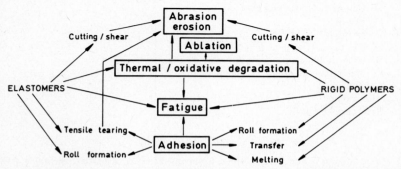

Fig. 1. Summary of wear processes for polymers. (Crown copyright. Reproduced by permission of the Controller, Her Britannic Majesty's Stationery Office.)

(Ratner and Farberova, 1960). The second is that a very considerable degree of interrelationship exists among the different processes that lead to failure. For example, adhesion influences the stress distribution around a localized contact and may thus influence fatigue; alternatively, thermal/oxidative degradation may change the mechanical properties of the surface layer and so influence subsequent abrasion or fatigue. An attempt to summarize the various failure mechanisms that can occur, together with the interrelationships among the conventional wear processes, is given in Fig. 1. The extent to which this is valid will emerge during the subsequent discussion.

II. Materials

There are several ways in which to classify the widely differing types of polymers currently available (Billmeyer, 1971; Brydson, 1975). The most conventional is to describe them as thermoplastic, thermosetting, or elastomeric. The generic term *plastic* usually refers to a thermoplastic or a thermosetting polymer containing one or more of a variety of additives. The materials most widely used in tribological applications are listed in Table I but comprise only a minute fraction of the many thousands of known polymers.

A. Thermoplastic Polymers

These materials are characterized by their ability to soften or melt at elevated temperatures. There are two general types, amorphous and crystalline, but neither of these descriptions is precise. In crystalline polymers, the degree of crystallinity is invariably less than 100% and often less than 50%, while in amorphous polymers there are often small-scale regions (domains) of ordered molecules (Geil, 1975). Amorphous thermoplastics are, at room temperature, below their glass transition temperature T_g and are thus relatively hard and brittle, i.e., glasslike. Crystalline thermoplastics consist of regions of crystallinity embedded in an amorphous matrix, which may be either above or below its T_g at room temperature, and it is this particular structure that is primarily responsible for their relatively high toughness and strength (Sharples, 1972). The majority of the thermoplastics listed in Table I are crystalline. Of these, the polyamides, polyacetals, and polyethylenes are well-established engineering polymers, but uses of the polyethylenes tend to be limited by their relatively low softening temperatures. PTFE is of major importance in tri-

TABLE I

POLYMERS, FILLERS, AND REINFORCEMENTS OF INTEREST FOR
TRIBOLOGICAL APPLICATIONS

Thermoplastic	Polyethylenes	High density polyethylene (HDPE), ultra-high-molecular-weight polyethylene (UHMWPE)
	Polyamides	Nylon 6, 6.6, 6.10, 11, 12
	Polyacetals	Homopolymer, copolymer
	Fluorocarbons	Polytetrafluoroethylene (PTFE) Copolymers of TFE with: hexafluoropropylene (FEP), ethylene (E-TFE), vinylidene fluoride (VF_2-TFE), fluorinated propylvinylether (PFA)
	High-temperature polymers	Poly(phenylene sulphide) (PPS), poly(ether sulphone) (PES), poly(p-hydroxy benzoic acid) (HBA)
Thermosetting		Phenolic, cresylic, polyester, epoxy, silicone, polyimide
Elastomeric		Natural rubber, styrene-butadiene (SBR), butadiene-acrylonitrile (nitrile), polyacrylate, fluorocarbons
Fillers and reinforcements	To improve mechanical properties	Asbestos, carbon, glass, textile, aromatic polyamide fibres, mica, metal oxides, carbon black, silica
	To reduce friction	Graphite, MoS_2, PTFE, mineral oils, silicones, fatty acids/amides
	To improve thermal conductivity	Bronze, carbon, silver

bology, as a matrix material suitably reinforced, as an additive to other polymers to reduce friction, as a thin film solid lubricant, or as a grease additive. The other thermoplastics are of more recent interest; the various fluorocarbons are more easily processable than PTFE, thermoplastic polyesters are very hard and dimensionally stable, while PPS, PES, and HBA possess high temperature stability (H. W. Hill *et al.*, 1974; Leslie *et al.*, 1975; Economy and Cottis, 1971).

B. Thermosetting Polymers

These materials undergo cross linking on curing to form infusible, three-dimensional structures which are relatively hard and brittle. They normally decompose on heating, without significant softening or melting. Of the materials listed in Table I, phenolic and cresylic resins are of most general application and are invariably filled or reinforced. Epoxies tend to be more expensive, but can be "flexibilized" and used in filled rather than in reinforced form (Pratt, 1967b). The outstanding characteristic of the silicone resins is their high temperature stability. The polyimides are normally classed as thermosetting because, although they are not necessarily cross linked, their T_gs are generally higher than their decomposition temperatures. Despite their relatively high cost, they are now becoming widely used for specialized applications at high temperatures (Sliney and Jacobson, 1975) and as binders for solid lubricants (Campbell and Hopkins, 1967).

C. Elastomeric Polymers

The characteristic feature of elastomers is their ability to undergo large reversible deformations. There are two basic requirements for this; the polymer must be above its T_g at the temperature of interest and must also be cross linked. A wide range of different elastomers is in common use for tribological applications. Natural rubber (essentially cis-1,4-polyisoprene) and styrene–butadiene rubber are both used in pneumatic tires, nitrile (butadiene–acrylonitrile), polyacrylate, and fluorocarbon rubbers are used for seals, and polyurethanes for abrasion-resistant coatings. Unfortunately, simple discussion of elastomers in the same terms as thermoplastics and thermosets is largely precluded because their properties are so highly dependent on compounding ingredients and the degree of cross linking induced during curing.

D. Properties Relevant to Tribology

In addition to the magnitudes of the coefficient of friction and the rate of wear, a wide range of other physical and mechanical properties of polymers is relevant to their applications in tribology. These include ultimate strength (compressive and tensile), elastic modulus, creep resistance, thermal conductivity and expansion coefficient, and environmental stability (to heat, radiation, vacuum, gases, fluids, etc.). Additives are frequently introduced into polymers to modify one or more of these properties and some of the most widely used are listed in Table I, together with their intended functions. It should be realized, however, that the action of any additive is seldom restricted solely to modifying one particular property. For example, graphite added to polymers to reduce friction may also improve mechanical properties, but can result in greatly increased wear in vacuum or inert gases. An increase in wear, and friction, may also occur when glass fibers are added to polymers in order to increase strength and stiffness. The selection of suitable additives for particular applications is therefore almost invariably a compromise.

In many applications, polymers are used, or are intended to be used, as direct replacements for lubricated metal components. Detailed information is therefore required concerning their relevant physical and mechanical properties. Such information for most of the polymers listed in Table I is readily available in the appropriate handbooks (e.g., Brydson, 1975) and data have also been published on many filled and reinforced products (e.g., O'Rourke, 1962; Theberge, 1971; ESDU, 1976). Rather than attempting to summarize these data here, it is probably more instructive to itemize the ways in which some of the physical and mechanical properties of polymers compare, in a general way, with those of metals.

(a) Polymers are viscoelastic and much more susceptible to creep than metals.

(b) Ultimate strengths and moduli are appreciably lower, typically by a factor of 10.

(c) Thermal expansion coefficients are relatively high, typically ten times greater than that of steel, and this may introduce problems of dimensional stability.

(d) Some polymers readily absorb fluids, including water from the environment (e.g., nylons), which may also affect dimensional stability.

(e) Thermal conductivities are very low, of the order of one-hundredth of that of steel, and the dissipation of frictional heat is therefore poor.

(f) Limiting temperatures associated with softening, melting, oxidation, thermal degradation are all relatively low ($<300°C$).

All these features are limitations to the use of polymers in tribology, but they are offset by a number of particular advantages.

(a) Physical and mechanical properties of polymers can be varied over a wide range by suitable choice of polymer type and/or fillers and reinforcements.

(b) Some materials, notably thermoplastics, are cheap and easy to fabricate into complex shapes.

(c) Many polymers, particularly fluorocarbons, are highly resistant to chemical attack by aggressive media, such as acids and alkalis.

(d) Coefficients of friction during unlubricated sliding against either themselves or metals are relatively low and typically within the range 0.1–0.4.

(e) Wear rates during sliding against smooth metal counterfaces are also relatively low, and polymers do not normally exhibit scuffing or seizure.

(f) Periodic maintenance, i.e., lubrication by fluids, can often be dispensed with.

(g) When fluid lubricants are present, polymers undergo elastohydrodynamic lubrication more readily than metals.

The significance of some of these aspects will be discussed in more detail later.

III. Friction

The mechanisms of both friction and wear of polymers have many features in common, and it is therefore pertinent at this stage to describe briefly some of the main factors affecting friction. More complete reviews have been given by Lancaster (1972d) and Tabor (1974). The two principal components contributing to friction arise from adhesion and deformation at the localized areas of contact.

A. Adhesion

As with metals, the adhesive component of the frictional force is given by $F = A_r s$, where A_r is the area of real contact and s is the shear strength of the junctions resulting from adhesive interactions over this area. For polymers, these interactions are considered to arise mainly from Van der Waals forces (Israelachvili and Tabor, 1973; Lee, 1974), possibly supplemented by electrostatic forces (Tabor, 1974). Determination of the real area of contact for realistic surfaces of polymers in contact with themselves or metals is fraught with difficulty. The simplest assumption to make is that localized plastic deformation occurs, in which case $A_r = L/P_m$, where L is the total load and P_m is the flow pressure of the polymer. Hence, the coefficient of friction $\mu = s/P_m$. However, as shown later, most polymer surfaces usually undergo localized elastic rather than plastic deformation. In these circumstances it is only possible to derive the magnitude of the real area of contact from either calculation or experiments with simplified model surface topographies (Archard, 1953; Kraghelskii, 1965) or from calculations using known topographical parameters of real surfaces (Thomas, 1975).

The shear strength of adhesive junctions is conventionally assumed to be equivalent to the bulk strength of the polymer, and this implies that cohesive failure occurs during sliding, leading to transfer. A direct demonstration of transfer on an atomic level has been given by Buckley and Brainard (1974) using field ion microscopy of a tungsten filament in contact with PTFE and polyimide. On the more macroscopic level, there have been numerous observations of transfer of polymers to metals or glass surfaces during sliding (Shooter and Tabor, 1952; Eiss *et al.*, 1978; Rhee and Ludema, 1978; Tanaka and Miyata, 1977). Cohesive (subsurface) failure and transfer are most common for crystalline thermoplastics (Pooley and Tabor, 1972). For amorphous thermoplastics, thermosets, and elastomers, however, failure is more usually adhesive, i.e., at the sliding interface itself, and transfer does not occur directly (Briscoe, 1977). With the more rigid polymers, it is not possible to measure the strengths of adhesion directly from experiments involving normal loading and unloading because the release of elastic stresses during unloading causes the surfaces to peel apart (Merchant, 1968). Such measurements are possible, however, between relatively smooth surfaces of very low modulus materials (Roberts, 1976) where the adhesion exhibited resembles the phenomenon of "tack" (Counsell, 1973). In these circumstances, the force required to separate the surfaces can be calculated in terms of surface energetics (Johnson *et al.*, 1971; Roberts and Thomas, 1975).

B. Deformation and Viscoelasticity

When a hard, rigid asperity moves over the surface of a polymer, energy can be dissipated by either elastic or plastic deformation. The simplest situation for elastic deformation arises during the rolling of a rigid sphere over a polymer plane. In these conditions adhesive effects are usually small, except for some very low modulus elastomers (Kendall, 1975; Briggs and Briscoe, 1976), and the energy dissipation can be calculated in terms of the hysteresis losses occurring during each deformation cycle (Tabor, 1955). The values obtained are generally commensurate with the measured values of friction (Greenwood *et al.*, 1961). In addition, close correlations have been observed between the variations of rolling friction with speed and temperature and the corresponding variations in mechanical losses (Flom, 1960; Ludema and Tabor, 1966).

When sliding occurs, it is no longer possible to identify unambiguously the separate contributions of adhesion and deformation to friction; both processes appear to be influenced by the viscoelastic properties of polymers in a similar way. For elastomers, Grosch (1963) has shown that the variations in the coefficient of friction with speed and temperature during sliding over abrasive paper and glass can all be reduced to single master curves with the aid of viscoelastic transforms (Williams *et al.*, 1955). Differences between these master curves in dry and lubricated sliding are attributed to the extra contributions to dry friction arising from adhesive interactions and localized surface failures (wear). Similar attempts to derive master curves, relating the friction of the more rigid polymers to speed and temperature, have been less successful (Bueche and Flom, 1958; MacLaren and Tabor, 1963; Ludema and Tabor, 1966), although Bahadur and Ludema (1971) have obtained some master curves for polyethylene and polypropylene. These appear to correlate reasonably well with similar master curves obtained from the speed and temperature variations of shear strengths and moduli, and it was therefore suggested that the major component of friction resulted from adhesion, rather than deformation. Significant contributions from deformation are most likely to occur only at very heavy loads (Bahadur, 1974).

C. PTFE

The extremely low coefficient of friction of PTFE under certain conditions of sliding has prompted numerous investigations into its cause. The lowest values, of the order of 0.05 or less, are only obtained at high loads, low speeds, and moderate temperatures, and examples of these variations are shown in Fig. 2. The original explanation for the low friction of PTFE, first suggested by Shooter and Thomas (1949), was that the charge on the carbon atoms is screened by the relatively large fluorine atoms (Hanford and Joyce, 1946), thus reducing interchain forces and enabling the polymer chains to slip easily over each other. Subsequently, attempts were made to explain the low friction in terms of the bulk, rather than the molecular structure (Makinson and Tabor, 1964). The individual units in the bulk structure of PTFE are thin crystalline bands of oriented chains, separated by disordered or amorphous regions (Bunn *et al.*, 1958; Speerschneider and Li, 1962a). At low speeds, easy slip is postulated within the disordered regions, leading to the drawing out of thin oriented films from the crystalline bands. At higher speeds, the reduced time of contact becomes commensurate with the relaxation times of the molecules in the disordered regions; slip is then more difficult, friction increases, and larger and more irregular particles are detached from the surface. More recently, Pooley and Tabor (1972) have suggested a return to the original molecular ideas and now consider that the low friction of PTFE results primarily from the smooth molecular profile of the chains. High-density polyethylene also has a smooth molecular profile and exhibits generally similar frictional behavior to PTFE, although with somewhat

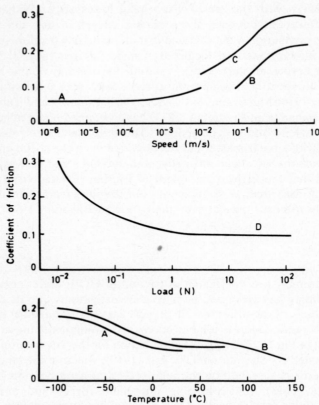

Fig. 2. Effect of sliding conditions on the coefficient of friction of PTFE. A (from Ludema and Tabor, 1966); B (from MacLaren and Tabor, 1963); C (from Tanaka *et al.*, 1973); D (from Shooter and Tabor, 1952); E (from King and Tabor, 1953).

higher coefficients of friction. In contrast, FEP and low-density polyethylene, which do not possess these smooth profiles, behave very differently.

IV.   Wear Testing

The ultimate technological objective of wear testing is to obtain data on the performance and reliability of specific components in service conditions. For this purpose, there is no wholly satisfactory alternative to full-scale testing of the component itself. Of more concern in the present context, however, are two intermediate objectives, which can be satisfied by the use of relatively simple laboratory equipment operating in conditions more clearly defined and controlled than those likely to be encountered in service: (a) to rate materials for wear resistance in conditions that simulate the essential features of an application; and (b) to obtain basic information about wear mechanisms as an aid to failure diagnosis or to the development of new materials.

In view of the wide variety of wear mechanisms that exist, it is evident that for both of the above purposes a range of different types of apparatus is required. For rating of materials against abrasive wear, there has been some degree of standardization of equipment and procedures via ASTM standards and details of the types of apparatus involved are given by Gavan (1967) and Harper (1961). Standardization of wear testing in nonabrasive conditions, however, is less well advanced.

It is seldom possible to obtain basic information about wear processes from standardized test apparatus and procedures. For this purpose, more versatile laboratory equipment is needed with which the major parameters influencing wear, such as load, speed, temperature, etc., can be defined accurately and varied over a wide range. There are two general types of such apparatus, involving conformal and nonconformal geometries, as illustrated in Fig. 3. For most purposes, the stationary member is a polymer, and the moving member is metal, except for abrasive wear conditions, in which the metal can be replaced by abrasive paper or a bonded-abrasive disk. Apparatus of the nonconformal type is often referred to as "pin-on-ring" (Fig. 3a) or "pin-on-disk" (Fig. 3b), and a major advantage is the ease with which very small amounts of wear can be determined via measurements of the wear scar diameters developed on suitably shaped pins. This method of wear measurement becomes increasingly inaccurate, however, when the elastic modulus of the polymer falls below about 300 MN/m^2 (Lancaster, 1969a). With conformal sliding arrangements, Fig. 3c, d, wear is usually measured from dimensional changes or weight loss. These techniques are relatively insensitive and problems can arise from temperature variations or moisture absorption. One way of overcoming this is to use a reference sample of the polymer that is always exposed to the same environmental conditions as the polymer being worn (Dowson et al., 1974). A more sensitive technique could be to use isotopic tracer labeled polymers, either stable or radioactive, but except in special cases, e.g., Cl-activated polychlorotrifluoroethylene (Eiss et al., 1976), this would involve the use of specially prepared materials.

A major difference between wear tests with conformal and nonconformal geometries is that in the former debris tends to be trapped within

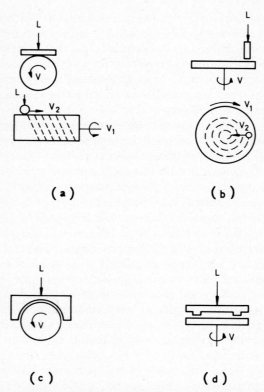

Fig. 3. Schematic arrangements of wear testing apparatus: (a,b) nonconformal; (c,d) conformal. (After Lancaster, 1972d. Crown copyright. Reproduced by permission of the Controller, Her Britannic Majesty's Stationery Office.)

the apparent contact area and can thus potentially influence the subsequent wear behavior, e.g., by abrasion from filler particles or by facilitating transfer film formation on one or both of the rubbing surfaces. With nonconformal contacts it is usually assumed that debris effects on wear are largely unimportant, but recent work suggests that this is not always correct. Play and Godet (1978) have shown from model experiments with chalk sliding on glass that debris particles aggregating in a small area near the trailing edge of a pin can significantly reduce the overall rate of wear. Lancaster (1975) has demonstrated that for carbons sliding on metals transfer of carbon to the metal from the leading edge of a pin modifies the wear subsequently occurring on the trailing edge. There is thus a small but significant dependence of the rate of wear on the apparent contact area of the pin and on its shape. The relevance of these effects to the wear of polymers, which are considerably more ductile than either chalk or carbons, remains to be examined, but somewhat similar effects of specimen size and shape on wear have already been reported for resin-bonded asbestos brake materials (Weintraub et al., 1974).

There are a number of ways in which the rate of wear of a polymer can be specified. In many situations it is observed that, following an initial period of "running in," the mass or volume of wear v becomes directly proportional to the time t or distance of sliding s, directly proportional to the applied load L, and almost independent of the apparent area of contact. It is thus possible to define a *specific wear rate* or *wear factor* $K = v/sL$. The units are m^2/N, but a greater degree of physical significance becomes apparent if these are expanded to mm^3/N m. For design purposes, where the depth of wear d, the mean pressure over the nominal contact area P, the speed V, and the time of operation are the major parameters of concern, $K = d/PVt$. The concept of a specific wear rate presupposes that variations in load (or pressure), speed, and time (or distance) of sliding are not accompanied by significant changes in any other variables that affect wear, such as temperature or the topography of the sliding surfaces. If such changes should occur, it is then only possible to relate the volume of wear to the operating conditions via empirical equations (Kar and Bahadur, 1974), and such equations have little or no fundamental validity. Two other criteria of wear should also receive brief mention. For friction (brake) materials, an *energy index* is widely used and is defined as the volume of wear per unit of work done during sliding, $K_i = v/\mu Ls = K/\mu$, where μ is the coefficient of friction and K is the specific wear rate as defined above. In the USSR, wear is often expressed as a dimensionless *wear intensity i*, defined as the volume of wear per unit distance of sliding per unit of real contact area, $i = v/A_r s$ (Kraghelskii, 1965). This has the merit of implicitly taking into account the mechanical properties of the materials and their topographical features since both of these parameters influence the magnitude of the real contact area A_r. However, as discussed previously, there is no convenient way of measuring the real contact area, and theoretical estimates are invariably required.

V. Abrasion

For rigid polymers, the simplest physical picture of abrasive wear is one in which hard surface asperities penetrate the polymer and remove material by shearing or cutting. This leads to the relationship $v/s = k(L/H) \times \tan \theta$, where H is the indentation hardness, θ the base angle of the indenting asperity, and the constant k expresses the fact that only a proportion of the material undergoing deformation appears as loose wear debris. An essential requirement for this type of wear is that deformation of the polymer be plastic rather than elastic, and two criteria are available to define this condition. Halliday (1955) suggests that the critical values of θ are given by tan $\theta = C(H/E)(1 - \nu^2)$, where ν is Poisson's ratio and the constant $C = 0.8$ for the onset of plasticity and $C = 2$ for full plasticity. Greenwood and Williamson (1966) introduce the concept of a "plasticity index" $(E/H)(\sigma/\beta)^{1/2}$, where σ is the standard deviation of the asperity heights and β is their average radius of curvature; plasticity begins when this ratio is equal to unity.

The onset of abrasive (cutting) wear in rigid polymers can best be illustrated by a model experiment involving conical indenters of various angles sliding over smooth, softer surfaces. Figure 4 shows how the vol-

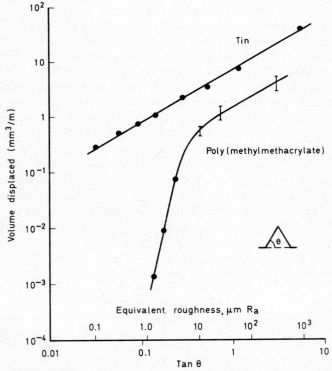

Fig. 4. Relationship between the volume of material displaced per unit sliding distance and the base angle of a conical indenter. (After Lancaster, 1969b. Crown copyright. Reproduced by permission of the Controller, Her Britannic Majesty's Stationery Office.)

ume of the groove produced per unit sliding distance varies with the base angle of the indenter for poly(methyl methacrylate) and a soft metal. With the latter, plastic deformation occurs at all angles and the relationship obtained is in agreement with the simple theory outlined above. For the polymer, however, there is a linear relationship between volume and tan θ only when θ exceeds about 30°. Approximate values of the equivalent surface roughness of abraded metal surfaces, corresponding to the different cone angles, are also shown on the abscissa of Fig. 4 and it is evident from these that plastic deformation and cutting of polymers are likely to occur only during sliding against surfaces of the order of 12 μm R_a. Such roughnesses are unlikely to be encountered on metals, but they are typical of the coarser abrasive papers; estimates of the R_a roughness of SiC paper range from 6 μm for 600 grade to 38 μm for 100 grade.

A. Wear against Abrasive Papers

Most investigations into the wear of polymers on abrasive papers have been made by Russian workers. Ratner (1967) has concluded that the wear rate of a polymer increases with increasing grit size, but that the relative wear resistance of different polymers is independent of grit size. In agreement with the theoretical predictions above, wear rates of polymers on abrasive papers are usually proportional to the applied load (Ratner and Farberova, 1960). Ratner et al. (1964) have also developed a simple theory of abrasive wear based on general considerations. They assume that there are three stages involved in the production of a wear particle: plastic deformation to give a real contact area of size inversely proportional to the hardness H, relative motion opposed by a frictional force $F = \mu L$, and disruption of material involving work equivalent to the area under the stress–strain curve at fracture. An approximate measure of the latter is the product of the breaking strength and the elongation at break se. The total wear is thus regarded as being proportional to the probability of completion of each stage, i.e., $v \propto \mu L / Hse$. A convenient way of assessing the validity of this relationship is to examine the variation of these parameters with temperature. Figure 5 shows the general way in which μ, H, s, and e change with temperature for an amorphous polymer, and, in combination, these variations predict a minimum in the wear rate at around the glass transition temperature of the polymer. Experimental verification does not appear to have been sought for wear against abrasive papers, but minima in wear have been observed during sliding against rough metal surfaces (Lancaster, 1969a). In the latter conditions, the product se appears to become the most important material parameter in-

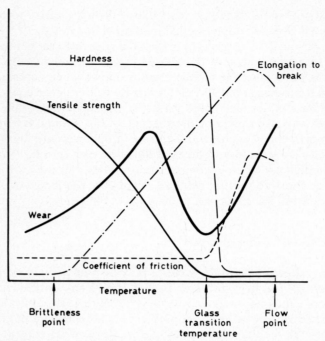

Fig. 5. Schematic variation of hardness, tensile strength, friction, and elongation to break with temperature for an amorphous polymer, and the resulting variation in wear. (After Ratner *et al.*, 1964.)

fluencing wear. The results shown in Fig. 6 demonstrate that there is a highly significant correlation between the wear rates of 18 polymers at room temperature and their individual values of *se,* as determined in conventional tensile tests at low strain rates. The strain rates involved in the wear process are of the order of 10^3 sec^{-1} and the existence of this correlation therefore implies that the magnitude of *se* is either independent of strain rate or varies with strain rate in a similar way for all polymers. There is insufficient information available at present to differentiate between these possibilities. The product *se* is also related to the impact strength or fracture toughness of a polymer and correlations have been sought between abrasive wear and these parameters. Ratner (1967) considers that the notched impact strength is likely to be the most significant because the strain rates involved approach those typical of the wear process; the observed relationship with polymer wear on abrasive papers, however, is nonlinear and exhibits considerable scatter. As a corollary, it should, fundamentally, be possible to relate the abrasive wear rates of

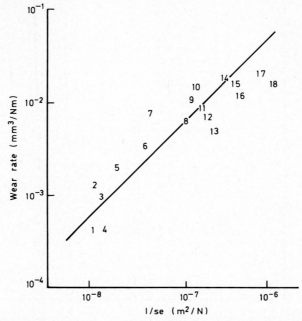

Fig. 6. Correlation between $1/se$ and wear rate during single traversals of polymers over rough mild steel (1.2 μm R_a). 1, (low-density) polyethylene; 2, polypropylene; 3, PTFE; 4, nylon 6.6; 5, nylon 11; 6, polyacetal; 7, methyl methacrylate–acrylonitrile copolymer; 8, polycarbonate; 9, acrylonitrile–butadiene–styrene copolymer; 10, poly(methyl methacrylate); 11, polysulphone; 12, polytrifluorochloroethylene; 13, poly(phenylene oxide); 14, epoxy resin; 15, polystyrene; 16, polyester resin; 17, poly(vinyl chloride); 18, poly(vinylidene chloride). (From Lancaster, 1973b. Crown copyright. Reproduced by permission of the Controller, Her Britannic Majesty's Stationery Office.)

polymers to their cohesive energies, i.e., to the energy of interaction between the polymer chains. A correlation of this type has been reported by Giltrow (1970) for wear on abrasive papers.

Although the simple theory of abrasive/cutting wear predicts an inverse relationship between wear rate and hardness, this relationship is much less well defined for polymers than for metals. The main reason is that the concept of "hardness" of a polymer has a different physical significance from that of a metal (Boor, 1960). During indentation of a polymer by a spherical or pyramidal indenter, a significant proportion of the load is supported elastically so that neither the depth of penetration nor the dimensions of the permanent impression correspond to full plasticity over the volume of the indentation. Despite these complications, the indentation hardnesses of polymers can be used as a very rough guide to

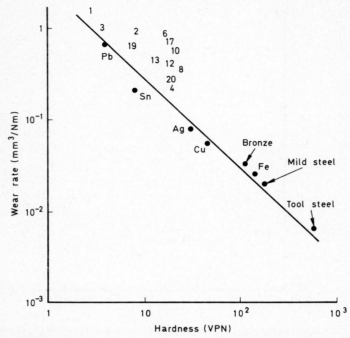

Fig. 7. Correlation between hardness and wear rate during abrasion on coarse (100 grade) carborundum paper. 1–18, as in Fig. 6; 19, FEP; 20, nylon 6. (From Lancaster, 1973b. Crown copyright. Reproduced by permission of Controller, Her Britannic Majesty's Stationery Office.)

compare their abrasive wear rates with those of metals in similar sliding conditions. Figure 7 shows such a comparison for wear against abrasive papers. It can be seen that all the rigid polymers examined exhibit higher rates of wear in these conditions than those of the common metals. This conclusion cannot, of course, be extended to include elastomers for which the deformation during both wear and indentation hardness measurements is almost entirely elastic.

B. Elastomers

As the elastic modulus decreases, the probability of wear occurring via plastic deformation and cutting also decreases. For elastomeric materials, cutting appears to occur only when sliding is against counterfaces with asperities of very small apex angles, such as sharp needles, etc. (Shallamach, 1952). Against more normal surfaces, two alternative wear processes can be identified—tensile tearing and fatigue. Only the first of these is discussed here; fatigue is deferred until later.

During sliding of a rough rigid counterface over a low modulus polymer, a bulge is formed ahead of each indenting asperity and a tensile stress develops at the rear (Shallamach, 1952; Ratner and Farberova, 1960). Localized failure can then occur by tearing in a direction at right angles to the direction of sliding. A similar process may occur during sliding against smooth counterfaces, provided that the interfacial adhesion is sufficiently high, but in this case the detached fragment can be deformed into a roll (Reznikovskii and Brodskii, 1967). Roll formation has also been observed with other, less ductile, polymers in certain conditions of sliding (Aharoni, 1973). Attempts have been made to correlate the wear of elastomers against rough surfaces with their mechanical properties, for example, tensile strength and hardness (Buist and Davies, 1946) and rebound resilience or hysteresis loss (Ratner and Klitenik, 1959). Shallamach (1958) has shown that the variation in wear rate with speed for an unfilled styrene–butadiene rubber is exactly paralleled by the variation in tensile strength.

One characteristic feature often associated with the abrasion of elastomers is the formation of a series of more or less uniformly spaced ridges on the elastomer surface, transverse to the direction of motion. These are called "abrasion patterns" and appear to result from the repeated passage of counterface asperities over stretched lips on the elastomer that are produced during previous encounters. A description of the various factors involved in the production of abrasion patterns has been given by Shallamach (1963) and by Southern and Thomas (1978).

C. Erosion

Two types of erosion may be distinguished, resulting from either normal or oblique impact (Bitter, 1963). In normal impact, the degree of damage depends on the severity of impact and on the mechanical properties of the material. If the conditions are sufficiently severe to cause plastic deformation, even a single impact can produce damage, but for elastic deformation multiple impacts are required before surface failure occurs via localized fatigue. Erosion during oblique impact arises from the fact that there is a velocity component parallel to the surface which can give rise to abrasive, cutting-type wear. At the same time, however, the normal force decreases so that there is a critical angle for maximum erosion; for most materials this is of the order of 20°. For oblique-impact erosion, the relationships between wear and mechanical properties are broadly similar to those already discussed for abrasion/cutting. Rigid polymers, in general, exhibit greater rates of erosion by abrasive particles than metals because of their lower values of the product *se* (Davies, 1966). However, unfilled elastomers, and particularly polyurethanes, are greatly superior to metals. The introduction of fillers into elastomers tends to increase their moduli, and decrease the product *se,* and thus usually leads to an increase in erosion (Marei and Izvozchikov, 1967).

A special case of erosion is that occurring due to impact with rain at high speeds and this is of some relevance in the present context because of the widespread use of polymers and their composites for radomes on aircraft (Fyall, 1970). Langbein (1965) has suggested that the rate of rain erosion of polymers is proportional to δ/e_i, where δ is the fraction of energy absorbed from the drops and e_i is the increase in internal energy of the polymer per unit volume following impact. The latter parameter is related to toughness or impact strength, and ultimately to the cohesive energy of the polymers. Correlations between rain erosion and these quantities have been sought, but in general, they appear to be valid only for a very limited range of materials (Hoff and Langbein, 1966). A correlation has also been reported between the erosion rates of a series of elastomers and their Shore hardnesses (Hammitt *et al.,* 1974). A more sophisticated theory of rain erosion has recently been developed by Springer (1976), based on fatigue considerations, which leads to complex analytical expressions relating the rate of erosion to impact severity and to the mechanical properties of the materials. These expressions appear to correlate reasonably well with existing experimental data and correctly predict the relative ratings of various material groups. In general, rain erosion resistance decreases in the following order: metals, elastomers, unfilled thermoplastics, filled/reinforced plastics composites, and foam/honeycomb structures with fiber-reinforced plastic skins. The very poor erosion resistance of the last named can be improved significantly by coatings of elastomers, and particularly polyurethanes. For polyurethanes there appears to be an optimum molecular weight for minimum erosion (Schmitt, 1967).

VI. Fatigue

Localized elastic deformation of a polymer during sliding becomes increasingly significant with decreasing elastic modulus and/or counterface roughness, and in these conditions it has already been noted that fatigue processes begin to play a major role in wear. This trend is illustrated schematically in Fig. 8, which emphasizes the important point that there can

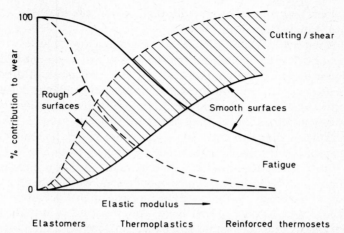

Fig. 8. Schematic variation of abrasion (cutting/shear) and fatigue wear with elastic modulus. (From Lancaster, 1973b. Crown copyright. Reproduced by permission of the Controller, Her Britannic Majesty's Stationery Office.)

be no rigid boundary between fatigue and abrasive/cutting wear. The fatigue properties of a polymer can be characterized approximately by the Wöhler relationship, $n \propto (\sigma_0/\sigma)^t$, where n is the number of cycles to failure, σ is the applied stress, σ_0 is the ultimate failure stress, and t is a material constant. Since the rate of wear will be inversely proportional to n, it is necessary to derive relationships among the applied stress σ, the surface topography causing this stress, and the mechanical properties of the polymer. The simplest case assumes indentation of a rigid hemispherical asperity of radius r into a smooth polymer surface under an applied load L. From elasticity theory, the contact radius $a \propto (rL)^{1/3}$ and the maximum stress $\sigma \propto r^{-2/3}L^{1/3}$. Assuming that the wear particles removed after n repeated contacts are equiaxed and of dimensions comparable to the contact diameter, the volume removed per unit distance of sliding is $v/s \propto a^2/n$. Hence $v/s \propto r^{-2(t-1)/3}L^{(t+2)-3}$. Values of t, derived from conventional fatigue data range from about 1.5 to 3.5 for elastomers and from about 3 to 10 for rigid thermoplastics and thermosets (Kraghelskii and Nepomnyaschii, 1965; Hollander and Lancaster, 1973). It can therefore be seen that the wear rates resulting from fatigue are likely to be very dependent on the topography of the counterface (r) and, unlike the situation in abrasive (cutting) wear, increase more rapidly than proportionately with load. More sophisticated theories of fatigue wear have been proposed by Reznikovskii (1960) and Kraghelskii and Nepomnyashchii

(1965), which lead to somewhat different power relationships, although the general trends remain the same.

It is claimed by Ratner (1967) that a convenient way to assess the wear properties of polymers in fatigue conditions is to use a geometrical sliding arrangement in which the counterface is a fine wire mesh of dimensions such that the "asperities" (mesh intersections) induce elastic deformation only. From experiments of this type it is possible to determine the index in the power relationship relating wear rate to load, $v/s \propto L^\alpha$. For rigid thermoplastics α ranges from 1 to 3, and in terms of the simple fatigue theory above, these values correspond to t ranging from 2 to 7. For elastomers, Klitenik and Ratner (1967) have concluded that α increases with increasing molecular polarity, cohesive energy density, and filler content; α decreases, however, when swelling occurs in the presence of fluid. Shcherbakov and Kaplan (1974) have also reported a reduction in the fatigue strengths of a polyamide and polypropylene with increased swelling, and this in turn implies a reduction in α.

An alternative view of fatigue wear emerges from work by Ratner and Lure (1966b). It implies that the wear rate is directly proportional to the rate at which the primary bonds in the polymer chain are broken, and that the activation energy for bond rupture is reduced by the applied mechanical stresses. This leads to a relationship of the form: (wear rate) $\propto e^{-(U_0 - kL)/RT}$, where U_0 is the activation energy, L the load, and the constant k involves the material properties. Wear rates of a number of thermoplastics sliding against wire gauze at different temperatures support the above relationship, and the activation energies derived appear to be in reasonable agreement with those obtained from other work on oxidative degradation. It follows from this view of fatigue wear that the introduction of oxidation inhibitors into a polymer might reduce the rate of wear and this has been confirmed by Ratner and Lure (1966a) in experiments with polyacetal and polyamide 6.8. The reductions in wear, however, are also accompanied by an increase in the index α, relating wear rate to load; the effects of oxidation inhibitors on wear are therefore most apparent at light loads. Reductions in wear by the incorporation of oxidation inhibitors have also been reported for rubbers (Brodskii and Reznikovskii, 1967).

Ratner (1967) further suggests that the fatigue wear process during sliding against a wire gauze is essentially the same as that when polymers slide against rough metals, provided that the asperities are insufficiently sharp to induce plastic deformation and cutting. Support for this view emerges from experiments in which various thermoplastics were worn in single traversals over steel drums of varying roughness (Lancaster, 1969a). Figure 9a shows the variations in wear rate with R_a roughness. From a computer analysis of the profiles of the surface topography of the

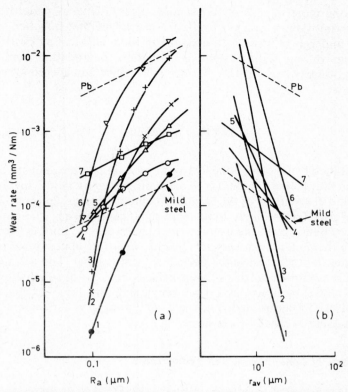

Fig. 9. Variation of wear rates during single traversals over mild steel with (a) surface roughness R_a; (b) average radius of curvature of the asperities r_{av}. 1, nylon 6.6; 2, poly-acetal; 3, poly(methyl methacrylate); 4, polyethylene; 5, polypropylene; 6, polystyrene; 7, PTFE. (Crown copyright. Reproduced by permission of the Controller, Her Britannic Majesty's Stationery Office.)

steel, the average radius of curvature r_{av} of the asperities can be derived, and Fig. 9b shows the wear rates plotted against these values of r_{av}. The slopes of the lines enable the values of t in the fatigue relationship to be derived, and these are similar in magnitude to values obtained in conventional fatigue tests (Hollander and Lancaster, 1973).

Although the above work provides strong circumstantial evidence in favor of localized fatigue wear processes in polymers, direct microscopic evidence for fatigue failures is difficult to find. Brown *et al.* (1976) have published micrographs of surface cracks in ultrahigh molecular weight polyethylene, which appear to occur in conjunction with a sudden increased wear rate after a period of prolonged sliding. These may, however, correspond more to macro- rather than microfatigue. A major difficulty in observing fatigue cracks results from the fact that polymer surfaces usually undergo very marked changes in their physical and mechanical properties during sliding, such as oxidation, degradation, and plastic flow (Egorenkov *et al.*, 1973). This aspect will be considered in more detail below.

VII. Adhesion

It is difficult to isolate adhesive wear as a unique mechanism because adhesion plays a part in almost all wear processes by modifying the magnitude and distribution of the localized contact stresses. In general, adhesion and transfer are likely to be most significant in wear when polymers slide repeatedly over themselves or over relatively smooth metal surfaces. In these circumstances, the specific wear rate almost invariably becomes constant after an initial period of "running in," and it is then reasonable to assume that the sliding surfaces have reached equilibrium or steady-state conditions. To avoid prejudging the issue of the precise mechanism of wear occurring in these conditions, it is convenient to refer to this regime as *steady-state wear*.

A. *Effects of Load, Speed, and Temperature*

Simple theories of wear (e.g., Archard, 1953) based on adhesion predict a direct proportionality between wear rate and load. This is frequently observed experimentally, as in Fig. 10a, but it should be emphasized that proportionality will be observed only when changes in load do not induce changes in any other variable affecting wear, such as temperature or the topography of the polymer and its counterface. The rapid rise in wear rate, shown in Fig. 10a for polyethylene and poly(methyl methacrylate) above a critical load, is attributable to thermal softening of the polymer. For low-density, high-density, and ultrahigh molecular weight polyethylenes, it has been reported that this critical load increases following γ irradiation, which introduces cross linking and inhibits melting (Matsubara and Watanabe, 1967; Shen and Dumbleton, 1974). The influence of temperature on the wear of thermoplastics is more complex than that of load, and some examples are shown in Fig. 10b. The initial decrease in wear rate with increasing temperature is a consequence of a reduction in the elastic modulus of the polymer; in terms of a fatigue wear process, a reduction in

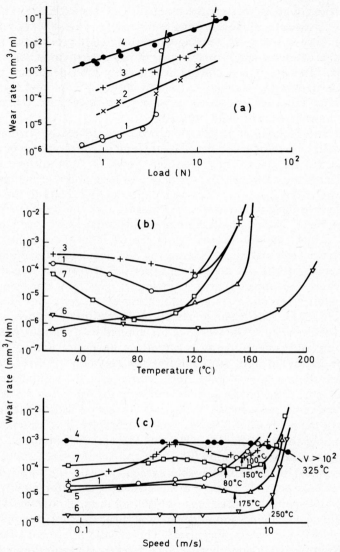

Fig. 10. Variation of steady-state wear rates against mild steel (0.15 μm R_a) with (a) load; (b) temperature; (c) sliding speed. 1, (low-density) polyethylene; 2, nylon 6; 3, poly(methyl methacrylate); 4, PTFE; 5, polyacetal; 6, nylon 6.6; 7, polypropylene. $\uparrow$, speed at which the calculated flash temperature reaches the melting/softening point of the polymer. (After Lancaster, 1972d. Crown copyright. Reproduced by permission of the Controller, Her Britannic Majesty's Stationery Office.)

modulus leads to an increase in the number of cycles to failure. The rapid increase in wear rate above a critical temperature is again a consequence of overall thermal softening, leading to extrusion and gross flow. Finally, the influence of sliding speed on the wear of some thermoplastics is shown in Fig. 10c. The maxima occurring at relatively low speeds are qualitatively consistent with viscoelastic (rate of strain) effects, and the subsequent minima and rapid increases in wear again result from temperature increases associated with high speeds.

Tanaka and Uchiyama (1974) have observed generally similar variations of wear rate with speed for polyethylene on steel and several other thermoplastics on glass. They consider, however, that surface melting occurs even at very low speeds, 0.1 m/sec or less, and that the rapid rise in wear rate is the result of reaching a depth and viscosity in the molten surface layer sufficient to permit gross extrusion of the polymer from the sliding interface. The main difficulty in deciding the exact speed at which localized contact melting begins lies in the fact that there is no reliable way of either calculating or measuring the localized temperatures—"flash" temperatures—at contacting asperities. However, one theoretical approach that appears to give estimates generally consistent with experimental observations is a modification of Jaeger's (1943) analysis (Lancaster, 1971). This assumes plastic deformation of a single asperity contact, which supports the whole of the applied load, and the calculated temperature must therefore be regarded as an upper limit. Figure 11 shows the predicted flash temperatures for polymers of three different hardnesses sliding against mild steel, where θ is the temperature rise above ambient, μ the coefficient of friction, L the load, and V the speed. The derivation of these curves includes a correction for the fact that the hardness of the polymer also varies with temperature. The critical speeds, corresponding to the softening points of the various polymers, derived from these curves are shown by the arrows adjacent to each curve in Fig. 10c, and appear to be in reasonable agreement with the onset of rapid wear. It is also possible to derive the critical speeds for melting when sliding occurs against counterfaces of differing thermal conductivity, and Fig. 12 shows calculated and experimental measurements for polyacetal sliding against steel, glass, and itself. The change in critical speed with counterface thermal conductivity is very marked, and this could well explain observations that polymer/polymer combinations often exhibit much higher wear rates than polymer/metals [see, e.g., Theberge *et al.* (1974)]. It is evident from Fig. 12 that when frictional heating is insufficient to cause surface melting, the wear rate of polyacetal on itself is very little different from that on glass or steel.

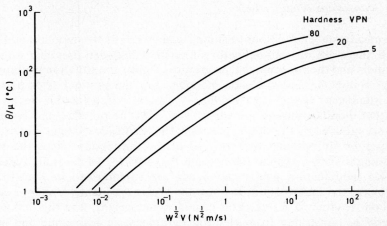

Fig. 11. Variation of θ/μ with $W^{1/2}V$ for polymers of different hardnesses. θ is the flash temperature, μ the coefficient of friction, W the load, and V the sliding speed. (Crown copyright. Reproduced by permission of the Controller, Her Britanic Majesty's Stationery Office.)

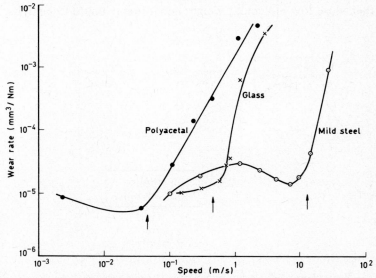

Fig. 12. Effect of thermal conductivity of the counterface K on wear rate/sliding speed relationships for polyacetal. Counterfaces as indicated. Polyacetal, $K = 0.2$ W/m °C; glass, $K = 1.5$ W/m °C; mild steel, $K = 46$ W/m °C. ↑, speed at which the calculated flash temperature reaches the melting point of the polymer (175°C). (Crown copyright. Reproduced by permission of the Controller, Her Britannic Majesty's Stationery Office.)

B. Counterface Modifications

In steady-state conditions of sliding against relatively smooth metal surfaces, there are no well-defined relationships between the wear rates of polymers and their mechanical properties. Unlike the situation in abrasive wear, neither hardness, elastic modulus, nor the product *se* is particularly significant (Lancaster, 1969a). The main reason for this lies in the fact that modifications to the surfaces of both the polymer and counterface are induced by the sliding process. These surface changes can be formalized by introducing the concept of a "third body" into the two-component sliding system (Godet and Play, 1976). The various types of surface modification which can occur are summarized in Fig. 13. In polymers, the molecules in the surface layers frequently undergo chain scission, leading to a reduction in molecular weight, and this has been directly demonstrated from electron paramagnetic resonance and mass spectrometry measurements on a number of thermoplastic polymers by Belyi and Sviridyonok (1974). Reductions in the molecular weight of wear debris have also been reported by Ratner and Lure (1966a) for several thermoplastics and by Arkles and Schireson (1976) for PTFE. It is possible that these low molecular weight constituents could "plasticize" the

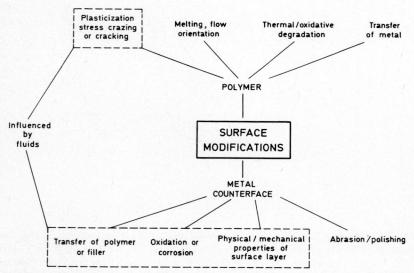

Fig. 13. Types of surface modifications induced during sliding of polymers against metals. (Crown copyright. Reproduced by permission of the Controller. Her Britannic Majesty's Stationery Office.)

surface layers of a polymer, or of transferred debris, leading to a lower elastic modulus and/or an increase in the elongation to break. The formation of low modulus layers on elastomers after sliding is well established (Shallamach, 1963).

With some crystalline thermoplastics, e.g., PTFE and high-density polyethylene, sliding induces a preferred orientation of the molecular chains (Makinson and Tabor, 1964; Pooley and Tabor, 1972). More extreme modifications have been reported by Tanaka and Uchiyama (1974), who observed the total disappearance (to a depth of the order of 10 μm) of the spherulitic structure in the surface layers of several crystalline thermoplastics. Thermal and oxidative degradation of surface layers are also common features of thermosetting polymers at high ambient temperatures or at high sliding speeds (Bark *et al.*, 1975). Oxidation of the surface layers will reduce their ductility, which can lead to enhanced abrasion (cutting) as a factor in the wear process. At even higher temperatures, pyrolysis may become significant (Liu and Rhee, 1976).

A further factor affecting the surfaces of both polymer and metal counterfaces during sliding is transfer of metal to the polymer. This phenomenon has been observed on a number of occasions (Rabinowicz and Shooter, 1952; James, 1958; Brainard and Buckley, 1973; Tanaka and Uchiyama, 1974) and occurs presumably when the subsurface strength of the metal falls below that of the adhesive or mechanical interactions at the interface. This situation could arise in several ways. The metal could be weakened locally by repeated contacts during sliding, as recently described by Suh (1973), and it is important to note that conventional surface-finishing treatments may generate locally weakened asperities even before sliding begins. Alternative mechanisms which may contribute to localized weakening of the metal are interactions with surface active agents, produced either by oxidation or degradation of the polymer (Gorokhovskii, 1965; Gorokhovskii and Agulov, 1966), or present in the surrounding medium (Rehbinder *et al.*, 1941), and selective migration— the preferential removal of one metallic constituent from the surface layer of an alloy (Egorenkov *et al.*, 1973). Any metal fragments transferred to the polymer will tend to become embedded in the surface layers and are then able to modify the counterface topography during subsequent sliding. Abrasion or polishing of the wear tracks on metal counterfaces during sliding against polymers is commonly observed, but whether any of the above mechanisms are primarily responsible remains an open question. More mundane possibilities are the presence of abrasive impurities or abrasive fillers in the polymer, and the ingress of abrasive particles from the environment. Further discussion of these aspects is deferred until later.

C. Transfer

It is generally recognized that the presence of thin films of transferred PTFE on a metal counterface significantly affects both friction and wear, and usually leads to a reduction in both. The mechanisms by which these films reduce wear appear to be twofold. First, the effective surface roughness of the counterface is reduced, which can lead to considerable reductions in wear. Second, sliding now occurs between oriented molecular chains of PTFE on both surfaces, the friction is reduced, the localized contact stresses are reduced, and so, in turn, is the rate of wear.

There is some uncertainty about the mechanism of adhesion of PTFE transfer films to metals. Mechanical interlocking effects within the surface depressions undoubtedly play some part, but there is also evidence for chemical effects. Mitchell and Pratt (1957) first showed that additions of bronze to PTFE greatly improved its wear properties, and that the reduced wear was associated with the formation of a well-defined transfer film on the counterface. Buckley and Johnson (1963) suggested that the adhesion of PTFE to metals is facilitated by thermal degradation induced by sliding and the formation of fluoride ions, and Pratt (1964, 1967a) has concluded that bronze catalyzes a reaction between PTFE degradation products and lead. A comprehensive review of possible chemical reactions influencing polymer wear has been given by Richardson (1971). More recently, Pocock and Cadman (1976) have examined a wide range of two- and three-component PTFE–metal and metal oxide mixtures by differential scanning calorimetry but, surprisingly, failed to find unequivocal evidence for reaction between Cu, Pb, and PTFE. A reproducible exothermic reaction was only observed with Cu–Sn–PTFE after prolonged heating at 430°C, leading to the formation of SnF_4. It is evident that the part played by chemical reactions in transfer film formation and wear is still rather obscure.

Although transfer film formation is particularly important for PTFE, it can also affect the wear behavior of other polymers. With the more ductile thermoplastics, such as polyethylene, polyacetal, and nylons, fragments transferred to the counterface can readily be deformed during repeated contacts and this ultimately leads to a reduction in counterface roughness and a reduced rate of wear. Relatively hard and brittle polymers, however, such as polyesters and some epoxies, can transfer large irregular fragments which increase the effective counterface roughness and hence, the wear (Lancaster, 1969a). Chemical effects can again play a part and Sviridyonok et al. (1973) have made extensive investigations into transfer film formation, using a variety of techniques for surface examination such as infrared and mass spectroscopy and electron paramagnetic resonance. They suggest that for highly polar polymers such as polyamides and poly(methyl methacrylate), transfer is controlled by the formation and interactions of free radicals generated during the sliding process.

VIII.   Composites

Although fillers and reinforcing fibers significantly improve many physical and mechanical properties of polymers, it is not valid to assume that they always improve their friction and wear properties also. For example, in abrasive conditions, i.e., wear against rough surfaces, the wear rates of filled or reinforced polymers can sometimes be greater than those of the matrix polymer alone (Lancaster, 1972a). The reason is that although the fillers generally increase the breaking strength s, they almost invariably reduce the elongation to break e; the product se can therefore be lower than for the unfilled polymer. However, in conditions of steady-state wear during sliding against relatively smooth metals, the effects of fillers on wear are usually beneficial and often dramatic.

A.   PTFE

Composites based on PTFE are more widely used and have received more attention than any other group of materials. The wear rate of unfilled PTFE is extremely high, but additions of almost any inorganic, and some organic, fillers can reduce wear by a factor of 100 or more. Some examples are shown in Fig. 14, based on data from pin and ring tests, and similar results for other sliding arrangements have also been reported (White, 1956; Mitchell and Pratt, 1957; Arkles *et al.*, 1974). The precise choice of filler or reinforcement depends on which particular combination of composite properties is most important for the intended application—friction, wear, dimensional stability, heat dissipation, stiffness, cost, etc. As already mentioned, some compromise is invariably necessary, and only a relatively small number of fillers is widely used commercially—graphite, carbon, glass fiber, and bronze (Wells, 1971). Wear rates against steel usually decrease rapidly with increasing filler contact to a minimum at around 40 vol % (Arkles *et al.*, 1974); most commercially available materials, however, contain typically 15–25% by volume of filler.

Carbon and graphite fillers are of particular interest because these materials are not only capable of exhibiting low friction and wear in their

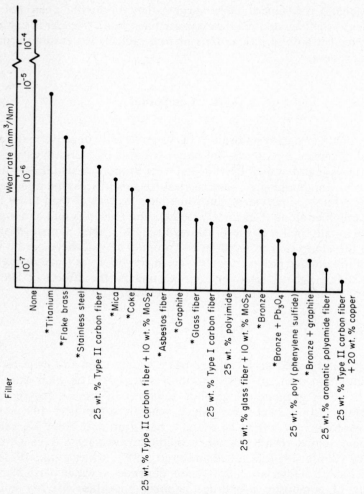

Fig. 14. Steady-state wear rates of PTFE composites against smooth steel (approximately 0.15 μm R_a).*—commercially available composites. (Crown copyright. Reproduced by permission of the Controller, Her Britannic Majesty's Stationery Office.)

own right but also reduce expansion coefficients and increase strength and thermal conductivity. Istomin and Khrushchov (1973) have emphasized the importance of the type of graphite used as a filler, and, in particular, its basal plane orientation with respect to the sliding interface. With thin flakes, highest wear (and lowest friction) is obtained when the basal planes are oriented parallel to the sliding surfaces, and lowest wear (but highest friction) when oriented normal to the surfaces. However, for graphite particles, which are more nearly equiaxed, there is little effect of basal plane orientation on wear. Generally similar results have been obtained by Briscoe *et al.* (1977), using oleophilic and oleophobic graphites as fillers. These effects of orientation have some practical significance because the usual method of fabricating filled PTFE composites—compression molding—can induce a preferred orientation of the filler (Griffin, 1972). In view of the extreme sensitivity of the friction and wear properties of carbons and graphite to environmental conditions, there has been a number of investigations into environmental effects on carbon–

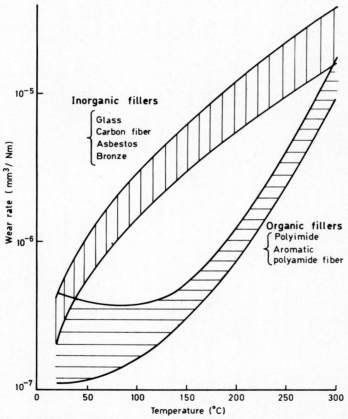

Fig. 15. Effect of temperature on the wear rates of PTFE composites sliding against stainless steel (0.15 μm R_a). (Crown copyright. Reproduced by permission of Controller, Her Britannic Majesty's Stationery Office.)

TABLE II

EFFECT OF FILLER-TYPE AND ENVIRONMENT ON THE WEAR BEHAVIOR OF PTFE COMPOSITES
(Adapted from Arkles *et al.*, 1974)

Environment	Ranking Low wear rate → High wear rate	Reference
Vacuum	glass fiber glass fiber/MoS₂; coke flour; graphite; silver copper	Buckley and Johnson (1963)
Air	bronze; carbon/graphite; glass fiber; MoS₂	Pratt (1964) Mitchell and Pratt (1957) Lewis (1968)
	bronze/MoS₂; glass fiber/MoS₂; carbon/graphite; glass fiber	O'Rourke (1965)
Air (wet, 5 × 10³ ppm)	bronze; carbon/graphite; carbon glass fiber bronze	Fuchsluger and Taber (1971)
	metal oxide; glass fiber; carbon/graphite	Schubert (1971)
Air (dry, 30 ppm)	glass fiber; carbon; bronze; carbon/graphite glass fiber	Fuchsluger and Taber (1971)
	metal oxide; carbon/graphite; bronze	Schubert (1971)

Environment	Materials	Reference
Nitrogen	bronze; MoS$_2$; glass fiber/MoS$_2$	O'Rourke (1965)
Nitrogen (wet, 5 × 10^3 ppm)	carbon	Fuchsluger and Taber (1971)
	carbon/graphite; glass fiber; bronze	Schubert (1971)
	glass fiber; carbon/graphite; bronze; metal oxide	
Nitrogen (dry, 2 × 10^2 ppm)	carbon/graphite; carbon; bronze; glass fiber	Fuchsluger and Taber (1971)
	glass fiber; carbon/graphite; bronze; metal oxide	Schubert (1971)
Nitrogen (dry, 30 ppm)	bronze; glass fiber; carbon/graphite	Fuchsluger and Taber (1971)
	carbon; glass fiber; carbon/graphite	Schubert (1971)
	carbon/graphite; glass fiber; bronze	
Nitrogen (dry, 2 ppm)	carbon; bronze; glass fiber; carbon/graphite	Fuchsluger and Taber (1971)
	carbon/graphite; bronze; glass fiber	Schubert (1971)
Oxygen (wet, 5 × 10^3 ppm)	glass fiber; carbon/graphite; bronze	Schubert (1971)
Helium	glass fiber; carbon	Hart (1966)
	carbon/graphite; glass fiber; bronze; metal oxide	Schubert (1971)
Water	carbon/graphite; glass fiber/MoS$_2$; bronze/MoS$_2$; glass	O'Rourke (1965)

PTFE composites. The results are conflicting. Hart (1966) claims that carbon-filled PTFE shows a tenfold increase in wear in dry gases, but Fuchsluger and Taber (1971) claim that the wear rate of a similar material is lower in dry nitrogen than in wet air. Similar confusion appears to exist in the reported environmental effects on wear of PTFE containing other types of fillers. A collection of data abstracted from the literature by Arkles *et al.* (1974) is given in Table II, and this shows clearly that the relative rankings of different PTFE composites for wear change significantly in different environments; there is no discernible pattern of behavior.

Filler additions to PTFE not only reduce wear rates under nominal room-temperature conditions but can also affect the way in which the wear rate varies with temperature. Figure 15 shows a selection of results from recent work by Evans (1977) that clearly demonstrates two general types of behavior. For mineral and inorganic fillers, the wear rates increase rapidly with temperature over the range 20–150°C, but for organic-filled PTFE composites, the wear rates remain virtually independent of temperature up to 150°C and then increase rapidly. Explanations are suggested in terms of the differing effects of the two types of filler on transfer film formation on the counterface, mineral and inorganic fillers being relatively abrasive as will be discussed later.

B. Other Polymers

In general, most filled or reinforced polymers other than PTFE and related fluorocarbons require solid lubricant additions to reduce friction and heat generation to an acceptable level for most dry sliding applications. PTFE itself is a very common additive, and the wear properties of a range of glass-fiber-reinforced thermoplastics containing PTFE powder have been evaluated by Theberge and Arkles (1973). It is difficult to draw any general conclusions from these results other than that friction, wear, and composite strengths all tend to decrease with increasing PTFE content. The unlubricated wear properties of some of the newer high-temperature-resistant polymers also benefit by PTFE additions, e.g., polyimides (Giltrow, 1973), poly(phenylene sulfide) (West and Senior, 1973), and poly(*p*-hydroxybenzoic acid) (Economy and Cottis, 1971). In many instances, mixtures of other additives with PTFE lead to even further reductions in wear, e.g., PbO + graphite + PTFE in poly(phenylene sulfide) (Graham and West, 1976), but optimum compositions have so far been developed entirely empirically; there appear to be no consistent explanations of why some mixtures are better than others.

The other common solid lubricant additions to polymer composites to reduce friction are graphite and MoS_2. In general they seem to be less effective than PTFE at concentrations below about 15 vol %, but graphite has the merit that it provides some degree of mechanical reinforcement as well as reducing friction. A greater degree of reinforcement is possible by using graphite, or carbon, in fiber form and Giltrow and Lancaster (1968, 1971) have shown that such reinforcements can lead to very low wear with both thermoplastic and thermosetting polymers. As with PTFE-based composites, low wear is only obtained when a coherent transfer film is established on the counterface surface, and some of the factors involved in transfer film formation from carbon-fiber-reinforced polymers have been discussed by Giltrow and Lancaster (1970). The effects of fiber orientation on the wear of carbon-fiber-reinforced polymers appear to be broadly similar to those described earlier for PTFE containing flake graphite; lowest wear occurs when the fibers are oriented normally to the direction of sliding (Lancaster, 1968b), and similar results have been obtained with composites containing other types of fibers (Tsukizoe and Ohmae, 1977). However, differences in orientation tend to become less significant with increasing fiber content and are relatively small beyond about 50 vol % (Giltrow and Lancaster, 1971; Brown and Blackstone, 1974). One rather curious observation is that in fretting conditions with carbon-fiber-reinforced polymers, minimum wear has been found to occur with the fibers lying in the plane of the surface and parallel to the direction of sliding (Ohmae *et al.*, 1974).

C. Mechanism of Action of Fillers

A variety of explanations has been proposed in the literature to account for the way in which various fillers and filler combinations reduce the wear of specific polymers. Chemical effects, resulting from reaction between Cu, Pb, and PTFE, have already been mentioned, and have also been proposed for CuO, Pb_3O_4, and high-density polyethylene sliding on steel (Briscoe *et al.*, 1974). On a more mechanistic level, Arkles *et al.* (1974) suggest that the filler particles in PTFE reduce wear by preventing the drawing out of long thin filaments characteristic of the wear of the unfilled polymer. For fiber-reinforced polymers, Lancaster (1968b) suggests that the fibers support the load preferentially, but this is only likely to occur when the fibers are relatively long and oriented mainly normally to the sliding surface. A more general explanation emerges from considering the effects of fillers and reinforcements on transfer to and modification of the counterface surface.

Solid lubricant fillers such as graphite, MoS_2, and PTFE appear to transfer preferentially to the surface of a metal counterface, reducing both

TABLE III

ABRASIVENESS OF PTFE AND PHENOLIC RESIN COMPOSITES

Composite	Abrasiveness (wear rate of bronze) $(10^{-6} \text{ mm}^3/\text{N m})$
PTFE	<0.01
+30% Glass fiber	62
+25% Asbestos fiber	10
+25% Carbon fiber (type II)	8
+30% Mica	3.1
+25% Coke	1.7
+25% Carbon fiber (type I)	0.2
+40% Bronze	0.08
+33% Graphite	0.05
Phenol-formaldehyde	—
Mineral filled	73
Asbestos paper reinforced	7.5
Wood filled	1.7
Paper reinforced	0.26
Cotton cloth reinforced	0.18

friction and wear, as discussed previously. Chemical effects may facilitate the development of these films. However, many fillers are abrasive to a metal counterface and can either prevent transfer film formation or cause wear of the metal itself. The relative abrasiveness of different fillers in two polymers is illustrated by the results in Table III, derived from a test in which a bronze ball was oscillated over the surface of a rotating composite disk (Lancaster, 1968a). The degree of surface damage to the counterface by the filler will also depend on the relative hardnesses (Lancaster 1972a) and on the shape of the filler particles (Speerschneider and Li, 1962b). Very soft counterfaces such as aluminum alloys are readily damaged by glass fibers, whereas harder counterfaces such as tool steels are merely polished. This may be one reason why the wear rates of polymer composites are generally observed to decrease with increasing counterface hardness, provided that the initial surface roughness remains the same (Lewis, 1967; Arkles *et al.*, 1974). A high counterface hardness also appears to be necessary to cause mechanical degradation of some fillers such as high-strength carbon fibers to a sufficiently fine state of subdivision in order to form a transfer film (Giltrow and Lancaster, 1970).

An example of the differing effects on the wear of a polymer caused by incorporating fillers of differing degrees of abrasiveness is shown in Fig. 16. As noted in Table III, high-strength carbon fibers (type II) are considerably more abrasive than the high-modulus variety (type I) and this dif-

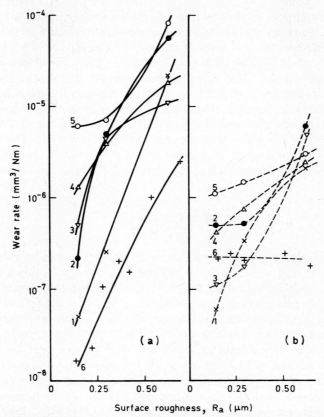

Fig. 16. Effect of counterface roughness on the wear rates of carbon fiber composites sliding against stainless steel. (a) type-I carbon fibers; (b) type-II carbon fibers. Thermoplastics (30 vol %). 1, poly(phenylene oxide); 2, PTFE; 3, nylon 6.6; 4, polyacetal; 5, (high-density) polyethylene. Thermoset (60 vol %). 6, epoxy. (Crown copyright. Reproduced by permission of the Controller. Her Britannic Majesty's Stationery Office.)

ference is reflected in the variations of the wear rates of composites containing these two fibers on steel surfaces of various initial surface roughnesses. The abrasiveness of the type II fibers is sufficient to modify the counterface topography of the steel during sliding so that the wear rates of these composites are much less sensitive to the magnitude of the initial roughness. The importance of the initial counterface topography on the wear of polymers and composites, in general, is still a somewhat controversial topic. It is often suggested that there is an optimum roughness for minimum wear, but the experimental evidence for this is conflicting. Lewis (1967) reports a minimum wear rate for PTFE against low-carbon steel at a roughness of about 0.2 μm R_a, but no minimum for an aromatic polyimide. Buckley (1974) shows a pronounced minimum in the wear of polyethylene on stainless steel at around 0.4 μm R_a, but Dowson et al. (1978) find only a shallow minimum for similar materials at about 0.1 μm R_a. No minima in wear have been observed for the carbon fiber-reinforced polymers in Fig. 16, or for an acetal copolymer (Clerico and Rosetto, 1973). In general, it would appear that an optimum counterface roughness for minimum wear *may* exist, but only for those polymers that rely on transfer film formation in order to exhibit low wear. The situation regarding polymers which do not transfer themselves but which contain lamellar solid lubricants such as graphite and MoS_2 is still unresolved.

IX. Lubrication

Fluid lubrication is a common feature of many tribological applications of polymers, and particularly of elastomers. Several reviews have recently been published dealing with both the fundamental aspects of lubrication of elastomers (Roberts, 1977a) and practical aspects relating to seals, tires, and wiper blades (Moore, 1975a; Grosch and Shallamach, 1977; Roberts, 1977b). For the more rigid polymers, most work has been concerned with the effects of fluids on friction. For polymers sliding against themselves, reductions in friction by conventional boundary lubricants, such as fatty acids, tend to be relatively small and this has been attributed to the inability of the lubricants to form a fully condensed adsorbed monolayer on polymers (Bowers et al., 1954; Fort, 1962). When polymers slide against metals, however, reductions in friction by boundary lubricants can be much greater because the lubricant can then interact with the metal (Matveevskii, 1961; Vinogradov and Bezborodko, 1962). For relatively simple fluids, attempts have been made to relate their lubricating ability (reductions in friction) to surface energies and wettability, but with little obvious success (Senior and West, 1971).

Before discussing the effects of fluids on the wear of polymers, two specific aspects of polymer/lubricant behavior should be mentioned. The first is that some polymers appear to have the ability to "retain" small amounts of lubricant on their surfaces for long periods of sliding. Booser et al. (1957) have measured the "dwell time" for various materials during sliding over a steel disk prelubricated by a known amount of mineral oil, the dwell time being the period of sliding before measurable wear occurred. For a nylon, this time exceeded 18 hr, but was only of the order of 10 min for conventional white-metal bearing materials. The same trend

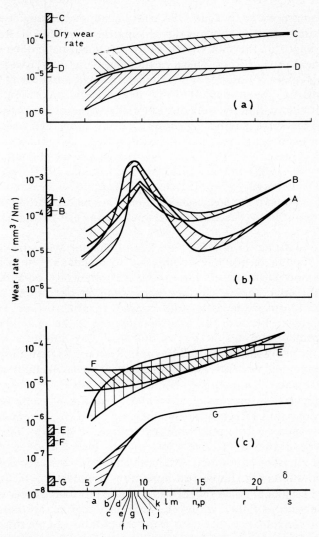

Fig. 17. Effects of fluids on the wear of polymers in conditions of boundary lubrication; δ, solubility parameter of liquid. A, poly(phenylene oxide); B, poly(methyl methacrylate); C, PTFE; D, polyacetal; E, PTFE/25 wt % type-I carbon fiber; F, PTFE/polyimide; G, epoxy/60 vol % type-I carbon fiber. a, silicone fluid; b, *n*-hexane; c, *n*-hexadecane; d, cyclohexane; e, tricresylphosphate; f, toluene; g, carbon tetrachloride; h, diethylhexylsebacate; i, benzene; j, acetone; k, *n*-octanol; l, *n*-propanol; m, dimethylformamide; n, methanol; p, ethylene glycol; r, formamide; s, water. (After Evans, 1978a and Lancaster, 1972c. Crown copyright. Reproduced by permission of the Controller, Her Britannic Majesty's Stationery Office.)

has also been reported by Pratt (1967b) from somewhat similar tests, who found, in addition, that polyacetal was superior to nylon 6.6. Reasons for the latter, based on thermodynamic considerations, have been suggested by Butterfield et al. (1971). These results show that effective lubrication of some polymers is possible with extremely small amounts of fluid, and this finding has been exploited by deliberately incorporating small amounts of fluid within the bulk structure of polymers. The fluid becomes exposed at the surface either by diffusion or as a result of wear. Various fluids have been examined, including fatty acids and amides (see, e.g., Allen, 1959; Briscoe et al., 1972), low molecular weight fluorocarbons (Bowers et al., 1965), mineral oils (see, e.g., Ikeda, 1975), and silicones (Arkles and Theberge, 1973; M. P. L. Hill et al., 1974; Fearon and Smith, 1974). Reductions in the wear rates of polymers incorporating fluids have also been reported by Pascoe and Dzhanokmedov (1978) and Abouelwafa et al. (1978).

The second aspect of polymer/lubricant behavior arises from the relatively low moduli of polymers compared to metals. In the presence of fluids, low-modulus materials can exhibit elastohydrodynamic film formation, leading to complete surface separation in conditions that, with metals, would ensure solid-to-solid contact (Cudworth and Higginson, 1976). Moreover, because the low modulus limits the magnitude of the localized contact stresses developed during sliding, lubricant viscosities cannot increase significantly, and the coefficient of friction in the elastohydrodynamic regime thus remains relatively low (Archard and Kirk, 1963; Hooke and O'Donohugue, 1972).

Meaningful measurements of lubricated wear are possible only when elastohydrodynamic effects are negligible and the surfaces remain in complete solid contact within the boundary regime. In general, this restricts the sliding conditions to relatively low speeds, heavy loads, and/or very small apparent areas of contact. Attempts have been made to define these limiting conditions for simple test apparatus (Lancaster, 1972b; Moore, 1975b). The results of some recent work on the lubricated wear of several polymers in various fluids, within the boundary regime, are illustrated in Fig. 17. For convenience, the wear rates are plotted against the solubility parameters of the various fluids, this being an approximate measure of the intermolecular cohesion within the fluid (Hansen and Beerbower, 1971). For the crystalline polymers, polyacetal and PTFE, Fig. 17a shows that most fluids slightly reduce the wear rate below its unlubricated value. Many fluids also reduce the wear rates of the amorphous polymers, poly(phenylene oxide) and poly(methyl methacrylate) (Fig. 17b), but very high wear occurs in those fluids that are strongly absorbed, leading to stress cracking and crazing. Finally, Fig. 17c shows that the wear rates of

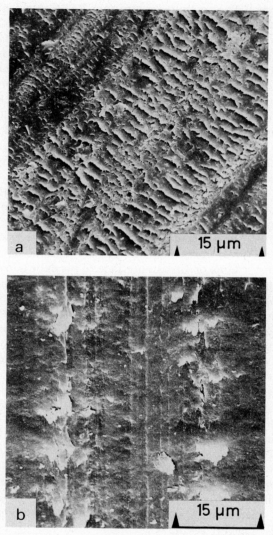

Fig. 18. Scanning electron micrographs of worn surfaces of polymers after sliding in liquids against stainless steel: (a) poly(phenylene oxide) in *n*-hexane; (b) polyacetal in *n*-propanol. (Crown copyright. Reproduced by permission of the Controller, Her Britannic Majesty's Stationery Office.)

an epoxy–carbon fiber composite and two PTFE composites are all in-creased by the presence of all the fluids.

The appearance of the surface layers on some polymers after sliding in fluids is, in general, very different from that obtained during dry sliding. Figure 18 shows typical examples for PPO and polyacetal, and the patterns of transverse ridges are very reminiscent of the abrasion patterns often developed on elastomers after the surface layers have been degraded and reduced in modulus by repeated sliding (Shallamach, 1958). Similar features on polyacetal have also been reported by Shen and Dumbleton (1976). The appearance of the worn surfaces strongly suggests that the surface layers have been extensively plasticized by the fluid, the degree of plasticization presumably being increased as a result of the contact stresses. In this connection, it is interesting to note that Zuer et al. (1974) have observed a significant increase in the degree of swelling of a rubber during sliding in oil.

It is clear from Fig. 17c that water is particularly deleterious to the wear rates of those polymer composites that, in dry conditions, rely on the formation of transfer films on the counterface to produce a low rate of wear. A similar conclusion for PTFE composites has been reported by Craig (1964). The main reason is that all fluids, but particularly water, prevent the formation of a transfer film on the counterface; the surface therefore remains relatively rough, and the wear rate of the composite remains high. This suggests that the abrasive action of fillers on counterface polishing, as discussed earlier, could well exert a dominant influence on wear in water-lubricated conditions. The effects of deliberate additions of small amounts of finely divided abrasives to two polymer composites in wet and dry conditions are illustrated in Fig. 19. It can be seen that the abrasive additions greatly reduce the wear rates of the composites in water to values that are not very different from those in dry conditions either with or without abrasives. It thus appears to be possible, by choosing fillers with a sufficient degree of abrasiveness, to formulate polymer composites whose wear properties remain insensitive to the presence of fluids. This aspect is of significance for those applications, such as dry bearings, in which contamination by fluids often occurs (Evans and Lancaster, 1976).

X. Applications

The largest tribological application of polymers is in bearings. These may be dry or lubricated and the materials used cover the whole range of those discussed above. Some of the most widely used products have been

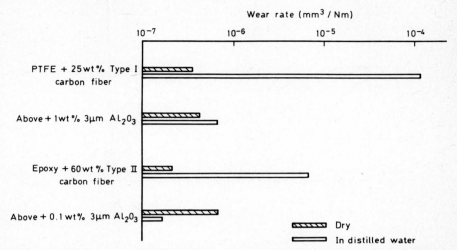

Fig. 19.   Effect of water on the wear rates of carbon fiber composites, with and without added abrasive, sliding against stainless steel (0.15 μm R_a). (After Evans, 1978b and Lancaster, 1972b. Crown copyright. Reproduced by permission of the Controller, Her Britannic Majesty's Stationery Office.)

reviewed by Pratt (1967b) and Lancaster (1973a). The selection of the most appropriate material for a specific dry bearing application is not always easy. The conventional way of rating performance of a material is in terms of its $P-V$ properties—the maximum allowable bearing pressure at a particular speed—and these may take two forms: the limiting PV product above which the wear rate, or the coefficient of friction, begins to increase rapidly, and the relationship between P and V for a specified wear rate. Some examples of $P-V$ relationships for a number of polymer composite bearings sliding against steel are shown in Fig. 20. The main difficulty with both of these approaches is that no account is taken of the variation in the strength of the material, and its wear rate, with temperature. To incorporate this aspect, a modified design procedure for dry bearings has recently been introduced (ESDU, 1976). The surface temperature of the bearing is first calculated by approximate methods and suitable materials are then selected from the known relationships between allowable pressure and temperature. Wear rates of these materials, and in turn bearing lives, are derived from the known specific wear rates obtained in moderate conditions of sliding where frictional heating is negligible, suitably corrected for temperature, pressure, speed, type of motion, and the initial counterface roughness. The specific wear rate of every material is, of course, different, but a general idea of the values typical of various groups of polymers and composites is shown in Fig. 21.

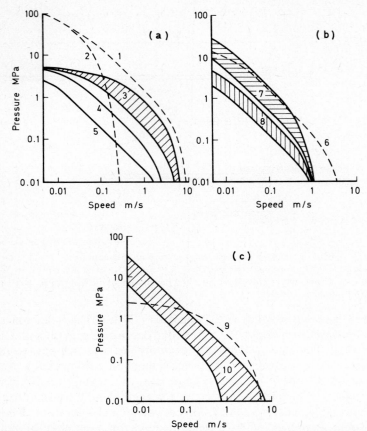

Fig. 20. *P–V* relationships for dry bearing materials at a depth-wear rate of 25 μm/100 hr. (a) PTFE composites: 1, porous bronze + PTFE + Pb; 2, woven PTFE fiber/glass fiber + thermosetting resin; 3, many fillers, including graphite, carbon, ceramic, glass fiber; 4, mica filled; 5, unfilled. (b) thermoplastics: 6, polyimide + graphite; 7, nylons and polyacetals with graphite, MoS_2, glass fiber, PTFE, or bronze fillers; 8, unfilled nylons and polyacetals. (c) filled or reinforced thermosetting resins: 9, epoxy with graphite, MoS_2, or PTFE fillers; 10, polyimides, silicones, epoxies, phenolics. (After Lancaster, 1973a. Crown copyright. Reproduced by permission of the Controller, Her Britannic Majesty's Stationery Office.)

Data on the lubricated wear of polymer bearings are less easy to find in the literature since such bearings are designed, wherever possible, to operate in fully hydrodynamic conditions. Information on water-lubricated bearings has been reviewed by Hother-Lushington (1976), while Smith (1967) has described the wear of polymer-based bearings in marine applications. A special case of lubricated bearing operation occurs with prosthetic joints. There is still some controversy as to whether the lubrication regime of natural joints is hydrodynamic (squeeze film) or boundary (Unsworth *et al.*, 1975), but for polymer–metal prostheses, the latter is more likely to occur. The potential of numerous thermoplastics for prostheses has been examined during sliding against stainless steel or cobalt–chrome alloy in water, saline solution, and synovial fluid, but ultrahigh molecular weight polyethylene (UHMWPE) is generally recognized to give lowest wear (Dumbleton and Shen, 1976). Extensive wear measurements have been made on this material using apparatus of the pin and disk type (see, e.g., Dowson *et al.*, 1974; Rostoker and Galante, 1976), or on complete prostheses in apparatus simulating the walking cycle (Beutler *et al.*, 1975; Trent and Walker, 1976). The low wear rates obtained in such tests appear to be in general agreement with those deduced from x-ray measurements on prostheses in situ and on those "withdrawn from service" (Atkinson *et al.*, 1978).

In addition to the uses of bulk polymers for bearings, there are many applications of polymers as thin film solid lubricant coatings. This topic has been comprehensively reviewed by Benzing (1964) and Campbell (1969). PTFE films can be formed on metals from thin self-adhesive tape, by powder spraying, by electrophoretic deposition, or by rf sputtering. PTFE is also sometimes used in conjunction with thermosetting resin binders to avoid the high temperatures needed for sintering ($\approx 325°C$). Thermosetting resin binders are usually an essential ingredient of thin

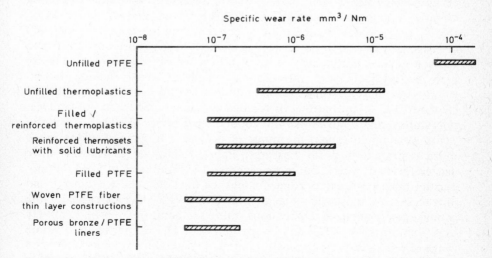

Fig. 21. Order of magnitude values for the specific wear rates of various groups of polymers and composites as dry bearing materials. (Crown copyright. Reproduced by permission of the Controller, Her Britannic Majesty's Stationery Office.)

films containing other types of solid lubricants such as graphite and MoS_2, and the materials involved include phenolics, vinyl butyral, acrylics, epoxies, silicones, and more recently, polyimides, polybenzothiazoles, and polybenzimidazoles. Binder ratios vary from about 25 to 50 vol %. Resin-bonded MoS_2 films are typically from 5 to 25 μm in thickness, and wear *per se* is seldom measured; the usual criterion of performance is the life to failure. However, Finkin (1970) has derived approximate values for the specific wear rates of such films, and they appear to be generally similar to those of bulk reinforced thermosetting polymers with solid lubricant additions (Lancaster, 1973a). While there are very significant differences in the performance of MoS_2 films containing different types of resin binders (e.g., Stupp, 1958), it has so far proved impossible to define the optimum combination of physical and mechanical properties required in an ideal binding resin (Kinner *et al.,* 1976).

A special case of thin film solid lubrication occurs when polymers are used as third components to deposit a lubricating film continuously by transfer to one or another of two contacting metal surfaces. This is described as "transfer lubrication" or, in the Russian literature, "rotaprint lubrication" (see, e.g., Troyanovskaya, 1974). The main application of the technique is in rolling element bearings with retainers fabricated from PTFE/glass fiber/MoS_2 (Sitch, 1973). In dry operation, PTFE (and/or MoS_2) is transferred from the retainer pockets to the balls and, in turn, to the races. Design information has recently become available relating bearing life to load, speed, and temperature for different bearing sizes with retainers of both PTFE/glass fiber/MoS_2 and polyimide/MoS_2 (NCT, 1977). Even in conditions of conventional lubrication by oils and greases, polymer retainers for rolling element bearings are now tending to replace metals, particularly in applications where rotational speeds are high or lubrication is intermittent; materials used include fabric-reinforced thermosetting resins, thermoplastics such as polyacetal (Kunkel, 1977), and porous or foamed polymers, including polyamides, polyimides, and a nitrile–acrylic copolymer (Christy, 1975).

For gear applications, the range of polymers in general use is much more restricted than for bearings. Fabric-reinforced thermosets have served for many years, but more recently the advantages of filled thermoplastics have started to become evident. Of the latter, the polyamides and polyacetals are most widely used, with UHMWPE showing considerable promise for specialized applications where temperatures remain relatively low (Rankin, 1970; Yelle and Poupard, 1975). The properties of a whole range of reinforced thermoplastics potentially suitable for gears have been described by Theberge (1971). Target and Nightingale (1970) conclude that for lubricated plastic gears the primary cause of failure is generally tooth fracture arising from fatigue, but for unlubricated gears wear is the most serious problem. Design information relating to gear tooth design, strength, and fatigue properties is readily available (see, e.g., Shanley and Lamont, 1976, 1977), but information on wear is less so. In only a few instances have comparative wear measurements been made on gears (see, e.g., Accaries and Ribollet, 1970), and it is usually necessary for design purposes to interpolate from wear data obtained in other conditions of sliding.

Conventional friction (brake) materials are composites of resin-bonded asbestos fibers and a variety of other additives such as metals, rubbers, graphite, oxides, and other inorganic compounds. Measurements of their wear properties are normally made either in dynamometer tests simulating the intended application (Starkey *et al.*, 1971) or in road vehicles themselves. In these circumstances, the wear-controlling parameters such as load, speed, and temperature all vary continuously with time. Only comparatively recently have wear measurements been made in controlled, isothermal conditions (Begelinger and de Gee, 1974). In a series of tests of the latter type, Rhee (1970, 1971, 1974) and Liu and Rhee (1976) have derived an empirical relationship, relating wear to the sliding conditions: wear $\propto P^a V^b t \; e^{-E/RT}$, where P is the pressure, V the speed, t the time, T the surface temperature, and a and b are constants. At low temperatures ($<230°C$) wear is considered to result from a combination of abrasive and adhesive processes, but at higher temperatures pyrolysis of the resin binder becomes the rate-controlling factor. Attempts to relate the wear of brake materials to their detailed composition and structure are seldom divulged in the open literature, but one exception is given by Weintraub *et al.* (1974), who have examined the effects on wear of the types of asbestos fiber and resin, the duration of resin cure, and the incorporation of a large number of different inorganic additives.

There are numerous other tribological applications of polymers in which wear is a significant problem, e.g., flooring materials (Gavan, 1967), face seals (Mayer, 1972), elastomeric lip seals (Brown, 1977), piston rings (Summers-Smith, 1971), dental restoratives (Lee *et al.*, 1974), and machine tool slides (Hemingray, 1973). Much of the data available in the literature on these topics, however, tend to be highly specialized and are difficult to summarize within the context of the present review. It is unfortunate that in many practically oriented publications, wear data are seldom presented in a form suitable for comparison with the general body of information derived from controlled laboratory tests, i.e., in terms of a specific wear rate. If specific wear rates were more widely quoted, considerable cross fertilization among different types of applications could lead, ultimately, to more rapid solutions of wear problems.

References

Abouelwafa, M. N., Dowson, D., and Atkinson, J. R. (1978). *Proc. Leeds–Lyon Symp. Tribol., 3rd,* p. 56. Mech. Eng. Publ. Ltd., London.

Accaries, M., and Ribollet, J. (1970). Paper presented at GAMI Conf., Paris, September.

Aharoni, S. M. (1973). *Wear* **25**, 309.

Allen, A. J. G. (1959). *J. Colloid Sci.* **14**, 206.

Archard, J. F. (1953). *J. Appl. Phys.* **24**, 981.

Archard, J. F., and Hirst, W. (1956). *Proc. R. Soc. London Ser. A* **236**, 397.

Archard, J. F., and Kirk, M. T. (1963). *Proc. Inst. Mech. Eng. Lub. Wear Convent.* p. 181.

Arkles, B. C., and Theberge, J. (1973). *Lub. Eng.* **29**, 552.

Arkles, B. C., and Schireson, M. J. (1976). *Wear* **39**, 177.

Arkles, B. C., Gerakaris, S., and Goodhue, R. (1974). *In* "Advances in Polymer Friction and Wear" (L.-H. Lee, ed.), Polymer Science and Technology Symposia Series, Vol. 5B, p. 663. Plenum Press, New York.

Atkinson, J. R., Charnley, J., Dowling, J. M., and Dowson, D. (1978). *Proc. Leeds–Lyon Symp. Tribol., 3rd,* p. 127. Mech. Eng. Publ. Ltd., London.

Bahadur, S. (1974). *Wear* **29**, 323.

Bahadur, S., and Ludema, K. C. (1971). *Wear* **18**, 109.

Bark, L. S., Moran, D., and Percival, S. J. (1975). *Wear* **34**, 131.

Begelinger, A., and de Gee, A. W. J. (1974). Erosion, Wear and Interfaces with Corrosion, p. 316. ASTM STP 567.

Belyi, V. A. and Sviridyonok, A. I. (1974). *In* "Advances in Polymer Friction and Wear" (L.-H. Lee, ed.), Polymer Science and Technology Symposia Series, Vol. 5B, p. 745. Plenum Press, New York.

Benzing, R. J. (1964). *In* "Modern Materials" (B. W. Gonser and H. H. Hausner, eds.), Vol. 4, p. 244. Academic Press, New York.

Beutler, H., Lehmann, M., and Stahli, G. (1975). *Wear* **33**, 337.

Billmeyer, F. W. (1971). "Textbook of Polymer Science," 2nd ed. Wiley (Interscience), New York.

Bitter, J. G. A. (1963). Pt. I, *Wear* **6**, 5. Pt. II, *Wear* **6**, 169.

Boor, L. (1960). *ASTM Bull.* **244**, 43.

Booser, E. R., Scott, E. H., and Wilcock, D. F. (1957). *Proc. Inst. Mech. Eng. Lub. Wear Conf.* p. 366.

Bowers, R. C., Clinton, W. C., and Zisman, W. A. (1954). *Ind. Eng. Chem.* **46**, 2416.

Bowers, R. C., Jarvis, R. C., and Zisman, W. A. (1965). *Ind. Eng. Chem. Prod. Res. Develop.* **4**, 86.

Brainard, W. A., and Buckley, D. H. (1973). *Wear* **26**, 75.

Briggs, G. A. D., and Briscoe, B. J. (1976). *Nature (London)* **260**, 313.

Briscoe, B. J. (1977). *PRI Conf. How long Do Polym. Last in Service, London, February,* Paper 9.

Briscoe, B. J., Mustafaev, V., and Tabor, D. (1972). *Wear* **19**, 399.

Briscoe, B. J., Pogosian, A. K., and Tabor, D. (1974). *Wear* **27**, 19.

Briscoe, B. J., Steward, M. D., and Groszek, A. J. (1977). *Wear* **42**, 99.

Brodskii, G. I., and Reznikovskii, M. M. (1967). *In* "Abrasion of Rubber" (D. I. James, ed.), p. 81. MacLaren, London.

Brown, J. M. (1977). *ASLE Trans.* **20**, 161.

Brown, K. J., Atkinson, J. R., Dowson, D., and Wright, V. (1976). *Wear* **40**, 255.

Brown, R. D., and Blackstone, W. R. (1974). *Conf. Composite Mater. Test. and Design, 3rd* p. 457. ASTM STP 546.

Brydson, J. A. (1975). "Plastics Materials," 3rd ed. Butterworth, London.

Buckley, D. H. (1974). *In* "Advances in Polymer Friction and Wear" (L.-H. Lee, ed.), Polymer Science and Technology Symposia Series, Vol. 5B, p. 601. Plenum Press, New York.

Buckley, D. H., and Brainard, W. A. (1974). *In* "Advances in Polymer Friction and Wear" (L.-H. Lee, ed.), Polymer Science and Technology Symposia Series. Vol. 5A, p. 315. Plenum Press, New York.

Buckley, D. H., and Johnson, R. L. (1963). NASA Tech. Note D-2073.

Bueche, A. M., and Flom, D. G. (1958). *Wear* **2**, 168.

Buist, J. M., and Davies, O. L. (1946). *Trans. IRI* **22**, 68.

Bunn, C. W., Cobbold, A. J., and Palmer, R. P. (1958). *J. Polym. Sci.* **28**, 365.

Burwell, J. T. (1957). *Wear* **1**, 119.

Butterfield, R., Farmer, D., and Scurr, E. M. (1971). *Wear* **18**, 243.

Campbell, M. E., and Hopkins, V. (1967). *Lub. Eng.* **23**, 288.

Campbell, W. E. (1969). *In* "Boundary Lubrication" (F. F. Ling, E. E. Klaus, and R. S. Fein, eds.), Chapter 10. American Society of Mechanical Engineers, New York.

Christy, R. I. (1975). *Lub. Eng.* **31**, 84.

Clerico, M., and Rosetto, S. (1973). *Meccanica* **8**, 174.

Counsell, P. J. C. (1973). "Aspects of Adhesion," Vol. 7, p. 202. Transcripta, London.

Craig, W. D. (1964). *Lub. Eng.* **20**, 456.

Cudworth, C. J., and Higginson, G. R. (1976). *Wear* **37**, 299.

Davies, G. R. (1966). IME Paper ADP9/66. Institution of Mechanical Engineers, London.

Dowson, D. Atkinson, J. R., and Brown, K. J. (1974). *In* "Advances in Polymer Friction and Wear" (L.-H. Lee, ed.), Polymer Science and Technology Symposia Series, Vol. 5B, p. 533. Plenum Press, New York.

Dowson, D., Challen, J. M., Holmes, K., and Atkinson, J. R. (1978). *Proc. Leeds –Lyon Symp. Tribol., 3rd,* p. 98. Mech. Eng. Publ. Ltd., London.

Dumbleton, J. H., and Shen, C. (1976). *Wear* **37**, 279.

Economy, J., and Cottis, S. G. (1971). *Encycl. Polym. Sci. Technol.* **15**, 292.

Egorenkov, N. I., Rodnerkov, V. G., and Belyi, V. A. (1973). *Sov. Mater. Sci.* **9**, 279.

Eiss, N. S., Warren, J. H., and Doolittle, S. D. (1976). *Wear* **38**, 125.

Eiss, N. S., Warren, J. H., and Quinn, T. F. J. (1978). *Proc. Leeds –Lyon Symp. Tribol., 3rd,* p. 18. Mech. Eng. Publ. Ltd., London.

ESDU (1976). Eng. Sci. Data Item 76029. Institution of Mechanical Engineers, London.

Evans, D. C. (1977). RAE Tech. Rep. TR 77070.

Evans, D. C. (1978a). *Proc. Leeds –Lyon Symp. Tribol., 3rd,* p. 47. Mech. Eng. Publ. Ltd., London.

Evans, D. C. (1978b). *ASLE 2nd Int. Conf. Solid Lub.,* p. 202. ASLE SP-6.

Evans, D. C., and Lancaster, J. K. (1976). Patent Appl. No. 14892/76.

Fearon, F. W. G., and Smith, R. F. (1974). *In* "Advances in Polymer Friction and Wear" (L.-H. Lee, ed.), Polymer Science and Technology Symposia Series, Vol. 5B, p. 481. Plenum Press, New York.

Finkin, E. F. (1970). *J. Lub. Technol.* **92**, 274.

Flom, D. G. (1960). *Anal. Chem.* **32**, 1550.

Fort, T. (1962). *J. Phys. Chem.* **66**, 1136.

Fuchsluger, J. H., and Taber, R. D. (1971). *J. Lub. Technol.* **93**, 423.

Fyall, A. A. (1970). *In* "Radome Engineering Handbook" (J. D. Walton, ed.), Chapter 8, p. 461. Dekker, New York.

Gavan, F. M. (1967). "Testing of Polymers" (J. V. Schmitz and W. E. Brown, eds.), Vol. 3, p. 139. Wiley (Interscience), New York.

Geil, P. H. (1975). *Ind. Eng. Chem. Prod. Res. Develop.* **14**, 59.

Giltrow, J. P. (1970). *Wear* **15**, 71.

Giltrow, J. P. (1973). *Tribology* **6**, 253.

Giltrow, J. P., and Lancaster, J. K. (1968). *Proc. Inst. Mech. Eng.* **182** (3N), 147.

Giltrow, J. P., and Lancaster, J. K. (1970). *Wear* **16**, 359.

Giltrow, J. P., and Lancaster, J. K. (1971). *Proc. Int. Conf. Carbon Fibers, Their Compos. and Appl.* p. 251. Plastics Inst., London.

Godet, M., and Play, D. (1978). *Proc. Leeds–Lyon Symp. Tribol., 3rd,* p. 77. Mech. Eng. Publ. Ltd., London.

Gorokovskii, G. A. (1965). *Sov. Mater. Sci.* **1**, 365.

Gorokovskii, G. A., and Agulov, I. I. (1966). *Sov. Mater. Sci.* **2**, 78.

Graham, I. D., and West, G. H. (1976). *Wear* **36**, 111.

Greenwood, J. A., and Williamson, J. P. B. (1966). *Proc. R. Soc. London Ser. A* **295**, 300.

Greenwood, J. A., Minshall, H., and Tabor, D. (1961). *Proc. R. Soc. London Ser. A* **259**, 480.

Griffin, G. J. L. (1972). *ASLE Trans.* **15**, 171.

Grosch, K. A. (1963). *Proc. R. Soc. London Ser. A* **274**, 21.

Grosch, K. A., and Shallamach, A. (1977). *Rubber Chem. Technol.* **49**, 862.

Halliday, J. S. (1955). *Proc. Inst. Mech. Eng.* **169**, 777.

Hammitt, F. G., Timm, E. E., Hwang, J. B., and Huang, Y. C. (1974). Erosion, Wear and Interfaces with Corrosion, p. 197. ASTM STP 567.

Hanford, W. E., and Joyce, R. M. (1946). *J. Am. Chem. Soc.* **68**, 2082.

Hasen, C. M., and Beerbower, A. (1971). *Encycl. Chem. Technol. Suppl. Vol.* p. 889.

Harper, F. C. (1961). *Wear* **4**, 461.

Hart, R. R. (1966). *Proc. Inst. Mech. Eng.* **181** (Pt. 1), 1.

Hemingray, C. P. (1973). *Proc. Mach. Tool Des. Res. Conf.* (S. A. Tobias and F. Koenigsberger, eds.), p. 99. Macmillan, New York.

Hill, M. P. L., Millard, P. L., and Owen, M. J. (1974). *In* "Advances in Polymer Friction and Wear" (L.-H. Lee, ed.), Polymer Science and Technology Symposia Series, Vol. 5B, p. 469. Plenum Press, New York.

Hill, H. W., Werkman, R. T., and Carrow, G. E. (1974). *Adv. Chem. Ser.* **134**, 149.

Hoff, G., and Langbein, G. (1966). *Kunstoffe* **56**, 2.

Hollander, A. E., and Lancaster, J. K. (1973). *Wear* **25**, 155.

Hooke, C. J., and O'Donohugue, J. P. (1972). *J. Mech. Eng. Sci.* **14**, 34.

Hother-Lushington, S. (1976). *Tribology* **9**, 257.

Ikeda, H. (1975). Japanese Patent 75 101, 441 [*Chem. Abstr.* **83** 207230 (1975)].

Israelachvili, J. N., and Tabor, D. (1973). *Progr. Surface Membrane Sci.* **7**, 1.

Istomin, N. P., and Krushchov, M. M. (1973). *Mech. Sci.* **1**, 97.

Jaeger, J. C. (1943). *Proc. R. Soc. NSW.* **76**, 203.

James, D. I. (1959). *Wear* **2**, 183.

Johnson, K. L., Kendall, K., and Roberts, A. D. (1971). *Proc. R. Soc. London Ser. A* **324**, 301.

Kar, M. K., and Bahadur, S. (1974). *Wear* **30**, 337.

Kendall, K. (1975). *Wear* **33**, 351.

King, R. F., and Tabor, D. (1953). *Proc. Phys. Soc.* **B66**, 728.

Kinner, G. H., Pippett, J. S., and Atkinson, I. B. (1976). RAE Tech. Rep. TR 76026.

Klitenik, G. S., and Ratner, S. B. (1967). *In* "Abrasion of Rubber" (D. I. James, ed.), p. 64. MacLaren, London.

Kraghelskii, I. V. (1965). "Friction and Wear." Butterworth, London.

Kraghelskii, I. V., and Nepomnyaschii, E. F. (1965). *Wear* **8**, 303.

Kunkel, H. (1977). *Ball Bear. J.* **191**, 1.

Lancaster, J. K. (1968a). *Tribology* **1**, 240.

Lancaster, J. K. (1968b). *Brit. J. Appl. Phys. J. Phys. D Ser. 2* **1,** 549.
Lancaster, J. K. (1969a). *Proc. Inst. Mech. Eng.* **183** (3P), 100.
Lancaster, J. K. (1969b). *Wear* **14,** 223.
Lancaster, J. K. (1971). *Tribology* **4,** 82.
Lancaster, J. K. (1972a). *Tribology* **5,** 249.
Lancaster, J. K. (1972b). *Wear* **20,** 315.
Lancaster, J. K. (1972c). *Wear* **20,** 335.
Lancaster, J. K. (1972d). *In* "Polymer Science" (A. D. Jenkins, ed.), Chapter 14. North-Holland Publ., Amsterdam.
Lancaster, J. K. (1973a). *Tribology* **6,** 219.
Lancaster, J. K. (1973b). *Plast. Polym.* **41,** 297.
Lancaster, J. K. (1975). *J. Lub. Technol.* **97,** 187.
Langbein, G. (1965). *Proc. Rain Eros. Conf., Meersburg* (A. A. Fyall and R. B. King, eds.), p. 81. RAE Farnborough, U. K.
Lee, L.-H. *In* "Advances in Polymer Friction and Wear" (L.-H. Lee, ed.), Polymer Science and Technology Symposia Series, Vol. 5A, p. 31. Plenum Press, New York.
Lee, L.-H., Orlowski, J. A., Glace, W. R., Kidd, P. D., and Enabe, E. (1974). *In* "Advances in Polymer Friction and Wear" (L.-H. Lee, ed.), Polymer Science and Technology Symposia Series, Vol. 5B, p. 705. Plenum Press, New York.
Leslie, V. J., Rose, J., Rudkin, G. O., and Feltzin, J. (1975). *Chem. Tech.* **5,** 426.
Lewis, R. B. (1967). *In* "Testing of Polymers" (J. V. Schmitz and W. E. Brown, eds.), Vol. 3, p. 203. Wiley (Interscience), New York.
Lewis, W. D. (1968). *Lub. Eng.* **24,** 122.
Liu, T., and Rhee, S. K. (1976). *Wear* **37,** 291.
Ludema, K. C., and Tabor, D. (1966). *Wear* **9,** 329.
MacLaren, K. G., and Tabor, D. (1963). *Proc. Inst. Mech. Eng. Lub. Wear Convent.,* p. 210.
Makinson, K. R., and Tabor, D. (1964). *Proc. R. Soc. London Ser. A* **281,** 49.
Marei, A. I., and Izvozchikov, P. V. (1967). *In* "Abrasion of Rubber" (D. I. James, ed.), p. 274. MacLaren, London.
Matsubara, K., and Watanabe, M. (1967). *Wear* **10,** 214.
Matveevskii, R. M. (1961). *Wear* **4,** 300; *Russ. Eng. J.* **40,** 27.
Mayer, E. (1972). "Mechanical Seals," 2nd ed. Iliffe, London.
Merchant, M. E. (1968). "Interdisciplinary Approach to Friction and Wear" (P. M. Ku, ed.), p. 181. NASA SP-181.
Mitchell, D. C., and Pratt, G. C. (1957). *Proc. Inst. Mech. Eng. Lub. Wear Conf.* p. 416.
Moore, D. F. (1975a). "The Friction of Pneumatic Tires." Elsevier, Amsterdam.
Moore, D. F. (1975b). *Wear* **35,** 159.
NCT (1977). Self-lubricating Bearings—a Performance Guide. Nat. Centre of Tribology, Warrington, UtK.
OECD (1969). Friction, Wear and Lubrication—Terms and Definitions. Res. Group on Wear of Eng. Mater., Paris.
Ohmae, N., Kobayashi, K., and Tsukizoe, T. (1974). *Wear* **29,** 345.
O'Rourke, J. T. (1962). *Mach. Des.* **34**(No. 21), 172.
O'Rourke, J. T. (1965). *Mod. Plast.* **42**(No. 1), 161.
Pascoe, M. W., and Dzhanokmedov, A. K. (1978). *Proc. Leeds–Lyon Symp. Tribol., 3rd,* p. 60. Mech. Eng. Publ. Ltd., London.
Play, D., and Godet, M. (1978). *Proc. Leeds–Lyon Symp. Tribol., 3rd,* p. 221. Mech. Eng. Publ. Ltd., London; *Wear* **42,** 197 (1977).
Pocock, G., and Cadman, P. (1976). *Wear* **37,** 129.
Pooley, C. M., and Tabor, D. (1972). *Proc. R. Soc. London Ser. A* **329,** 251.

Pratt, G. C. (1964). *Trans. Plast. Inst.* **32**, 255.

Pratt, G. C. (1967a). *Proc. Inst. Mech. Eng.* **181** (30), 58.

Pratt, G. C. (1967b). *In* "Lubrication and Lubricants" (E. R. Braithewaite, ed.), Chapter 7. Elsevier, Amsterdam.

Rabinowicz, E., and Shooter, K. V. (1952). *Proc. Phys. Soc.* **B65**, 671.

Rankin, I. M. (1970). Pt. I, *Eng. Dig.* **31**(10), 97; Pt. II, *Eng. Dig.* **31**(11), 73.

Ratner, S. B. (1967). *In* "Abrasion of Rubber" (D. I. James, ed.), p. 23. MacLaren, London.

Ratner, S. B., and Farberova, I. I. (1960). *Sov. Plast.* (No. 9), 51; *also in* "Abrasion of Rubber" (D. I. James, ed.), p. 297. MacLaren, London.

Ratner, S. B., and Klitenik, G. S. (1959). *Zav. Lab.* **25**, 1375.

Ratner, S. B., Farberova, I. I., Radyukevich, O.V., and Lure, E. G. (1964). *Sov. Plast.* (No. 7); 37; *also in* "Abrasion of Rubber" (D. I. James, ed.), p. 145. MacLaren, London.

Ratner, S. B., and Lure, E. G. (1966a). *Vysokomol. Soed.* **8**, 88; *also in* "Abrasion of Rubber" (D. I. James, ed.), p. 161. MacLaren, London.

Ratner, S. B., and Lure, E. G. (1966b). *Dokl. Akad. Nauk. SSSR* **166**, 909; *also in* "Abrasion of Rubber" (D. I. James, ed.), p. 155. MacLaren, London.

Rehbinder, P., Lichtmann, V. I., and Maslenikov, V. M. (1941). *C. R. Acad. Sci. URSS* **32**, 125.

Reznikovskii, M. M. (1960). *Sov. Rubber Technol.* **19**, 32; *also in* "Abrasion of Rubber" (D. I. James, ed.), p. 119. MacLaren, London.

Reznikovskii, M. M., and Brodskii, G. I. (1967). *In* "Abrasion of Rubber" (D. I. James, ed.), p. 14. MacLaren, London.

Rhee, S. H., and Ludema, K. C. (1978). *Proc. Leeds –Lyon Symp. Tribol., 3rd,* p. 11. Mech. Eng. Publ. Ltd., London.

Rhee, S. K. (1970). *Wear,* **16**, 431.

Rhee, S. K. (1971). *Wear* **18**, 471.

Rhee, S. K. (1974). *Wear* **29**, 391.

Richardson, M. O. W. (1971). *Wear* **17**, 89.

Roberts, A. D. (1976). *Tribology* **9**, 75.

Roberts, A. D. (1977a). *Tribology* **10**, 115.

Roberts, A. D. (1977b). *Tribology* **10**, 175.

Roberts, A. D., and Thomas, A. G. (1975). *Wear* **33**, 45.

Rostoker, W., and Galante, W. O. (1976). *J. Biomed. Mater. Res.* **10**, 303.

Schmitt, G. F. (1967). *Proc. Meersburg Conf. Rain Eros. Allied Phenomena, 2nd* (A. A. Fyall and R. B. King, eds.), p. 329. R. A. E., Farnborough, UK.

Schubert, R. (1971). *J. Lub. Technol.* **93**, 216.

Senior, J. M., and West, G. H. (1971). *Wear* **18**, 311.

Shallamach, A. (1952). *J. Polym. Sci.* **9**, 385.

Shallamach, A. (1958). *Wear* **1**, 384.

Shallamach, A. (1963). *In* "The Chemistry and Physics of Rubber-like Substances" (L. Bateman, ed.), Chapter 13. MacLaren, London.

Shanley, R., and Lamont, L. (1976). Pt. I, *Mach. Des.* December 9, 125.

Shanley, R., and Lamont, L. (1977). Pt. II, *Mach. Des.* January 6, 75.

Sharples, A. (1972). *In* "Polymer Science" (A. D. Jenkins, ed.), Chapter 4. North-Holland Publ., Amsterdam.

Shcherbakov, S. V., and Kaplan, M. B. (1974). *Russ. Eng. J.* **59**, 40.

Shen, C., and Dumbleton, J. H. (1974). *Wear* **30**, 349.

Shen, C., and Dumbleton, J. H. (1976). *Wear* **38**, 291.

Shooter, K. V., and Tabor, D. (1952). *Proc. Phys. Soc.* **B65**, 661.

Shooter, K. V., and Thomas, P. H. (1949). *Research (London)* **2**, 533.

Sitch, D. (1973). *Tribology* **6**, 262.

Sliney, H. E., and Jacobsen, T. P. (1975). *Lub. Eng.* **31**, 609.

Smith, W. V. (1967). "Testing of Polymers" (J. V. Schmitz and W. E. Brown, eds.), Vol. 3, p. 221. Wiley (Interscience), New York.

Southern, E., and Thomas, A. G. (1978). *Proc. Leeds–Lyon Symp. Tribol., 3rd*, p. 157. Mech. Eng. Publ. Ltd., London.

Speerschneider, C. J., and Li, C. H. (1962a). *J. Appl. Phys.* **33**, 1871.

Speerschneider, C. J., and Li, C. H. (1962b). *Wear* **5**, 392.

Springer, G. S. (1976). "Erosion by Liquid Impact." Wiley, New York.

Starkey, W. L., Foster, T. G., and Marco, S. M. (1971). *Trans. ASME Ser. B* **93**, 1225.

Stupp, B. C. (1958). *Lub. Eng.* **14**, 159.

Suh, N. P. (1973). *Wear* **25**, 111.

Summers-Smith, D. (1971). *J. Lub. Technol.* **93**, 293.

Sviridyonok, A. I., Belyi, V. A., Smurugov, V. A., and Savkin, V. G. (1973). *Wear* **25**, 301.

Tabor, D. (1955). *Proc. R. Soc. London Ser. A* **229**, 198.

Tabor, D. (1974). *In* "Advances in Polymer Friction and Wear" (L.-H. Lee, ed.), Polymer Science and Technology Symposia Series, Vol. 5A, p. 1. Plenum Press, New York.

Tanaka, K., and Uchiyama, Y. (1974). *In* "Advances in Polymer Friction and Wear" (L.-H. Lee, ed.), Polymer Science and Technology. Symposia Series, Vol. 5B, p. 499. Plenum Press, New York.

Tanaka, K., and Miyata, T. (1977). *Wear* **41**, 383.

Tanaka, K., Uchiyama, Y., and Toyooka, S. (1973). *Wear* **23**, 153.

Target, J., and Nightingale, J. E. (1970). *Proc. Inst. Mech. Eng. Conf. Gear.* **184**, 30.

Theberge, J. E. (1971). *ASLE Proc. Int. Conf. Solid Lub.* p. 166. ASLE SP-3.

Theberge, J. E., and Arkles, B. C. (1973). *Am. Chem. Soc. Div. Org. Coat. Plast. Chem. Pap.* **33**, 188.

Theberge, J. E., Arkles, B. C., and Cloud, P. (1974). *Mach. Des.* October 31, 60.

Thomas, T. R. (1975). *Wear* **33**, 205.

Trent, P. S., and Walker, P. S. (1976). *Wear* **36**, 175.

Troyanovskaya, G. I. (1974). *Russ. Eng. J.* **54**, 50.

Tsukizoe, T., and Ohmae, N. (1977). *Proc. ASTM Int. Conf. Wear.* p. 518.

Unsworth, A., Dowson, D., and Wright, V. (1975). *J. Lub. Technol.* **97**, 369.

Vinogradov, G. A., and Bezborodko, M. D. (1962). *Wear* **5**, 467.

Weintraub, M. H., Anderson, A. E., and Gaeler, R. L. (1974). *In* "Advances in Polymer Friction and Wear" (L.-H. Lee, ed.), Polymer Science and Technology Symposia Series, Vol. 5B, p. 623. Plenum Press, New York.

Wells, D. M. (1971). *Engineering* December, 1029.

West, G. H., and Senior, J. M. (1973). *Tribology* **6**, 269.

White, H. S. (1956). *J. Res. Nat. Bur. Std.* **57**, 185.

Williams, M. L., Landel, R. F., and Ferry, J. D. (1955). *J. Am. Chem. Soc.* **77**, 3701.

Yelle, H., and Poupard, M. (1975). *Polym. Eng. Sci.* **15**, 90.

Zuer, Yu. S., Borshchevskaya, A. Z., and Kamenskaya, G. G. (1974). *Sov. Mater. Sci.* **10**, 414.

The Wear of Carbons and Graphites

J. K. LANCASTER

Ministry of Defence, Procurement Executive
Materials Department
Royal Aircraft Establishment
Farnborough, Hants, England

I. Introduction

It is important to recognize from the outset that carbons and graphites comprise a wide range of materials whose structure and properties vary enormously, depending on the raw materials involved and the parameters associated with their processing. For simplicity, the generic term *carbon* will be used to describe all these materials, unless specific attention is being drawn to individual types, as defined later. The main tribological applications of carbons are in sliding electrical contacts (brushes), bearings and seals for predominantly hostile environments, and more recently, high-performance aircraft brakes. This chapter reviews the more fundamental aspects of the wear behavior of carbons relevant to these applications. Information on wear is widely scattered throughout the literature,

Copyright © 1979 by Academic Press, Inc.
All rights of reproduction in any form reserved
ISBN 0-12-341813-5

and only one general survey has previously been published (Badami and Wiggs, 1970). Discussions on wear in books by Holm (1958) and Shobert (1965) concentrate primarily on carbon brushes in electrical machines, while Mayer (1972) provides some information on the wear of carbons in seals.

II. Materials

A. *Graphite*

The crystalline structure of graphite is now well characterized (Bernal, 1924). The most common form is hexagonal, containing four atoms per unit cell, in which sheets of hexagonally arranged carbon atoms—basal planes—are stacked successively on top of each other. Within the planes, the carbon atoms are relatively close together, 1.417 Å, and bonding is covalent. The interatomic spacing between planes, however, is much greater, 3.35 Å, and bonding is by Van der Waals forces supplemented by π-electron interactions. Successive planes are displaced with respect to each other so that only alternate planes come into exact register—the so-called *ABAB* . . . stacking sequence. A less common structural modification is rhombohedral graphite, containing six atoms per unit cell, in which three layers are successively displaced to give an *AB-CABC* . . . stacking sequence. This modification can arise from mechanical deformation of the hexagonal form (Friese and Kelly, 1963). In addition to these differences in stacking sequence, defects can occur in which successive planes undergo an annular rotation with respect to each other—a turbostratic structure.

B. *Manufactured Carbons*

Crystalline graphite, approximating to the ideal structure, can be prepared by pyrolytic deposition (Klein *et al.,* 1962), but most manufactured products deviate very significantly from the ideal. General information about the various manufacturing processes involved has been given by Hutcheon (1970). Briefly, various carbonaceous constituents are mixed, together with a viscous binder, pressed into blocks, and then heat treated. Figure 1 shows some of the more common raw materials. The pressure during processing influences density and porosity, and heat treatment progressively volatilizes impurities. At temperatures above about 2000°C, structural reordering begins toward a graphitic structure. The extent to which graphitization occurs depends on both the temperature and the

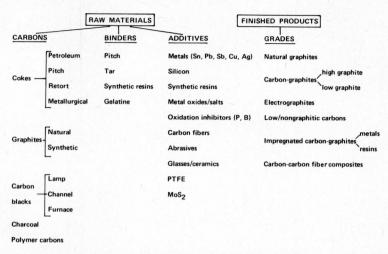

Fig. 1. Typical constituents in manufactured carbons and graphites. (Crown copyright. Reproduced by permission of the Controller, Her Britannic Majesty's Stationery Office.)

type of raw materials involved. Petroleum and pitch cokes graphitize more readily than carbon blacks, and some forms of carbon chars from cellulose or synthetic resins are virtually nongraphitizable. The term *electrographite* is often used to denote materials that have been graphitized by heating with an electric current passing through the structure. Following heat treatment, the residual porosity in the materials can be impregnated with resins, metals, inorganic salts, etc. to improve particular properties.

It is convenient to divide the general range of manufactured carbons into eight groups, as listed in Fig. 1, although the dividing lines among them are often blurred. In addition to these groups there are specialty materials, such as pyrolitic carbon and graphite, glassy carbons, and carbon fibers. Only the last named has so far made any significant contribution in tribology.

C. Properties

It is evident from the above brief and necessarily oversimplified description that the various types of carbon are likely to have a wide spectrum of physical and mechanical properties. Table I gives some *approximate* values for the eight general groups already mentioned, and these can be put into perspective by comparison with similar properties for mild steel. Several characteristic features of carbons may be noted.

TABLE I

Approximate Properties of the General Types of Manufactured Carbons and Graphites

Group	Specific gravity	Compressive strength (MPa)	Tensile strength (MPa)	Compressive modulus (GPa)	Expansion coefficient $\times10^{-6}$/°C	Thermal conductivity (W/m °C)
Natural graphite (pitch-bonded)	1.65	30	5	5	4	50
Electrographite (petroleum coke)	1.7	65	15	7	5	50
Carbon-graphite	1.65	85	20	9	3	35
High graphite	1.6	150	25	11	3	10
Low graphite	1.6	200	25	11	4	7
Low/nongraphitic carbon (petroleum coke)						
Impregnated carbon-graphite						
Resin (phenolic)	1.8	175	40	14	5	10
Metal (Cu-Pb)	2.7	250	50	16	5	25
Carbon–carbon fiber composites	1.5	500	200	100	4	5
Mild steel	7.85	400	550	180	11	45

(a)   Tensile strengths are very much less than compressive strengths, sometimes by as much as a factor of 10.

(b)   Strengths tend to decrease with increasing graphite content or with degree of graphitization.

(c)   Elastic moduli are typically less by factors of 10 to 20 than those of steel.

(d)   Coefficients of thermal expansion are one-quarter to one-half those of steel.

(e)   Thermal conductivity increases with increasing degree of graphitization, and becomes comparable to that of steel for the most graphitic carbons.

Carbon–carbon fiber composites are a comparatively recent development and involve densification of fiber mat or cloth by resin impregnation and subsequent carbonization, or by chemical vapor deposition of carbon from organic vapors. The properties given in Table I are even more approximate than those for the other carbons because they depend greatly on the type of fiber, its physical form (mat or cloth), the fiber concentration, the densification route chosen (resin impregnation or CVD), and the final heat-treatment temperature. They suffice to show, however, that it is possible to produce materials with strengths and stiffnesses roughly comparable to those of mild steel.

Carbons do not exhibit any significant degree of plasticity in their stress–strain behavior. The elastic modulus usually decreases with increasing strain, and failure occurs by brittle fracture at strains typically within the range 0.1–2%. There are no large changes in mechanical properties with temperature and, in contrast to most other materials, strengths and moduli actually increase slightly with temperature in vacuum or inert gas atmospheres. In air, however, some degradation in mechanical properties can result from oxidation, which begins to become significant at around 350–500°C, depending on the type of carbon. In general, the greater the degree of graphitic order, the lower the rate of oxidation at any one temperature.

III.   Friction

Both friction and wear are manifestations of the interaction of small localized areas of contact during sliding, and an examination of the mechanisms of friction thus provides a relevant introduction to the topic of wear. Bragg (1928) first suggested that the low friction of graphite could be a consequence of its anisotropic crystal structure and proposed the idea of easy shear between basal planes. It was soon found, however, that the friction of graphite remains low only in the presence of water or other condensable vapors in the environment (Dobson, 1935; Savage, 1945). In vacuum or inert gas atmospheres, the friction increases typically by a factor of 10 and the wear rate by factors of between 100 and 1000. Bryant *et al.* (1964) have also demonstrated directly that the interlamellar binding energy of graphite in vacuum is relatively high and only decreases when oxygen or water is present. They consider that water might reduce the binding energy by a "stress etch" of the π-electron bonds during cleavage. In contrast, Rowe (1960) suggests that oxygen or water might intercalate between planes and again reduce π-electron bonding. However, x-ray diffraction has failed to reveal any evidence of an increased interlayer spacing, which would necessarily result from intercalation (Arnell and Teer, 1968). It follows that graphite is not an intrinsically low-friction solid such as MoS_2, and the concept of easy shear—or cleavage—is insufficient alone to account for its low coefficient of friction.

An alternative explanation begins by assuming that the contacts occurring during sliding are mainly between basal plane surfaces. This is in agreement with the observation that most worn surfaces of graphite exhibit a preferred orientation of basal planes. It was originally thought that the planes were oriented approximately parallel to the surface (Jenkins, 1934), but later work showed that instead they were tilted at a small angle with their normals pointing against the direction of motion of the opposing surface (Porgess and Wilman, 1960; Midgley and Teer, 1961; Quinn, 1963). Direct observation in the electron microscope of worn surface topographies has also revealed a pattern of overlapping tilted flakes consistent with the orientation evidence from electron diffraction (Fulham and Savage, 1948; Bradley *et al.*, 1956). The basal planes are low-energy surfaces with values of the order of 40 mJ/m² (Larsen *et al.*, 1971) to 80 mJ/m² (Roselman and Tabor, 1976), and in consequence, the adhesive forces between them must be relatively low. During sliding, shear will occur primarily at the interface between contacting basal planes, rather than beneath the surface, thus leading not only to low friction but also to low wear (Roselman and Tabor, 1976). The shear strength associated with basal plane interactions has been estimated at 21 MPa (Holm, 1958), 20 MPa (Skinner, 1971) or 1–5 MPa (Seldin and Nezbeda, 1970). Skinner *et al.* (1973) have shown that the friction coefficient of a tungsten stylus sliding over the basal plane of single-crystal graphite is very low (0.005–0.02) even in high vacuum, provided that there is no surface damage. Once damage occurs, edge sites are exposed and the friction increases by a factor of 10 or so. The shear strength associated with edge-site interactions has been estimated at around 200 MPa, i.e., 10–100 times greater than that between basal planes (Skinner, 1971). It follows that for circumstances in which significant numbers of edge-site interactions can occur, the adhesion between these sites could well dominate the magnitude of the observed friction, as originally suggested by Deacon and Goodman (1958). Evidence for increased friction on graphite oriented to expose mainly edge sites during sliding has been reported by Campbell and Kozak (1948) and Pike and Thompson (1969); Arnell *et al.* (1966), however, failed to find a significant difference in friction with oriented pyrolytic graphite.

Shobert (1965) considers that the adsorbed film of water vapor normally present on carbons in air functions as a hydrodynamic fluid film in reducing friction. However, if it is the edge-site interactions that dominate the frictional behavior, it seems more reasonable to attribute the effects of adsorbed gases and vapors to reduced adhesion between these sites. Savage (1951) first showed that the amount of water vapor required to maintain low friction was only sufficient for about 10% monomolecular coverage over the whole surface, thus implying selective adsorption on preferred sites. There is, however, conflicting evidence. First, it has been concluded that the water vapor responsible for low friction is adsorbed physically, rather than chemisorbed (Campbell and Kozak, 1948; Savage, 1948); on edge sites, however, chemisorption is more probable (Bobka, 1961). Second, while polar molecules such as water are preferentially adsorbed on edge sites, nonpolar organic molecules are more likely to be adsorbed onto the basal planes (Groszek, 1969). Some nonpolar hydrocarbons, e.g., *n*-heptane, are considerably more effective than water in maintaining low friction and wear (Savage and Schaefer, 1956). It is therefore evident that there is still some confusion about the precise role played by water and other vapors in maintaining a low friction for graphite.

As discussed earlier, most manufactured carbons are multiphase materials and often include various components with differing degrees of graphitic order. Even the most highly graphitic phases may contain dislocations, vacancy and interstitial atoms, and interlayer stacking faults. Explanations for the low friction of these materials, based on observations with single crystals of graphite, are thus subject to a considerable degree of extrapolation and uncertainty. One fundamental question still unresolved is whether or not there is any relationship between friction and the degree of graphitic order in carbons. Results by Longley *et al.* (1964), which indicate that the friction coefficients of carbons of high and low graphiticity are very similar, appear at first sight to discount the idea of such a relationship. Unambiguous interpretation of their results, however, is complicated by the fact that all types of carbons undergo drastic surface modifications during sliding. Apart from the development of a preferred orientation of basal planes, there is usually gross movement of material via transfer, or the aggregation of loose wear debris into surface films on one or both of the sliding surfaces (Clark *et al.*, 1963). It is the properties of these surface films that determine the magnitude of the dynamic coefficient of friction; the structure and properties of the bulk material are relevant only insofar as they determine the type and quality of the surface films. Examples of film formation on the surfaces of three different types of carbons are shown in Fig. 2.

It is intuitively obvious that the degradation of carbons into finely divided wear debris during sliding will be more difficult for strong nongraphitic carbons than for weaker graphitic materials. Thus, consolidated debris films from low or nongraphitic carbons can only be formed during sliding against hard counterfaces, whereas films from graphitic carbons form on either hard or soft counterfaces (Clark and Lancaster, 1963; Giltrow and Lancaster, 1970). Little is known about the structure and properties of these films beyond the facts that they are all relatively soft, exhibit some degree of graphitic order irrespective of that of the bulk carbon, and show no discernible structural features greater than about 20 nm. The mechanism of debris film formation is equally uncertain. It is possible that hydrogen bonding occurs between adsorbed water molecules on the edge sites of debris particles, and this is consistent with the observation that film formation almost invariably decreases as the temperature increases, and is not observed at all in dry atmospheres or in vacuum.

In general, the magnitude of the coefficient of friction between surfaces involves two components arising from adhesion and deformation. It is seldom possible, except in highly simplified laboratory experiments, to isolate and quantify each of these contributions. Deformation (interpenetration of asperities, displacement of material and fracture) will be most significant in the early stages of sliding between freshly prepared surfaces that are relatively rough. It is during these early stages that carbon debris film formation occurs and as it proceeds, the surfaces gradually become smoother; the deformation component of friction then presumably decreases in importance. An initial decrease in friction with time of sliding for carbon–carbon or carbon–metal combinations is very frequently observed. Following this initial decrease in friction, the magnitude of the friction coefficient is determined by the properties of the surface films generated, and adhesion probably becomes the most important component. However, a further complication can now ensue. The total (adhesion) frictional force $F = As$, where A is the real area of contact and s is the shear strength per unit area of the regions over which the adhesive

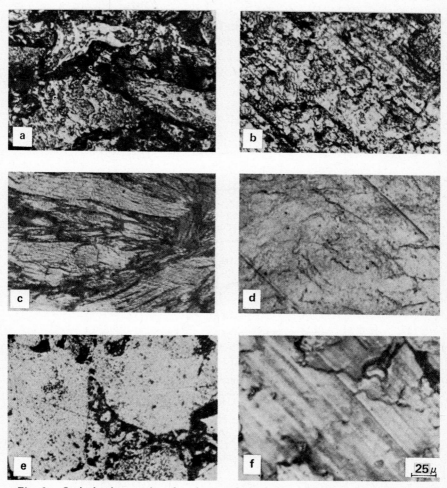

Fig. 2. Optical micrographs of carbon surfaces before and after sliding: (a,b) low-graphitic carbon; (c,d) natural graphite; (e,f) electrographite. (From Clark and Lancaster, 1963.)

forces operate. Changes in either of these parameters will affect friction, and it is not always possible to conclude which is most significant in any particular situation. Two examples will suffice to illustrate the ambiguities involved.

It is well known that the coefficient of friction of graphitic carbon brush materials sliding on metals decreases when current flows across the interface (Barlow, 1924; Hayes, 1947). Some typical results are shown in Fig. 3a. Holm (1962) attributes the fall in friction to a reduction in the specific shear strength s caused by the localized temperature rise consequent on the passage of current. In contrast, Lancaster and Stanley (1964) consider that the friction decreases because of a reduction in the real area of con-

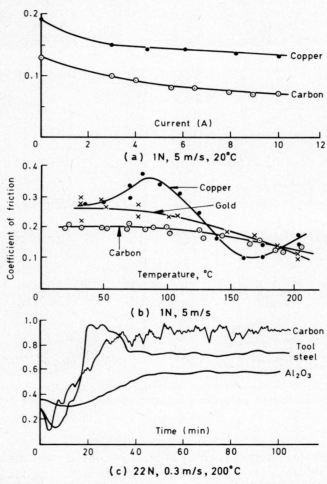

Fig. 3. Variations of friction with (a) current—electrographite; (b) temperature—electrographite; (c) time—low-graphitic carbon. The counterface material is shown adjacent to each curve. (Crown copyright. Reproduced by permission of the Controller, Her Britannic Majesty's Stationery Office.)

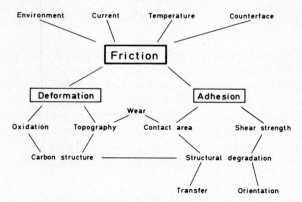

Fig. 4.   Summary of factors influencing the friction of carbons. (Crown copyright. Reproduced by permission of the Controller, Her Britannic Majesty's Stationery Office.)

tact A caused by the fact that the localized temperature rise induces selective oxidation in the contact areas and increases their degree of subdivision. Figure 3b shows that reductions in friction are also observed with increasing temperature in the absence of current, at least on unreactive counterfaces such as gold or carbon. Similar explanations can again be invoked. On copper, however, the variation of friction with temperature is more complex and effects caused by oxidation of the copper itself begin to influence the friction mechanism (Stanley, 1966).

The second example concerns the way in which the coefficient of friction of low or nongraphitic carbons increases with time of sliding, as shown in Fig. 3c. The generally accepted explanation is that of a rise in the real area of contact associated with the development of a very smooth surface topography (Longley *et al.*, 1964; Lancaster, 1977). However, the low coefficients of friction shortly after the onset of sliding are associated with the formation of debris films, which may reduce friction by the conventional mechanism of solid lubrication—a low shear strength film on a harder substrate. These films are gradually removed during prolonged sliding, and a part of the subsequent increase in friction could thus result from an increase in the specific interfacial shear strength.

In summary, it is apparent that the friction of carbons is influenced by a number of complex interrelationships among material properties, the conditions of sliding, and the way in which the surface layers are modified during sliding. Figure 4 shows the most important parameters known, or believed, to influence the magnitude of the coefficient of friction and some of the ways in which they can interact.

IV. Wear

The basic mechanisms of wear of graphite have been studied micro-scopically by Savage (1959a,b,c, 1960) in an elegant series of experiments involving fine wires sliding over the basal plane of single crystals. Several distinct processes were identified including punching of particles out of the basal plane, delamination in front of the slider, and peeling of lamellae at right angles to the direction of sliding. Considerable imagination is re-quired, however, to extrapolate these elementary processes to the wear of polycrystalline, multiphase, partially graphitic, manufactured carbons.

A. Abrasion

One of the most significant properties of a manufactured carbon in rela-tion to wear is its relatively low elastic modulus. Application of Halliday's (1955) criterion to calculate the minimum asperity angles at which the elastic limit is first exceeded gives values of around $7°–15°$; these are an order of magnitude larger than those typical of metals, $0.5°–2°$. In conse-quence, abrasive wear is likely to occur for carbons only during sliding against extremely rough counterfaces, or in the presence of hard, sharp, abrasive particles. Several investigations have been made of the wear of carbons against abrasive paper. Figure 5 shows that in these conditions, the wear rates of different carbons are approximately inversely propor-tional to their indentation hardness, a result in general agreement with that for metals (Krushchov and Babichev, 1953). It can also be seen from Fig. 5 that the abrasive wear rates of carbons are significantly higher than those of polymers and soft metals sliding in the same conditions. The main reason for this is the lack of ductility in carbons. Porgess and Wilman (1960) have shown that the proportion of an abrasion groove vol-ume resulting in loose wear debris is about four times greater for carbons than for metals. Carbons suffer extensive fragmentation at the groove edges, whereas metals merely undergo plastic flow. A further dif-ference between the abrasive wear properties of carbons and metals is in the relationship between the wear rate and the coefficient of friction on abrasive papers of differing grit size. For metals, proportionality is usually observed (Goddard et al., 1959), but for carbons the wear rate in-creases much more rapidly than proportionately (Porgess and Wilman, 1960). The latter is again attributed to extensive fragmentation of the carbon at the abrasion groove edges.

Many manufactured carbons contain hard, nongraphitic constituents, and almost all contain metallic oxide impurities, particularly those grades

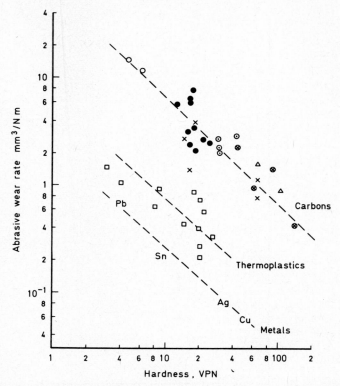

Fig. 5. Relationships between abrasive wear rates on SiC paper (100D) and hardness. Load is 10 N; speed is 0.02 m/sec: O, natural graphites; ×, impregnated carbon graphites; ●, electrographites; ⊗, low/nongraphitic carbons; ⊙, carbon graphites; △, carbon–carbon fiber composites. (Crown copyright. Reproduced by permission of the Controller, Her Britannic Majesty's Stationery Office.)

based on natural graphite. If either, or both, of these are released during the sliding process, the possibility arises of three-body abrasion. The extent to which this might influence wear of the carbon itself is speculative, but of potentially greater significance is the possibility of wear and surface damage to the metal counterface. Figure 6 shows the results of experiments in which various types of carbons were worn against copper rings for periods sufficiently long to cause measurable wear of the copper. The mean wear rate of the copper was assessed from surface profiles, and it can be seen that there is a general increase with increasing impurity content in the carbons. The significance of counterface damage and roughening in relation to the wear of the carbon itself is discussed in more detail later.

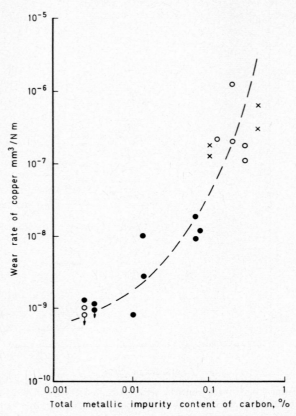

Fig. 6. Variation of wear rate of copper with impurity content in carbons: ○, natural graphites; ●, electrographites; ×, low/nongraphitic carbons. (From Lancaster, 1963a.)

B. Fatigue

It is often supposed that the initially high rates of wear of carbons sliding against metals during "running in" are abrasive in type, but this is not necessarily so. Figure 7a shows a comparison of the mean wear rates of an electrographite and a soft metal (Cd) during sliding in single traversals over lubricated mild steel surfaces of different roughnesses. For the metal, the wear rate increases relatively slowly with the R_a roughness and from computations of the mean base angle of the asperities associated with each roughness (Lancaster, 1964), it can be shown that the wear rate is proportional to tan θ as expected from simple abrasion theory (Rabinowicz, 1964). For the electrographite, however, the increase in wear with roughness is much more rapid. Even on the roughest surfaces θ does not exceed about 10°, and it is therefore probable that throughout the whole range of roughnesses, the deformation of the graphite remains elastic. If so, it is necessary to attribute wear to some mechanism other than abrasion. Several items of evidence have now accumulated to suggest that localized surface fatigue can play a major role in the wear of carbons. Model experiments involving crossed cylinders sliding against each other under nominal point contact stresses below the elastic limit have shown that surface disruption only begins after a finite number of repeated cycles (Clark and Lancaster, 1963). Figure 8 shows that the relationships between stress and cycles to failure bear a marked resemblance to con-

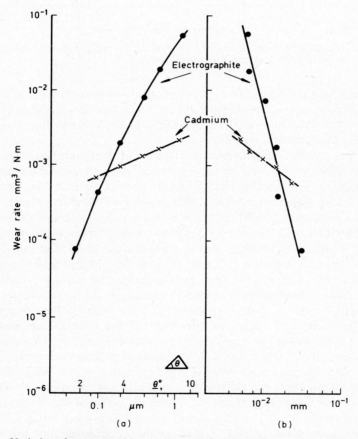

Fig. 7. Variation of wear rate with topographical parameters during single traversals over mild steel counterfaces: (a) counterface roughness, R_a; (b) average radius of counterface asperities. (From Hollander and Lancaster, 1973. Crown copyright. Reproduced by permission of the Controller, Her Britannic Majesty's Stationery Office.)

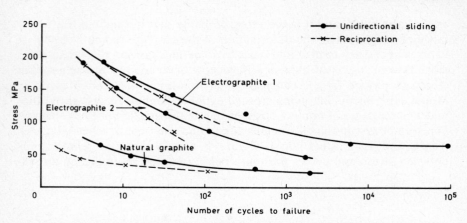

Fig. 8. Variation of cycles to failure with stress for elastically deformed, crossed cylinders of carbons. (From Clark and Lancaster, 1963.)

ventional fatigue curves. The introduction of stress reversals, via reciprocating sliding, accelerates failure. White (1969) has also reported evidence for localized fatigue on worn carbon surfaces from observations using the scanning electron microscope.

Indirect support for a fatigue mechanism of wear can be derived from the way in which the wear of carbons varies with counterface roughness, as in Fig. 7a. Several theories of fatigue wear (Reznikovskii, 1960; Kraghelskii and Nepomnyashchii, 1965) predict an inverse proportionality between wear rate and R^α, where R is the average radius of the counterface asperities, and the index α depends on the particular model of counterface topography invoked and is greater than unity. On the simplest fatigue theory, outlined in the chapter by Evans and Lancaster, $\alpha = \frac{2}{3}(t - 1)$, where t is the exponent in the fatigue relationship $n \propto (\sigma_0/\sigma)^t$; n the number of cycles to failure, σ_0 the failure stress, and σ the applied stress. Figure 7b shows the wear rates of Fig. 7a replotted against the average radius of curvature of the counterface asperities, derived from computer analyses of profilometer traces (Hollander and Lancaster, 1973). The index α is appreciably greater than unity, as predicted, and the calculated value of the exponent $t = 7.6$. While there are little or no conventional fatigue data available for carbons with which to compare this value, expressing the "fatigue" curve of Fig. 8 for the same electrographite 1 in terms of a power relationship gives $t = 8.3$.

The results given in Fig. 7 lead to the conclusion that the surface roughness of the mating counterface is one of the most important parameters influencing the wear rate of carbons. The initial wear rate between freshly prepared surfaces will be characteristic of their roughness and mode of preparation, but the final limiting wear rate in steady-state conditions must depend on the roughness generated by the sliding process itself. It is therefore pertinent at this stage to discuss briefly some of the factors that influence the counterface topography generated during sliding. In general there are two main components involved, abrasion/polishing and transfer. Abrasion by hard constituents and/or impurities has already been discussed, but graphite itself may also be intrinsically abrasive as a consequence of the anistropic mechanical properties resulting from its lamellar structure (Lancaster, 1966). Transfer films on metal counterfaces form most readily from graphitic carbons, mask the underlying topography, and lead, in general, to smoother surfaces. After transfer film formation, the limiting rates of wear of a graphitic carbon on abraded copper surfaces ranging in roughness from 0.1 to 8 μm R_a differ only by a factor of about 10, whereas before transfer film formation the initial wear rates differ by a factor of 1000 (Lancaster, 1963a). Transfer is also influenced by the degree of oxidation of the counterface metal. There is some evidence suggesting that oxidation facilitates the adhesion of graphite debris films to metals (Buckley and Johnson, 1964) and some suggesting the reverse (Lancaster, 1962). In addition, many oxides are harder than their parent metals and may be beneficial in generating a smoother counterface topography by polishing. It is thus clearly evident that the particular topography generated on metals during sliding against carbons can result from a variety of processes, some of which depend on properties of the counterface itself and others on properties of the carbon. Attempts to relate the wear rates of different types of carbons, in standardized arbitrary test conditions, to their conventional physical and mechanical properties are thus unlikely to be successful. Figure 9a confirms this conclusion, showing that there is no significant correlation between the steady-state wear rates and elastic moduli of 11 different types of carbons sliding on copper. However, if the variable topographical factor is eliminated by continuously generating a constant wear track on the copper with a heavily loaded subsidiary carbon, Fig. 9b shows that the wear rates of the different materials now exhibit a very significant inverse correlation with their elastic moduli. Curve 1 in Fig. 9b was obtained by generating a wear track on the copper with a nongraphitic carbon, causing wear of the copper, whereas curve 2 involved an electrographite which generated a transfer film on the copper.

C. Adhesion

Although the significance of adhesion as a factor in the friction of carbons is now widely recognized, its role in wear is much less clear. As

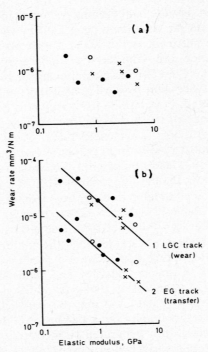

Fig. 9. Variation of wear rate of carbons on copper with their elastic moduli: (a) in individual wear tracks, load is 10 N; (b) in a constant wear track produced by an auxiliary carbon, load is 0.8 N. ○, natural graphites; ●, electrographites; ×, low/nongraphitic carbons. Speed is 14 m/sec. (From Lancaster, 1963a.)

already mentioned, adhesion will be greatest for those contacts involving edge-site interactions, and in these circumstances it is possible to envisage basal planes being removed by a process of progressive cleavage. Spreadborough (1962) has observed rolled-up sheets of basal planes on graphite surfaces after sliding, which might have been produced in this way. However, for manufactured carbons in which graphitic order is only poorly developed, it is difficult to envisage the above mechanism being significant. A more plausible physical picture is one in which adhesion operates in conjunction with some other failure process. For example, the stress distribution around a localized asperity contact will depend on the magnitude of the shear stress imposed by overcoming adhesive forces, and this, in turn, will influence the dynamics of fatigue wear. Alternatively, subsurface crack propagation could weaken the bulk material locally until even very weak interfacial adhesive forces suffice, either alone or in combination with mechanical deformation, to detach a wear fragment. This last hypothesis is substantially the same as the delamination theory of wear for metals, recently proposed by Suh (1973).

V.   Electrical Contacts

Since the publication of the original carbon brush patent in 1885, a voluminous literature has grown up dealing with the friction and wear of brushes. Much of the early work is reviewed by Holm (1946). Most experiments concerned with carbon brushes have been made in "realistic" conditions, i.e., on electrical machines either in service or in the laboratory, and in these circumstances it is seldom possible to identify precisely the wear processes involved. Among the various mechanisms that have been suggested are abrasion by the copper counterface (Shobert, 1965), abrasion by counterface oxides or corrosion products (Perrier, 1930; Hessler, 1937; Baker and Hewitt, 1937), abrasion by loose carbon debris (Mayeur, 1957), and thermal disruption of the brush surface by localized differential thermal expansion (Hessler, 1935).

For the present discussion, it is most convenient to examine first the wear of carbon brush materials sliding in the absence of current, and then to see how the introduction of current modifies the behavior. In general, an electrographitic carbon sliding on copper can exhibit two distinct regimes of wear, as illustrated in Fig. 10a, curve 1. By analogy with metallic wear processes, these may be described as "mild" and "severe" (Archard and Hirst, 1956). However, these two regimes are not always distinct and often merge gradually into each other, e.g., at lower speeds, as shown in Fig. 10a, curve 2. In the severe wear regime, there is appreciable transfer of carbon to the copper, oxidation of the copper is inhibited, the electrical contact resistance remains low, and the worn surfaces are relatively rough. In the mild wear regime, there is little transfer of carbon, appreciable oxidation of the copper, a high contact resistance, and the worn surfaces are relatively smooth. Figure 10b, curve 1, shows the effect of speed on wear. As the speed decreases, there is a gradual transition toward the mild wear regime associated with diminishing transfer of carbon and increasing oxidation and contact resistance. There is some controversy as to whether the extent of oxidation in the wear track on a copper ring depends primarily on the mean surface temperature or on the localized flash temperatures. The results in Fig. 10b, curve 1, together with others (Lancaster, 1962), support the former, as do calculations by Spry and Scherer (1961); the opposite viewpoint, however, is taken by Holm (1958) and Quinn (1962).

It follows from the above that, in general, low wear tends to be asso-

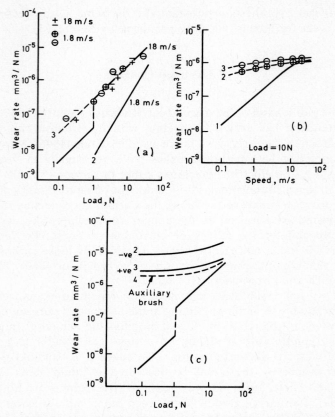

Fig. 10. (a) Variation of wear rate with load at different speeds. (b) Variation of wear rate with speed with and without current. (c) Effect of current on the wear rate–load relationship. Electrographite on copper. $\oplus$, Positive brush; $\ominus$, negative brush. (After Lancaster, 1962, 1963b.)

ciated with pronounced oxidation of the copper and a high electrical contact resistance. An inverse relationship of this type has been reported by Betz (1973), who also suggests that contact resistance measurements might, in fact, be used for the prediction of wear rates in service. However, when current is passed across the sliding interface, the protective oxide film characteristic of the mild wear regime must be disrupted. In this regime, therefore, wear tends to increase steadily with increasing current until a maximum wear rate is reached, corresponding to that of severe wear. Figure 10a, curve 3, and Fig. 10b, curves 2 and 3, clearly demonstrate the onset of severe wear at high current (10 A). There is a slight effect of brush polarity, the wear rate being highest on the negative brush, and similar results have been reported by Hessler (1935) and Holm (1958). In contrast, for natural graphite brushes on steel, Thompson and Turner (1962) observed greater wear on the positive brush. Much more serious effects of current on brush wear, however, become apparent when the sliding contacts are electrically interrupted, leading to arcing. Curves 2 and 3 in Fig. 10c show the effect of introducing insulated segments into a copper ring and it can be seen that the increases in wear rate are very large, particularly for the negative brush. Arcing introduces two extra components into the wear process. There is a direct loss of carbon by vaporization at the arc spot, and this is most pronounced for the negative brush, which undergoes positive ion bombardment. A second contribution, however, arises because the arc also roughens the surface of the copper and thus introduces an additional mechanical component of wear. Curve 4 in Fig. 10c shows the effect of roughening of the copper on wear, as assessed by an insulated subsidiary brush sliding in the same wear track as that undergoing arcing. It can be seen that for the positive brush, almost all the extra wear introduced by arcing is attributable to roughening of the copper by the arc; for the negative brush, however, there is a significant extra contribution from direct loss of carbon in the arc.

In practice, arcing can arise from a variety of causes. If brush loads are light and speeds high, there may be partial or complete aerodynamic film formation (Stanley, 1962). A practical solution in this situation is the use of very porous carbon brushes. Alternatively, brush inertia effects, combined with rotational eccentricities in a slip ring, may give rise to "bouncing." Solutions here are to improve concentricity and/or to reduce brush inertia (Sketch *et al.*, 1960). An interesting recent development has been a return to the original concept of a "brush" by using bundles of high-strength carbon fibers for current collection (Bates, 1973; Cook and Tomaszewski, 1976). These have all the advantages of low inertia, very uniform distribution of current transfer, and no tendency to aerodynamic film formation.

"DUSTING"

The onset of a regime of very high friction and wear of graphitic carbons first became a serious problem during the early 1940s on the carbon brushes of electrical equipment in aircraft flying at high altitudes (Ramadanoff and Glass, 1944). Similar phenomena were subsequently observed in vacuum (Savage, 1945) and in inert gases (Campbell and Kozak, 1948). The high wear rates result in large quantities of finely divided debris (Savage and Brown, 1948; Bobka, 1961)—hence the descriptive term "dusting"—and also lead to severe roughening on the opposing counterface. If the water vapor required to prevent dusting is physically adsorbed, as discussed earlier, the amount present on the surface will be controlled by the ratio P/P_0, where P is the partial pressure of water and P_0 is its saturation vapor pressure at the appropriate temperature. A critical amount of adsorbed vapor could thus be reached either by a reduction in P or by an increase in P_0 resulting from a rise in temperature. It follows that at sufficiently high ambient temperatures, loads, or speeds dusting might therefore occur in normal atmospheric environments. This has now been observed several times (Plavnik *et al.*, 1963; Plutalova and Panyusheva, 1970; Lancaster, 1975a). Levens (1976) has also reported instances of the onset of rapid wear for electrographite sliding on stainless steel in air, but attributes these to machining of the graphite, following the formation of work-hardened lumps on the steel, rather than to dusting.

A detailed study of the onset of dusting for graphitic carbons sliding against themselves in air has been reported by Lancaster (1975a). This work suggests that dusting begins when the combination of load, speed, and ambient temperature suffices to raise the total temperature of the individual asperity contacts to a critical value, of the order of 150–180°C. At this temperature, the ratio P/P_0 falls below that needed for adsorption of a sufficient quantity of water vapor. The critical value of $P/P_0 \approx 0.001$ is very much less than that required for water vapor in vacuum, $P/P_0 \approx 0.1$, because of the availability of atmospheric oxygen. This conclusion is also consistent with observations by Pardee (1967).

The attainment of a critical localized contact temperature depends on the interaction of two groups of factors. On one side are those which determine the rate of heat generation and dissipation (load, speed, coefficient of friction, thermal properties of the materials, and the ambient temperature), while on the other side are the factors which influence the number, size, and distribution of the localized contacts (surface roughness, debris film formation, running-in procedures, and the size of the apparent contact areas). In general, dusting occurs most readily for rough surfaces of small apparent contact area sliding against counterfaces of low thermal conductivity. A suggested sequence of events involved in the initiation of dusting is given in Fig. 11.

In addition to water vapor, several studies have been made of the effects of other gases and vapors in preventing dusting. Dry Ar, He, H_2, and N_2 are all ineffective (Ramadanoff and Glass, 1944; Savage and Van Brunt, 1944; Pardee, 1967), but both O_2 and CO_2 will inhibit dusting if present in sufficiently large amounts (Campbell and Kozak, 1948). There is also evidence to show that the required quantity of water depends on the atmosphere in which it is contained. Pardee (1967) has shown that 70 ppm will suffice in air, but 420 ppm are needed in He and 640 ppm in N_2.

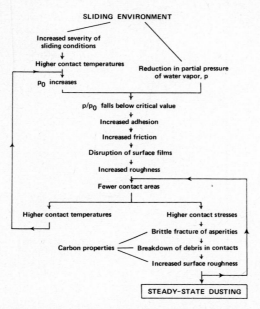

SLIDING ENVIRONMENT

Increased severity of
sliding conditions

Higher contact temperatures

p_0 increases

Reduction in partial pressure
of water vapor, p

p/p_0 falls below critical value

Increased adhesion

Increased friction

Disruption of surface films

Increased roughness

Fewer contact areas

Higher contact temperatures

Higher contact stresses

Brittle fracture of asperities

Carbon properties —— Breakdown of debris in contacts

Increased surface roughness

STEADY-STATE DUSTING

Fig. 11. Sequence of events involved in dusting of graphitic carbons. (Crown copyright. Reproduced by permission of the Controller, Her Britannic Majesty's Stationery Office.)

He suggests that the critical amounts of water are determined by the extent to which the adsorption of other gases inhibits or facilitates adsorption of water. The critical water vapor concentrations also appear to depend on the type of counterface; Moberly and Johnson (1959) claim that seven times more water is needed to prevent dusting on steel than on copper. This result, however, may be connected with the lower thermal conductivity of steel, leading to higher contact temperatures. The significance of the contact temperature on dusting does not seem to have been appreciated until comparatively recently.

Most organic vapors are considerably more effective than water in preventing dusting, although silicones in enclosed environments are reported to have disastrous effects on carbon brush wear, (Marsden and Savage, 1948; Moberly, 1960). Figure 12 shows some results by Savage and Schaefer (1956), from which it can be seen that there is a general decrease in the critical ratio of P/P_0 as the molecular size increases. These data have been interpreted by Cannon (1964) in terms of the formation of mobile liquid adsorbed films that are free to migrate to selected sites on the basal planes. Figure 12 demonstrates that the concentrations of the larger organic molecules required to inhibit dusting are only a few parts per mil-

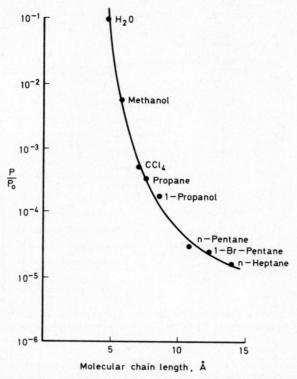

Fig. 12. Variation of the critical ratio of P/P_0 to prevent dusting with size of organic molecules. (From Savage and Schaeffer, 1956.)

lion, and in view of this it is hardly surprising that instances of dusting on electrical equipment in typical industrial service conditions are comparatively rare, and seldom reproducible. Because of their porous nature, manufactured carbons will readily absorb contaminant fluids such as oils, and vaporization of these fluids at elevated temperatures could easily provide the requisite vapor concentrations to inhibit dusting. Deliberate impregnation of porous carbons is widely employed in practice to produce special grades intended for operation in water-deficient environments. In addition to fluids and other, solid organic materials, a wide variety of inorganic compounds—"adjuvants"—are also known to be effective; these include halides of Pb, Cd, Ba, and Sr, P, and B, and compounds, MoS_2, and PTFE. A review of the patent literature on this topic is given by Paxton (1967). Various suggestions have been made to account for the action of some of these solid adjuvants in preventing dusting. Savage (1951) considers that halides, such as CdI_2, improve the retention of water vapor by the carbon; McCubbin (1957) suggests that BaF_2 is chemisorbed on the copper surface; and Lynn and Elsey (1949) postulate that BaF_2 catalyzes oxidation of copper to provide a more protective surface film. No all-embracing explanation has yet been proposed.

VI. Bearings and Seals

Because of their general lack of reactivity, except to oxidizing agents, carbons are particularly suitable for bearings and seals in hostile environments. The choice of a particular carbon for a specific application is usually made on the basis of experience and/or recommendation by the manufacturer; there are few general guide lines on wear, except for some of the most widely used grades (ESDU, 1976). When fluids are present, the ideal situation is one of complete hydrodynamic lubrication, and carbons are valuable in this respect because of their ability to generate very smooth surfaces during running in. Wear rates also tend to be relatively low in conditions of boundary lubrication. Khrishanova *et al.* (1970) consider that the wear of electrographite on metals in different fluids is related to the ability of the fluid to disperse, or peptize, loose wear debris. Wear in water or aqueous solutions is usually higher than in organic fluids because carbon wear debris aggregates in water and increases wear either directly by abrasion of the carbon or indirectly via abrasion and roughening of the metal counterface. Further work (Lancaster, 1972) has confirmed that the wear rate of an electrographite on stainless steel in water is higher than in most organic fluids, but attributes the difference to the inhibiting effect of water on the formation of a transfer film on the metal counterface. High wear rates, and an absence of transfer films, are also characteristic of the wear of carbons in cryogenic fluids, such as liquid H_2, He, or N_2, but wear can be reduced considerably by impregnation of the carbon with the types of additive found useful in preventing dusting in the absence of water vapor—halides, organic resins, and PTFE (Wisander and Johnson, 1960). For seal applications, the permeability of carbons is an important parameter, and the patent literature on impregnation treatments with resins, metals, inorganic salts, etc., to reduce porosity is extensive (Paxton, 1967). Apart from its effect on leakage in a seal, excessive porosity in carbons can also be detrimental to wear. Fluid may be forced into the pores under high stresses leading to subsurface failure and the formation of pits (Strugala, 1972).

In dry sliding applications, the limiting feature of carbons is usually temperature (Johnson *et al.*, 1956). Wear rates generally increase with increasing temperature, and Fig. 13 shows some examples which demon-

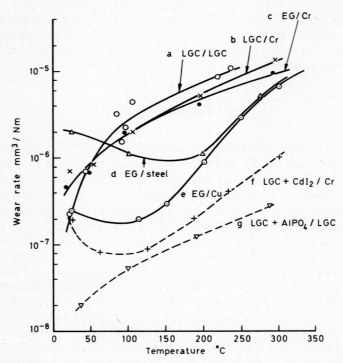

Fig. 13. Variation of wear rate with temperature for carbons on various counterfaces. Load is 10 N, speed is 0.5–2 m/sec; EG, electrographite; LGC, low-graphitic carbon. (Crown copyright. Reproduced by permission of the Controller, Her Britannic Majesty's Stationery Office.)

strate that the type of counterface also influences the results. On unreactive counterfaces—curves a–c—the increase in wear rate is relatively rapid during the first 100°C rise in temperature, and this is associated with a diminution in the extent of transfer film formation on the counterface. On more reactive metals, however, such as copper and steel, the wear rate only begins to increase rapidly above about 150–200°C. These results are consistent with the conclusion by Bisson (1964) that low wear (and friction) of carbons occurs only when either a transfer film of carbon or a coherent oxide film is present on the metal surface. If so, it follows that additives to carbons that either facilitate transfer, promote oxidation of the metal, or possibly react with the metal to form films other than oxides should be beneficial in reducing wear at high temperatures. A wide variety of such additives has been examined, mostly in connection with increasing the life of graphite solid lubricant films at high temperatures, and many inorganic salts and oxides have been found beneficial, e.g., PbO, CdO, Na_2SO_4, $CdSO_4$ (Peterson and Johnson, 1956); NaF (Abe *et al.*, 1971); and Na_2SiO_3, copper phthalocyanine, and numerous metal oxides (Cook, 1971). A further example is shown in curve f of Fig. 13 for carbon containing CdI_2.

There are three possible explanations for the general increase in wear with temperature; the introduction of an abrasive component of wear by debris either from the carbon itself and/or from metal oxides, a loss of adsorbed water vapor from the carbon surface (Giltrow, 1973), or an increasing loss of carbon by oxidation. The last is possibly the most plausible. McKee *et al.* (1972) have concluded that there is a general correlation between the wear of carbon brushes at high temperatures and their oxidation characteristics. The type of metal counterface also influences carbon oxidation, and in turn wear, because some metals, such as copper, are oxidation catalysts, whereas others, such as zinc, are inhibitors. A variety of carbon oxidation-inhibiting additives containing boron or phosphorus have been shown to be effective in reducing the wear of a graphitic carbon at high temperatures (Vasilev and Emelyanova, 1970). An example is shown in Fig. 13, curve g, for carbon containing $AlPO_4$.

VII.   Carbon–Carbon Combinations

Until the advent of high-performance aircraft brake disks made from carbon–carbon fiber composites (Weaver, 1972), there was little practical interest in the friction and wear of carbons sliding against themselves. For brake applications, a primary requirement is the maintenance of a stable coefficient of friction and a low rate of wear over a wide range of sliding conditions, particularly temperature. The tendency for graphitic carbon combinations to undergo dusting at elevated temperatures has already been noted. Low or nongraphitic carbons exhibit similar transitions to dusting, but the increases in friction and wear are much smaller, and the total contact temperatures involved are much higher, $\simeq 750°C$ (Lancaster, 1977). A more serious problem with nongraphitic carbon combinations is the gradual increase in friction with time to relatively high values as shown earlier in Fig. 3c. Microscopic examination of worn surfaces after various times of sliding has revealed several different ways in which wear can occur (Lancaster, 1975b). These are illustrated in Fig. 14 which shows:

(a)   Flaking of the surface films of debris found in the early stages of sliding when the coefficient of friction is low. Similar phenomena have

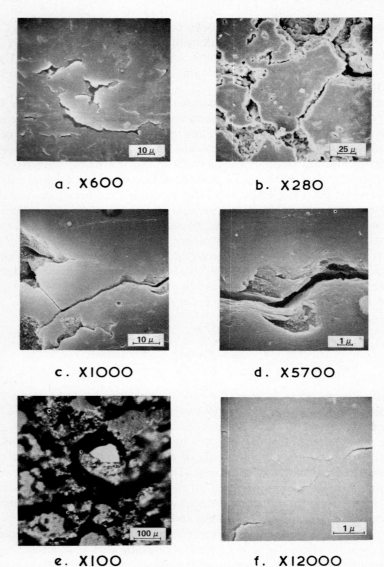

a. X600

b. X280

c. X1000

d. X5700

e. X100

f. X12000

Fig. 14. Worn surfaces of low-graphitic carbon after sliding against itself. (a–d), (f) scanning electron micrographs; (e) optical micrograph. (From Lancaster, 1975b. Crown copyright. Reproduced by the permission of the Controller, Her Britannic Majesty's Stationery Office.)

also been observed for graphitic carbons on metals where the debris films often undergo "blistering" prior to rupture (Swinnerton and Turner, 1966).

(b) Preferential removal of weakly bonded material at the grain boundaries as the coefficient of friction increases.

(c) Crack formation and propagation within coke grains.

(d) Chipping and delamination at the edges of coke grains.

(e) Subsurface fracture of whole grains; the smooth bright patch made visible by oblique illumination is inclined at about 15° to the general plane of the surface.

(f) Generation of very smooth surfaces on individual grains after prolonged sliding. This implies the existence of wear on a scale of the order of a few tens of nanometers or less, almost approaching the concept of "atomic" wear postulated by Holm (1958). In terms of its direct contribution to the total wear, this process may be quantitatively negligible; however, its indirect contribution, by generating smooth surfaces, increasing the real area of contact, and, in turn, the coefficient of friction, could well be of considerable significance. A high frictional stress over an individual grain might lead to subsurface fracture, as in (e), and when this process occurs it is likely to comprise the major component in the total wear. If so, it follows that there should, in general, be a direct, causative relationship between the coefficient of friction and the rate of wear, and this has recently been observed (Lancaster, 1978). For one particular carbon sliding in different conditions, the wear rate increases more rapidly than proportionately with friction, consistent with an increasing probability of subsurface grain fracture with increasing friction.

The concept of a causative relationship between friction and wear leads to a possible explanation of why the wear rates of nongraphitic carbon combinations are so readily reduced by a wide variety of inorganic additives such as metal oxides and salts (Lancaster, 1978). Many of these additives are mildly abrasive to carbons and prevent the development of a very smooth topography, thus reducing the real contact area, the coefficient of friction and, in turn the rate of wear. As an example, Fig. 15 shows the effect of incorporating resin-bonded abrasive particles of Al_2O_3 of different particle sizes into small holes machined into the surface of a nongraphitic carbon. The coefficient of friction is reduced and the smaller particle sizes also greatly reduce the rate of wear. The increase in wear for particle sizes beyond 0.3 μm, without any similar large increase in the coefficient of friction, is presumably attributable to an increasing component of abrasive wear. The wider relevance of this mechanism to the wear of additive-containing carbons on metals remains to be explored.

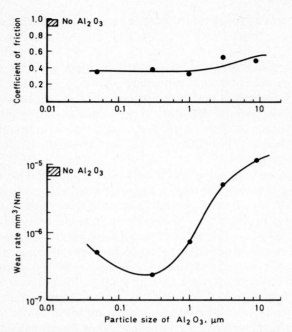

Fig. 15. Variation of friction and wear of low-graphitic carbon containing Al_2O_3 with particle size. Load is 22 N, speed is 2 m/sec, temperature is 200°C. (Crown copyright. Reproduced by permission of the Controller, Her Britannic Majesty's Stationery Office.)

VIII. Summary

Three main conclusions clearly emerge from the preceding discussion.

(i)   The wear rates of carbons are extraordinarily sensitive not only to environmental conditions (water, organic vapors, etc.) but also to the "purity" of the carbon itself (organic contamination, metal oxides, etc.).

(ii)   Wear may occur by means of several processes, operating together on different scales of size, and each of these processes is likely to be affected in different ways by the environment and the imposed conditions of sliding.

(iii)   The formation of surface films of wear debris, following structural breakdown of the carbon, plays a very significant role in wear as well as in friction.

As will have become very evident, it is often difficult, if not impossible, to define the wear processes occurring in carbons in terms of the conventional mechanisms, i.e., abrasion, adhesion, fatigue, and oxidation. All of

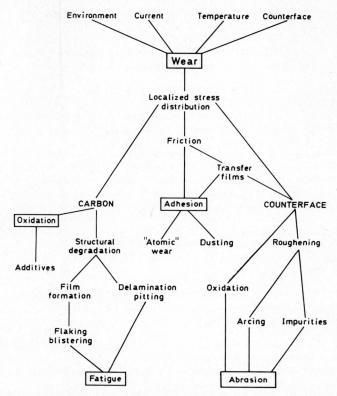

Fig. 16. Summary of factors influencing the wear of carbons. (Crown copyright. Reproduced by permission of the Controller, Her Britannic Majesty's Stationery Office.)

these interact with each other in complex ways, and an attempt to summarize the various interrelationships is shown in Fig. 16. Many aspects are still imperfectly understood, but perhaps the one most in need of further clarification is debris film formation. Little information is yet available about the mechanics of film formation and the physical and chemical processes involved in interparticle cohesion and adhesion. Only when these have been elucidated will it really become possible to derive more satisfactory explanations for the sensitivity of carbon wear to environmental vapors, contaminants, and additives.

References

Abe, W., Mutch, A., Terada, Y., and Segawa, T. (1971). *Proc. Int. Conf. Solid Lub.* ASLE SP-3, p. 1.

Archard, J. F., and Hirst, W. (1956). *Proc. R. Soc. London Ser. A* **236,** 397.

Arnell, R. J., and Teer, D. G. (1968). *Nature (London)* **218,** 1155.

Arnell, R. J., Midgley, J. W., and Teer, D. G. (1966). *Proc. Inst. Mech. Eng.* **179** (3j), 115.

Badami, D. V., and Wiggs, P. K. C. (1970). Friction and wear, *In* "Modern Aspects of Graphite Technology" (L. C. F. Blackman, ed.), Chapter VI. Academic Press, New York.

Baker, R. M., and Hewitt, C. W. (1937). *Trans. AIEE* **56,** 123.

Barlow, H. M. (1924). *J. Inst. Electr. Eng.* **62,** 133.

Bates, J. J. (1973). Carbon fiber brushes for electrical machines, *In* "Carbon Fibers in Engineering" (M. Langley, ed.), Chapter 6. McGraw-Hill, New York.

Bernal, J. D. (1924). *Proc. R. Soc. London Ser. A* **106,** 749.

Betz, D. (1973). *Proc. Holm Semin. Electr. Contact Phenomena, 19th, IIT* p. 164.

Bisson, E. E. (1964). "Advanced Bearing Technology," NASA-SP 38, Chapter 8, p. 203.

Bobka, R. J. (1961). *Proc. Bienniel Carbon Conf., 5th* p. 287. Pergamon, Oxford.

Bradley, D. E., Halliday, J. S., and Hirst, W. (1956). *Proc. Phys. Soc.* **B69,** 484.

Bragg, W. L. (1928). "Introduction to Crystal Analysis," p. 64. Bell, London.

Bryant, P. J., Gutshall, P. L., and Taylor, L. H. (1964). "Mechanisms of Solid Friction," p. 118. Elsevier, Amsterdam.

Buckley, D. H., and Johnson, R. L. (1964). *ASLE Trans.* **7,** 91.

Campbell, W. E., and Kozak, R. (1948). *Trans. ASME* **70,** 491.

Cannon, P. (1964). *J. Appl. Phys.* **35,** 2928.

Clark, W. T., and Lancaster, J. K. (1963). *Wear* **6,** 467.

Clark, W. T., Connolly, A., and Hirst, W. (1963). *Brit. J. Appl. Phys.* **14,** 20.

Cook, C. R. (1971). *Proc. Int. Conf. Solid Lub.* ASLE SP-3, p. 13.

Cook, R. J., and Tomaszewski, M. J. (1976). *Proc. London Int. Carbon Graphite Conf., 4th* SCI, p. 717.

Deacon, R. F., and Goodman, J. F. (1958). *Proc. R. Soc. London Ser. A* **243,** 464.

Dobson, J. V. (1935). *Electr. J. London* **32,** 527.

ESDU (1976). A Guide to the Design and Selection of Dry Rubbing Bearings. Eng. Sci. Data Item 76029. Institution of Mechanical Engineers, London.

Freise, E. J., and Kelly, A. (1963). *Phil. Mag.* **8,** 1519.

Fulham, E. G., and Savage, R. H. (1948). *J. Appl. Phys.* **19,** 654.

Giltrow, J. P. (1973). *ASLE Trans.* **16.** 83.

Giltrow, J. P., and Lancaster, J. K. (1970). *Wear* **16,** 359.

Goddard, J., Harker, H. J., and Wilman, H. (1959). *Nature London* **184,** 333.

Groszek, A. J. (1969). *Proc. R. Soc. London Ser. A* **314,** 472.

Halliday, J. S. (1955). *Proc. Inst. Mech. Eng.* **169,** 777.

Hayes, M. E. (1947). "Current Collecting Brushes in Electrical Machines." Pitman, London.

Hessler, V. P. (1935). *Trans. AIEE* **54,** 1050.

Hessler, V. P. (1937). *Trans. AIEE* **56,** 8.

Hollander, A. E., and Lancaster, J. K. (1973). *Wear* **25,** 155.

Holm, E. (1962). *J. Appl. Phys.* **33,** 156.

Holm, R. (1946). "Electric Contacts," 1st ed. Gerber, Stockholm.

Holm, R. (1958). "Electric Contacts Handbook," 3rd ed. Springer, New York.

Hutcheon, J. M. (1970). Manufacturing technology of baked and graphitized carbon bodies. *In* "Modern Aspects of Graphite Technology" (L. C. F. Blackman, ed.), Chapter II. Academic Press, New York.

Jenkins, R. O. (1934). *Phil. Mag.* **17,** 457.

Johnson, R. L., Swikert, M. A., and Bailey, J. M. (1956). NACA-TN-3595.

Khrisanova, L., Belogorskii, V., and Kontarovich, S. (1970). *Konstr. Mater. Osn. Grafita.* (5), 158 (FDT-HC-23-0991-72).
Klein, C. A., Straub, W. D., and Diefendorf, R. J. (1962). *Phys. Rev.* **125**, 468,
Kraghelskii, I. V., and Nepomnyashchii, E. F. (1965). *Wear* **8**, 303.
Krushchov, M. M. and Babichev, N. A. (1953). *Dokl. Akad. Nauk. SSSR* **88**, 445 (NSF Transl. Tr-15, 1953).
Lancaster, J. K. (1962). *Br. J. Appl. Phys.* **13**, 468.
Lancaster, J. K. (1963a). *Br. J. Appl. Phys.* **14**, 497.
Lancaster, J. K. (1963b). *Wear* **6**, 341.
Lancaster, J. K. (1964). *Proc. Inst. Mech. Eng. Lub. Wear Group Convent., 1st* p. 190.
Lancaster, J. K. (1966). *Wear* **9**, 169.
Lancaster, J. K. (1972). *Wear* **20**, 335.
Lancaster, J. K. (1975a). *ASLE Trans.* **18**, 187.
Lancaster, J. K. (1975b). *Wear* **34**, 275.
Lancaster, J. K. (1977). *ASLE Trans.* **20**, 43.
Lancaster, J. K. (1978). *Proc. ASLE Int. Conf. on Solid Lub., 2nd.,* ASLE SP-6, p. 176.
Lancaster, J. K., and Stanley, I. W. (1964). *Br. J. Appl. Phys.* **15**, 29.
Larsen, J. V., Smith, T. G., and Erickson, P. W. (1971). NOL-TR-71-165.
Levens, M. B. (1976). *Proc. London Int. Carbon Graphite Conf., 4th* SCI, p. 691.
Longley, R. I., Midgley, J. W., Strang, A., and Teer, D. G. (1964). *Proc. Inst. Mech. Eng. Lub. Wear Group Convent., 1st,* p. 193.
Lynn, C., and Elsey, H. M. (1949). *Trans. AIEE* **68**, 490.
Marsden, J., and Savage, R. H. (1948). *Trans. AIEE* **64**, 1196.
Mayer, E. (1972). "Mechanical Seals," 2nd ed. Iliffe, London.
Mayeur, R. (1957). *Rev. Gen. Electr.* **66**, 207.
McCubbin, W. L. (1957). RAE-TN-EI137.
McKee, D. W., Savage, R. H., and Gunnoe G. (1972). *Wear* **22**, 193.
Midgley, J. W., and Teer, D. G. (1961). *Nature London* **189**, 735.
Moberley, L. E. (1960). Eng. Proc. P-36, Electr. Contacts, p. 117. Penn State Univ., University Park.
Moberley, L. E., and Johnson, J. L. (1959). *Trans. AIEE* **PAS-78**, 263.
Pardee, R. P. (1967). *Trans. IEEE* **PAS-86**, 616.
Paxton, R. R. (1967). *Electrochem. Tech.* **5**, 174.
Perrier, M. (1930). *Bull. Soc. Fr. Electr.* **40**, 903.
Peterson, M. B., and Johnson, R. L. (1956). NACA-TN-3657.
Pike, E. C., and Thompson, J. E. (1969). *Wear* **13**, 247.
Plavnik, G. M., Plutalova, L. A., and Rovinskii, B. M. (1963). *Izv. Akad. Nauk. SSSR Otn. Tekh. Nauk. Mekh. Mashinostr.* **4**, 179.
Plutalova, L. A., and Panyusheva, Z. A. (1970). *Mashinov* (4), 112 (FDT-MT-24-293-72).
Porgess, P. V. K., and Wilman, H. (1960). *Proc. Phys. Soc.* **B76**, 513.
Quinn, T. F. J. (1962). *Br. J. Appl. Phys.* **13**, 33.
Quinn, T. F. J. (1963). *Br. J. Appl. Phys.* **14**, 107.
Rabinowicz, E. (1964). "Friction and Wear of Materials." Wiley, New York.
Ramadanoff, D., and Glass, S. W. (1944). *Trans. AIEE* **63**, 825.
Reznokovskii, M. M. (1960). *Sov. Rubber Technol. (Engl. Transl.)* **19** (9), 32.
Roselman, I. C., and Tabor, D. (1976). *J. Phys. D* **9**, 2517.
Rowe, G. W. (1960). *Wear* **3**, 454.
Savage, R. H. (1945). *Gen. Electr. Rev.* **48**, 13.
Savage, R. H. (1948). *J. Appl. Phys.* **19**, 1.
Savage, R. H. (1951). *Ann. N. Y. Acad. Sci.* **53**, 862.

Savage, R. H. (1959a). *Nature London* **183,** 315.

Savage, R. H. (1959b). *Nature London* **183,** 454.

Savage, R. H. (1959c). *Nature London* **184,** 45.

Savage, R. H. (1960). *Proc. Carbon Conf., 4th* p. 727. Pergamon, Oxford.

Savage, R. H., and Brown, C. (1948). *J. Am. Chem. Soc.* **70,** 2362.

Savage, R. H., and Shaeffer, D. L. (1956). *J. Appl. Phys.* **27,** 136.

Savage, R. H., and Van Brunt, C. (1944). *Gen. Electr. Rev.* **47,** 16.

Seldin, E. J., and Nezbeda, C. W. (1970). *J. Appl. Phys.* **41,** 3389.

Shobert, E. I. (1965). "Carbon Brushes: The Physics and Chemistry of Sliding Contacts." Chemical Publ. Co., New York.

Sketch, H. J. H., Shaw, P. A., and Splatt, R. J. K. (1960). *Proc. IEE* **107A,** 336.

Skinner, J. (1971). Ph.D. Dissertation, Univ. of Cambridge [Quoted in Roselman and Tabor (1976)].

Skinner, J., Gane, N., and Tabor, D. (1971). *Nature London* **232,** 195.

Spreadborough, J. (1962). *Wear* **5,** 18.

Spry, W. J., and Scherer, P. M. (1961). *Wear* **4,** 137.

Stanley, I. W. (1962). *Wear* **5,** 363.

Stanley, I. W. (1966). *Br. J. Appl. Phys.* **17,** 775.

Strugala, E. W. (1972). *Lub. Eng.* **28,** 333.

Suh, N. P. (1973). *Wear* **25,** 111.

Swinnerton, B. R. G., and Turner, M. J. B. (1966). *Wear* **9,** 142.

Thompson, J. E., and Turner, M. J. B. (1962). *Proc. IEE* **109A,** 235.

Vasilev, Yu. N., and Emelyanova, V. M. (1970). *Izv. Akad. Nauk. SSSR Neorgan. Mater.* **6,** 201.

Weaver, J. V. (1972). *Aero. J.* **76,** 695.

White, J. R. (1969). *Wear* **13,** 145.

Wisander, D. W., and Johnson, R. L. (1960). *Adv. Cryogen. Eng.* **6,** 210.

Scuffing

A. DYSON

Shell Research Limited
Thornton Research Centre
Chester, England

I.   Introduction

The objects of this chapter are to introduce the topic of scuffing, to describe its physical manifestations and the conditions under which it appears, to assess its importance in practice, and to review the various empirical criteria and theoretical interpretations that have been advanced to describe and to explain the observed facts. The treatment of scuffing in practical machinery is mainly in terms of gears, although contacts between cams and tappets and between piston rings and cylinders will also be discussed. Since it is difficult to do fundamental work and to measure important quantities such as friction in practical machinery, much of the experimental work that will be described has been undertaken with disk machines, in which most of the essential factors of real machinery may be reproduced. The rapid changes in conditions that occur during the operating cycles of most useful machine parts are usually absent in disk machines, however, and this absence may be important in limiting the application of research findings in practice.

The review and the references are not intended to be exhaustive, since an expansion by something approaching one order of magnitude would be required to attain this condition. A critical review inevitably reflects the personal opinions, experiences, and prejudices of the reviewer. An apology is due if these are found to be unduly obtrusive in the present review.

II.   Definitions

The terminology of the subject is somewhat confusing. In general, the definitions adopted by the Institution of Mechanical Engineers (1957) will be used in this chapter. According to these definitions, scuffing is "gross damage characterized by the formation of local welds between the sliding surfaces." Seizure is defined as "the stopping of a mechanism as the result of interfacial friction." This is rarely encountered in practice in the mechanisms discussed here, but important authorities, notably Blok in his earlier papers, use the term seizure in the sense of scuffing as defined above.

Scoring is a form of failure related to scuffing and defined as "scratching across the rubbing surfaces without modification of the general form." The relation between scoring and scuffing will be discussed later. The two terms are often used indiscriminately. A form of surface damage encountered in high-speed rolling bearings, called "smearing," appears to be similar in nature to scuffing or scoring. Perhaps the most graphic description of the scuffing phenomenon is the German term *fressen,* with its connotation of gobbling by animals. This describes truly the catastrophic wear rate that is one of the characteristics of severe scuffing.

III.   Physical Manifestations of Scuffing

A.   *Background*

1.   HISTORICAL

A piece of machinery is invented, designed, constructed, or evolved to do a certain job, and once this job is seen to be done effectively, the obvious temptation is to reduce the size, weight, and cost of the machinery while asking it to do the same job, or to increase the difficulty of the job done by the same piece of machinery. By the difficulty of the job is meant, say, the power transmitted by a set of gears, or the mass controlled and the frequency of operation of a valve gear. In this process, a series of performance barriers is commonly encountered. For example, the first gear teeth were probably wooden pegs with obvious limitations in strength. Stronger materials were evolved in the course of time and steel is now used in virtually all gears transmitting appreciable quantities of power except for the wheels of worm gears, for which bronze is used because of the severe sliding conditions.

2.   THE SEQUENCE OF EVENTS IN A SCUFFING TEST

Under mild conditions, at low loads and temperatures, the surfaces are completely separated by a film of lubricant and remain in their original state. In a typical scuffing test, the load is increased in stages. At each such increase the bulk temperature of the parts increases rapidly at first and then more slowly, and the minimum thickness of the lubricant film, which is controlled by the viscosity of the lubricant, decreases accordingly. At first the surfaces will remain undisturbed, but at some stage they will interact and become modified. At first this will be confined to the peaks of the highest asperities and will result in a reduction in their height and curvature. Some chemical reactions, e.g., oxidation, may also occur, and in general the coefficient of friction will start to increase slightly. This process is normal and desirable and it is an essential part of "running in" or "breaking in." But in some cases the interaction becomes too severe and the surfaces are damaged. The first form of damage noted with relatively soft gears is usually either pitting or mild adhesive wear, and both may be countered by through hardening and then case hardening, first of the pinion and then of the wheel.

3. SCORING AND SCUFFING

Further increases in load may produce scoring, a series of grooves and ridges in the surface oriented in the sliding direction, possibly accompanied by a metallurgical transformation. In the early stages, there may be a considerable area of very smooth, polished surface between the grooves. This is usually a progressive form of wear and damage, and it may be difficult to judge when it began. Finally, still further increases in load may produce the sudden failure known as scuffing.

The exact sequence of events may vary with the materials, the design, the lubricant, and with the running conditions. This variation in the course of the failure has been responsible for much confusion in the past, and in the author's opinion we are only just beginning to understand it.

B. *Immediate and Obvious Manifestations*

The first obvious signs of gear scuffing are sudden increases in noise and vibration, and in further operation these are followed by a progressive increase in operating temperature, which may cause smoke to be given off by the lubricant. Other dramatic effects, such as showers of sparks, may sometimes be seen. Examination of the surfaces reveals the typical scuffing marks illustrated in Fig. 1. The surfaces appear as if they had been welded together at discrete points, and this feature forms the basis of the definition of scuffing already noted. The surfaces are often completely nondirectional in appearance, there being no indication of the directions either of the original grinding marks or of the sliding motion.

Scuffing varies in severity, and this aspect has received comparatively little attention. Many authors report the occurrence during a scuffing test of initial scuffs that apparently heal and are followed at a higher load by a final severe scuff. A mild scuff by itself may heal after further running, and in this case the evidence that a scuff has ever occurred is provided only by visual examination, perhaps long after the event. A severe scuff may lead to tooth breakage, while Fig. 2 shows an intermediate stage in which the tooth is severely damaged but not broken. This result was obtained by loading the gears beyond the initial failure load as judged by visual inspection (Padmore and Rushton, 1964–1965).

Scuffing of cams and tappets in valve gear systems may lead to gross wear, to deviations of the valve motion from that prescribed by the designer, and hence to loss in engine performance. Scuffing of piston rings

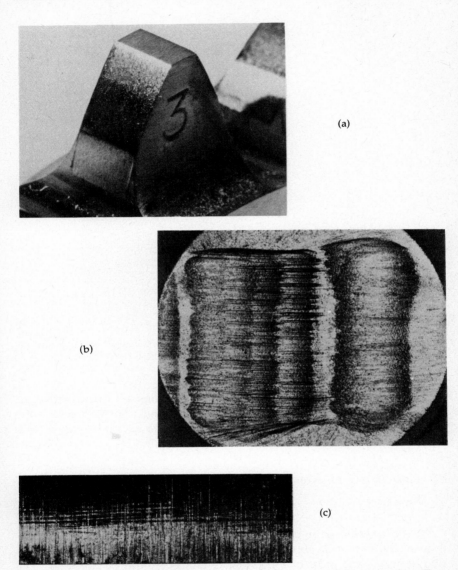

(a)

(b)

(c)

Fig. 1. Appearance of scuffed parts. (a) Gear tooth, (b) tappet, (c) piston ring. (From Dyson, 1975. Copyright IPC Science and Technology Press, Ltd. Reproduced by permission.)

Fig. 2. A moderately severe scuff. [Reproduced by permission of the Council of the Institution of Mechanical Engineers from Padmore and Rushton (1964–1965). *Proc. Inst. Mech. Eng. London,* **179,** 3J, 278–289.]

and cylinder liners leads to increased blow-by and oil consumption. It may be responsible for the breakage of piston rings, but in this case the evidence is often largely destroyed and it may be difficult to reach firm conclusions about the cause of the damage.

C. Less Obvious Physical Manifestations

1. FRICTION

Probably the most important of the less obvious physical manifestations of scuffing is the abrupt rise in friction. It is difficult to measure friction in gears, and most of the evidence comes from disk machines. The friction rises by a factor of something like 3, typically from a coefficient of about 0.07 to one of approximately 0.2. The rise is very abrupt and appears to coincide with the sudden increase in noise and vibration within the response time of the instruments. Some authorities see the abrupt rise in friction as a cause of scuffing (Hirst, 1976), but it seems likely that this is a "chicken-and-egg" situation. The severe wear, the marking of the surfaces, and the abrupt increase of friction are all symptoms of the complex phenomenon that we recognize as scuffing, and it does not yet seem possible to assign to it a single definite cause.

2. CONTACT RESISTANCE

The electrical resistance of the contact between the two rubbing bodies becomes very low at scuffing, and this is only to be expected from the occurrence of microwelding. The contact resistance may be measured by applying a low voltage to the contact, typically by a potential divider circuit. The effective resistance of the contact may be inferred from the potential difference across it, which may be displayed on an oscilloscope or averaged by a circuit with a suitably long time constant and displayed on a meter. In general, the potential difference displayed on an oscilloscope fluctuates between zero and the full applied potential. As conditions become more severe, the fraction of the time for which the potential difference is zero or very low increases and becomes unity at scuffing. There is rather a marked disagreement about the behavior just before scuffing. Some authorities take the absence of a detectable voltage over a sufficient length of time as an indication and even as a criterion of scuffing or scoring (Leach and Kelley, 1965; Blok, 1972); others find that machines will run for an appreciable time without scuffing but with no detectable contact resistance (Christensen, 1966; Bell and Dyson, 1972a); still others find the electrical contact method useful under some conditions but not under others (Staph and Ku, 1968).

3. WHITE LAYERS

Metallurgical features of scuffed ferrous surfaces have received much attention. A hard white etch-resistant phase is found on the surfaces of scuffed parts. The identity of the phase is still a matter of controversy, but it is commonly thought to be a mixture of austenite, martensite, and carbide. It is also thought that it has passed through an austenitic phase at some stage in its history. The presence of this transformed layer, often separated from the original untransformed bulk of the material by a tempered layer, suggests that the scuffed surfaces have been subject to very high temperatures and rapid cooling. It is often concluded that the temperature at the surface must have been at least the minimum value of 732°C required for the austenitic transformation, but high pressures could cause a reduction in this critical temperature by an unknown amount. In some cases these high temperatures have been confirmed by heat flow calculations, e.g., by Sakmann (1947). Padmore and Rushton (1964–1965) estimated the temperatures on their gear teeth by a hardness relaxation method. Their results ranged from a maximum of 235°C for an unscuffed tooth to a maximum of 590°C at one corner of a scuffed tooth. Appreciable portions of this tooth had experienced temperatures of about 500°C. These would be the temperatures of the untransformed underlying material, so the surface temperatures could have been over 700°C.

Two distinct white layers with different properties have been found on the surfaces of scuffed piston rings (Rogers, 1969, 1970). A review of the subject of these white layers has been given by Eyre and Baxter (1972), who suggest that the layers may have commercial applications.

IV.   The Practical Importance of Scuffing

A.   General

Scuffing is perhaps not commonly encountered in practice, but the problem cannot be dismissed as of no importance. A scuffed machine part will produce a noticeable drop in the performance of the machine as a whole, and in the worst cases the part must be replaced. Scuffing is therefore an expensive nuisance and manufacturers will go to great lengths to avoid it. They manage to avoid it only because they have a background of experience that tells them how far they can go along the road of increased output or decreased size before they run into trouble. This background of experience is entirely empirical and it would be an advantage if there were some degree of theoretical understanding that would allow this experience to be extrapolated to cover machinery of novel design. For example, the approximate scuffing limits of involute gears are known from experience, but those for Wildhaber–Novikov gears could not be deduced in advance and had to be established by further experience.

B.   Measures to Avoid Scuffing

Scuffing may be postponed or avoided by the incorporation into the lubricant of additives, the so-called "extreme-pressure" (EP) additives. These are usually compounds containing one or more of the elements sulfur, chlorine, and phosphorus. These presumably act by laying down a film of a nonmetallic reaction product on the metal surfaces, thereby preventing the welding that is an essential feature of scuffing. But since we do not know the mechanism of scuffing in detail, we do not know in detail how these additives work, and the selection of additives for use in any given situation is again entirely empirical. A whole range of materials is available, from the curiously named "mild EP" type to extremely active and corrosive ones. These additives have their disadvantages: They may promote pitting, sludging, oxidation, corrosion, or wear of other metals, e.g., bronze, and they may interfere with the action of other additives present in the lubricant. It is therefore desirable to temper the activity of the additive to the conditions of operation and to use no more EP activity than is required to avoid scuffing. But this degree of activity is not known in advance, and again must be found by experience.

Another method to avoid scuffing is to coat the metal parts with a solid film, usually produced by chemical reaction with the finished or partly finished part or by electrolytic deposition. Many commercial treatments are available: The films may contain phosphorus, sulfur, carbon, nitrogen, or other metals such as chromium, molybdenum, etc. A review has been given by Wilson (1973).

C. Gears

The scuffing limitation has its effect on the design of gears, notably in the choice of the pitch. Designers can choose between a small number of large teeth, which are strong but are liable to scuff because of the higher sliding speeds, and a large number of small teeth, which are weaker but less liable to scuff. In many cases the scuffing limitation dictates a finer pitch than would be chosen from other considerations.

An instructive example of the importance of scuffing in practical gears concerns the rear axle drive of road vehicles. In the conventional design, power from the engine is taken through the drive shaft and must then be transmitted through a right angle to the rear axle. The simplest arrangement is the bevel gear, in which the axes of the drive shaft and of the rear axle intersect. This arrangement is also the easiest to lubricate, but the geometry of the gear dictates an inconveniently high position for the drive shaft.

The hypoid gear permits the axis of the drive shaft to be some distance below that of the rear axle and thus avoids the inconvenience. This gear introduces a considerable amount of sliding, and scuffing is avoided only by the use of coatings on the surfaces and of very active extreme pressure agents in the oil. This is a case very suitable for the use of such additives since the quantity of oil involved is small and it may be isolated from other more sensitive materials.

Finally, the worm gear configuration arises when the drive shaft is taken to its lowest possible position. Here the degree of sliding introduced is so great that it is impossible to use the steel-on-steel material combination, even with the most effective surface coatings and oil additives. The wheel is usually made of the more expensive bronze, which is considerably softer than steel, so that the pressures transmitted are lower, and the physical dimensions must be greater than if hardened steel were to be used.

The experience of a designer of aircraft gears may be of some interest (Shotter, 1977). In his experience, scuffing is seen mainly as a running-in problem; it can sometimes occur in fully run-in gears, but usually there is some exceptional reason, e.g., poor oil supply, lack of alignment, etc. Scuffing in aircraft gears is a relatively infrequent happening, and when it does occur the resulting damage is usually so great that it is difficult to observe where it started. But occasionally some clues may be obtained from examples in which the failure was discovered and arrested at an early stage during overload testing on the bench.

In this designer's experience, the evidence of such examples suggests that scuffing is often provoked by an edge or end contact. At the start of the contact between two teeth there is no possibility of building up a hydrodynamic film to protect the surfaces, and this is seen as the most dangerous condition. The edge or end geometry is critical, and in some cases a phosphate coating may make matters worse because it produces a rougher edge or end surface. No cases have been found in which it has not been possible to cure scuffing by modifications to the geometry, the materials (including lubricant), the surface finish, and the surface treatment.

D.   Cams and Tappets

Scuffing of cams and tappets becomes more difficult to avoid as engine speeds increase. Steel is rather prone to scuffing, and at one time nearly all European designers used chilled cast-iron cams and tappets. These are very resistant to scuffing, but are liable to pit. American engine manufacturers use hardenable cast iron, which does not pit but is still quite liable to scuff, although less so than is steel. So lubricants for American engines must have quite a high degree of EP activity, but this could result in worse pitting if such an oil were used in European engines with chilled cast-iron materials. The job of the lubricant supplier is therefore a difficult one. The scuffing difficulty is one reason for the adoption of overhead camshafts for high-speed engines. This design reduces the mass of the reciprocating part of the valve train and therefore reduces the load that must be carried by the contact between cam and tappet. Even these overhead camshaft designs can give trouble. It is an increasingly common practice for a finger follower to be inserted between the cam and the valve, and wear or scuffing of both the cam and of the follower has been experienced. Information on cam and tappet problems has been surveyed by MIRA (1972) and by Abell (1977).

E. Piston Rings and Cylinders

In a review of piston ring and cylinder scuffing, Neale (1971) states that few engines suffer from this defect in service, but that it is encountered in the development of many types, and much effort is devoted to solving the problem before the engine reaches the market.

F. Conclusion

We have seen that although scuffing is not encountered frequently in practice, it is nevertheless always present in the background as a factor influencing both the design of sliding mechanisms and the materials of their construction. It is therefore important for the designer to understand the factors affecting scuffing. This aspect is discussed in the next section and leads naturally to a consideration of empirical criteria of scuffing in which these factors of importance may be expected to occur, explicitly or implicitly.

V. Factors Affecting Scuffing

A. Gears

Sliding is the one essential feature common to all systems that scuff. Systems in which there is no sliding do not scuff but fail by some other mechanism, e.g., pitting. Table I gives a list of operating factors affecting scuffing in gears or in disk machines operated under conditions intended to simulate those in gears. The effect of the velocities of the surfaces re-

TABLE I

FACTORS INFLUENCING SCUFFING OF GEARS AND DISKS

Factors promoting scuffing		Factors retarding scuffing	
Increasing:	Load	Increasing:	Rolling speed
	Sliding speed		Lubricant viscosity
	Surface roughness		Chemical activity
	Temperature		of environment

quires further discussion. The mean rolling speed $\overline{U}$ is defined as one-half of the vector sum of the velocities U_1 and U_2 of the two surfaces relative to their conjunction, while the sliding speed U_s is the vector difference between U_1 and U_2. In any one gear combination the maximum sliding speed is proportional to the mean rolling speed, i.e., the maximum slide/roll ratio $U_s/\overline{U}$ is constant. In tests under these conditions of constant slide/roll ratio the scuffing load decreases with increasing speed over the range of most practical interest, but Borsoff (1959) showed that the scuffing load eventually passes through a minimum and finally increases again at high speeds (10,000–20,000 rpm). A similar result is found with disks (Watson, 1957). There are two speed conditions in which scuffing is particularly likely to occur. One is when the two velocities U_1 and U_2 are equal in magnitude but opposite in direction, and the other is when the contact is stationary on one of the surfaces. Lubrication is known to be more difficult when the peripheral velocities of the discs U_1 and U_2 are in opposite directions, since one of the disks then drags hot oil from the conjunction back into the inlet.

The effect of chemical activity also merits further discussion. Tests both with gears (Baber *et al.*, 1968) and with disks (Bjerk, 1973) show that complete exclusion of atmospheric oxygen results in failure at very light loads. Failure may be postponed or avoided by an increase in chemical activity of the atmosphere, of the lubricant, or of the materials of the rubbing surfaces. Too much chemical activity gives excessive wear by corrosion, however (Goldman, 1969), and for any given system there is an optimum level of chemical activity giving minimum wear.

B.   Materials

The materials of the two rubbing surfaces may affect the incidence of scuffing in a number of different ways. Pairs having different mechanical, and particularly chemical, properties are better than pairs of identical materials, while increasing heterogeneity is also helpful (Hirst, 1976). The Hertzian stresses in contacts between gear teeth and between cams and tappets are so high as to exclude virtually all except hardened ferrous materials, and of these the heterogeneous materials, hardenable and chilled cast iron give much better performance in cam and tappet systems than do the more homogeneous steels. But considerations of impact and bending strength rule out the cast irons as gear materials, and steel must be used in this application. Austenite is very inert chemically, and the higher the austenite content, the more easily the steel will scuff. Thus alloying elements that tend to increase the austenite content, such as Ni, Mn, C, and N, should be avoided as much as possible, and the use of Cr,

W, Mo, Va, and Si, which reduce the austenite content, is to be encouraged (Niemann and Lechner, 1967).

C. Edge Effects

Scuffing of components in nominal line contact is very sensitive to edge effects. Scuffing often starts at an edge, and in one investigation (Bell *et al.*, 1975) the load was removed and the machine stopped as soon as sudden large increases in noise, vibration, and friction indicated a scuff. The time delay until the machine stopped was about 5 sec, and during that time the band of visible scuffing had progressed inwards from the edge of the contact towards the center by about 1–1.5 mm. A slight imbalance of 1%–2% in the loads carried by the two push rods of the split loading system was sufficient to cause the majority of the scuffs to start from this more heavily loaded edge. The accuracy of alignment of these disks was thought to be better than that in most industrial gears. Time-lapse photography of a scuffing test carried out in an IAE gear test machine with overhung gears showed scuffing starting at one edge. De Gruchy and Harrison (1963) found that scuffing started at random, sometimes at the edges, sometimes in the center of the track. Even with extreme precautions to promote uniformity of loading, scuffs frequently start at an edge (Miller and Mackenzie, 1973). If one of the disks is crowned, scuffing usually starts at the center of the track. Edge effects in gears have been discussed in Section IV,C.

D. Running In

Scuffing also depends critically on running in. A carefully run-in component will carry a much higher load than will an unrun component, and the demands on marine gears are rendered particularly arduous by the practice of verifying the maximum ship speed during the trials. The reverse flanks of marine gears are particularly susceptible to such damage because the first time they are run under load is often when the engines are run astern with forward way on the ship after it has been worked up to maximum speed. Thus the first load on the reverse flanks is often an overload, and severe scuffing may result.

E. Piston Rings and Cylinders

Factors affecting the scuffing of piston rings have been surveyed in a review by Neale (1971). The two most influential factors appear to be piston speed and bore surface finish, but the usually accepted center-line average (CLA) criterion of surface finish is not by itself an adequate measure of the characteristics of the bore surface. Piston ring and cylinder scuffing appears to be more responsive to oil viscosity and less responsive to the action of EP additives than is scuffing in gears or in cams and tappets.

The factors affecting scuffing in service have now been enumerated and these would be expected to appear either explicitly or implicitly in empirical criteria of scuffing, which will now be discussed. The importance to the designer of such criteria has already been explained.

VI. Empirical Criteria for Scuffing

A. *Total Contact Temperature—The Blok Postulate*

1. INTRODUCTION

Any discussion of criteria for scuffing must be dominated by the name of Blok and by the subject of flash temperatures. Blok (1939) put forward the hypothesis that a characteristic temperature at scuffing may be allotted to each combination of lubricant and rubbing surfaces, this characteristic temperature being nearly independent of load, speed, and bulk lubricant temperature. The relevant temperature is the total contact temperature, which is the sum of two components. One, the bulk temperature of the metal parts, may in principle be measured or estimated without difficulty. The other component is the instantaneous rise in the temperature of the surface as it passes through the heat source, which is represented by the area of contact between the rubbing surfaces. This component is known as the flash temperature and is difficult to measure since it is of very short duration and decays rapidly both with time after the surface has left the contact and with depth below the surface. An experimental estimate of a mean total contact temperature across the area of contact may be made if the two materials develop an appreciable thermal emf, but this is not always possible, and the thermal emfs will be very low if both materials are ferrous. Furthermore, scuffing will depend on the maximum temperature within the contact, whereas the thermal emf will give some sort of mean temperature. Thus in most cases this second component of the total contact temperature, the flash temperature, must be estimated. Blok (1937a,b) also developed theories by which this estimation of the flash temperature could be undertaken. The coefficient of friction enters into the theoretical expression for the flash temperature, but it is usually impossible to measure this quantity in practical machinery. This disadvantage is shared by most of the other empirical criteria that have been proposed.

2. RANGE OF APPLICABILITY

An important point concerning Blok's postulate of a constant scuffing temperature is that it is confined to straight mineral oils, i.e., those not containing EP additives. Although the constancy of the total contact temperature at scuffing was illustrated in the original paper (Blok, 1939) by reference to experimental results obtained with a straight mineral oil, EP oils were not specifically excluded. A later statement (Blok, 1972), however, made it clear that the postulate was intended from the outset to apply to nonadditive mineral oils in conjunction with the hard steels usual for heavy-duty gears.

3. EXPERIMENTAL EVIDENCE

a. Gears. Blok's hypothesis was originally based on results obtained with gears. Unfortunately, the experimental data were destroyed during the 1939–1945 war and have never been published. One disadvantage of work with gears is that the instantaneous coefficient of friction cannot be measured. An estimate of the mean coefficient may be obtained, e.g., from the power consumption of a "back-to-back" gear set, but the instantaneous coefficient will vary during the meshing cycle as a result of the changes in sliding speed. Furthermore, the normal loads between the gear teeth are influenced by dynamic effects, and they must be measured by strain gauges rather than estimated from the static loading.

Lemanski (1965) obtained an impressive correlation between the occurrence of scuffing in aircraft gears and the total contact temperature calculated on the basis of an assumed constant coefficient of friction. The same data with some additions were reanalyzed by Kelley and Lemanski (1967–1968) with individual estimates of the coefficient of friction for each case, and the correlation between scuffing and contact temperature then became considerably less impressive.

Ku *et al.* (1975) were able to predict the scoring loads of gears from the results of tests on disks by using the critical total contact temperature hypothesis, and their results offer the most impressive evidence in favor of the Blok postulate.

b. Disk Machines. Most of the other evidence was obtained with disk machines, usually intended to simulate gear operation. The position has been reviewed by Blok (1970). The Blok postulate has been confirmed experimentally by Kelley (1953) and by Leach and Kelley (1965). Meng (1960) presented his results in the form of a nearly constant average conjunction temperature. When Blok (1970) converted these average temperatures into maximum conjunction temperatures, the results agreed with the postulate and with the temperatures previously calculated by Blok from results obtained with a gear test machine, based on various assumptions about the coefficient of friction and the thermal properties of the materials.

Fleck (1963) used a machine in which the disks were in oscillatory motion. He reported, first, that the flash temperatures, as measured by the thermal emf between disks of two different materials, were about 30% higher than those calculated from the theory and, second, that the total conjunction temperature at scuffing was not constant. Blok (1970) has given reasons for doubting both. conclusions. O'Donoghue *et al.* (1967–1968) obtained results demonstrating the validity of the Blok hypothesis, but they covered only a narrow range of speeds and bulk temperatures. The most impressive evidence was given by Ku *et al.* (1975).

Many workers have found that the total contact temperature at scuffing was not constant, but that it was usefully correlated with other variables. Thus Fein (1960) found that although the total contact temperature at failure varied over a range of up to about 170°C, it was correlated with the ratio of load to sliding speed, and that this correlation could be used to predict failure conditions. Later work (Fein, 1965) showed deviations from the suggested correlation under certain condition, while still further work (Fein, 1967) made it clear that the apparent correlation depended on the test procedure. This later work (Fein, 1967) was especially noteworthy in that one of the lubricants (squalane) was nominally the same as one used by Leach and Kelley (1965), while the machine was of identical design. The conclusions of the two sets of investigators were diametrically opposed: Leach and Kelley (1965) found that the total contact temperature at scoring was constant within experimental error (about ±20°C), while Fein found that the temperature at scuffing varied by something like 200°C!

Carper *et al.* (1973) and Staph *et al.* (1973) found a useful correlation between the total contact temperature at scuffing and a parameter

$$\xi = \eta_0^2 |U_2^2 - U_1^2| / R^2 (\overline{p}_H)^2$$

where η_0 is the viscosity of the lubricant at atmospheric pressure and at the temperature of the surface of the disks, U_1, U_2 are the peripheral velocities of the two disks, R is the radius of relative curvature, and $\overline{p}_H$ is the mean Hertzian pressure in the contact.

Some caution is necessary in the use of this correlation for prediction of scuffing. The total contact temperature is highly correlated with the surface temperature, which in turn controls η_0, so that the history of any one test run shows a very strong correlation between the parameter ξ and the total contact temperature. The usefulness of the relation for prediction depends on the fact that the correlation line for points corresponding to scuffing failure is different from that for those describing the histories of test runs. Later results (Ku *et al.*, 1975) suggested that the total contact temperature alone was a better criterion.

Results inconsistent with the Blok postulate have been reported by Genkin *et al.* (1949), Cocks and Tallian (1971), Drozdov and Gavrikov (1968), Bjerk (1973), Czichos and Kirschke (1972), Pike and Spillman (1970), Ali and Thomas (1964), Bell and Dyson (1972a), Bell *et al.* (1975), and Appeldoorn (1970). Bell *et al.* (1975) gave an example in which the total contact temperature at failure was less than 60°C, whereas the same lubricant and materials could be run at total contact temperatures of over 300°C without failure. The only differences between the two cases were the rolling and sliding speeds. Failure at intermediate temperatures could be obtained by a suitable choice of speeds.

c. A Critical Experiment. Genkin *et al.* (1949) performed a critical experiment giving a direct proof that the total contact temperature at failure decreased with decreasing speed at constant slide/roll ratio. Under certain circumstances their failure load increased with increasing speed. They ran a loaded pair of disks until temperature equilibrium had been attained and then gradually reduced the speed at constant load and slide/roll ratio. The frictional traction increased as the speed decreased, but the product of sliding speed and frictional traction presumably decreased; it would be extremely unusual if it did not. Both the measured bulk temperature of the disks and the calculated flash temperature in the conjunction decreased with decreasing speed, and therefore the total contact temperature must have decreased. Nevertheless, the disk scuffed as the speed decreased. This shows directly that the critical total contact temperature at scuffing must have decreased with decreasing speed.

A similar conclusion may be drawn rather more indirectly from the many results available from scuffing tests both with disks (Watson, 1957; De Gruchy and Harrison, 1963; Genkin *et al.*, 1949) and with gears (Padmore and Rushton, 1964–1965; Borsoff, 1959; Niemann and Lechner, 1967; Ku and Baber, 1959), which show the existence of a high-speed region in which the scuffing load increases with increasing speed at constant slide/roll ratio. Again, the coefficient of friction decreases with increasing speed, but the product of coefficient of friction and sliding speed normally increases continuously. Both the surface temperature and the flash temperature therefore increase with speed at constant load;

therefore the total contact temperature increases with speed at constant load, and hence the critical total contact temperature at failure must increase even more rapidly with increasing speed and increasing load. Therefore the total contact temperature at failure cannot be constant as is required by Blok's postulate. It is only fair to add that similar objections may be raised against other criteria, e.g., the frictional power intensity, which will be discussed later.

d. Conclusion. There is obviously a serious conflict of evidence about the applicability of the Blok postulate of a constant total contact temperature at failure.

The most plausible conclusion would seem to be that the Blok criterion represents a useful approximation over certain ranges of experimental conditions. Thus Bell *et al.* (1975) showed that the total contact temperature at scuffing was indeed approximately constant over certain bands of rolling (1.6–6.2 m s^{-1}) and sliding (1.6–2.5 m s^{-1}) speeds, and that significant deviations were found when the experimental conditions were outside this band. Their results are shown in Fig. 3.

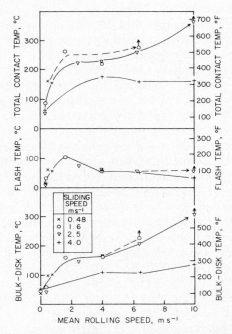

Fig. 3. Temperatures just before failure. (From Bell *et al.*, 1975. Copyright American Society of Lubrication Engineers. Reproduced by permission.)

The postulate is therefore of service for the purpose of design and prediction, provided that the conditions do not depart too far from those in which previous experience has been gained.

B. Other Temperature Criteria

Niemann and Lechner (1967), Niemann and Seitzinger (1971), and Michaelis (1973) concluded that the mean temperature of the surfaces of the gear teeth was more nearly constant than was the total contact temperature. Drozdov and Gavrikov (1968) and Drozdov (1972) favored a parameter that was equivalent to the flash temperature divided either by the load per unit transverse width or by the mean Hertzian pressure. Bailey and Cameron (1973) found that the total contact temperature at scuffing was constant at low bulk metal temperatures, but that the flash temperature was constant at higher bulk metal temperatures. Wydler (1973) suggested the flash temperature itself as a criterion.

C. Frictional Power Intensity

A criterion with a history even longer than that of the total contact temperature is the product of the mean Hertzian stress $\overline{p_H}$ and the sliding speed V_s (Almen 1935, 1942, 1948). Since the directions of the vectors of the normal pressure and the sliding speed are at right angles, the product of their magnitudes is of doubtful physical significance in this context, and a criterion of greater meaning is obtained by further multiplication by the coefficient of friction μ to give the product of the mean shear stress due to friction $\overline{\tau} = \mu\overline{p_H}$ and the sliding speed. This is the product of two parallel vectors, and the result is the rate of dissipation of energy by friction per unit area of nominal contact, the "specific power of friction" of Matveevsky (1965) or the "friction power intensity" of Bell and Dyson (1972a). By an argument similar to that used in connection with the total contact temperature criterion the friction power intensity must increase with increasing speed at constant slide/roll ratio in the high-speed regime in which the scuffing load increases with increasing speed.

Figure 4 shows the variation with mean rolling speed of the friction power intensity, calculated for conditions obtaining just before scuffing, according to various authors (Watson, 1957; Bell et al., 1975; Carper et al., 1973; Staph et al., 1973; Bjerk, 1973; Meng, 1960; Genkin et al., 1949; Drozdov and Gavrikov, 1965; Pike and Spillman, 1970; Matveevsky, 1956; Remezova, 1959; Vinogradov and Podolsky, 1960). All refer to hardened steel surfaces lubricated by mineral oils, and in many cases the

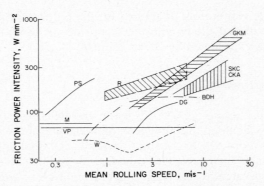

Fig. 4. Variation with mean rolling speed of friction power intensity just before scuffing. BDH, Bell *et al.* (1975), disks; CKA, Carper *et al.* (1972), disks; DG, Prozdov and Garrikov (1968), disks; Genkin *et al.* (1949), disks; M, Matveevsky (1965), four ball; PS, Pike and Spillman (1970), four ball; R, Remezova (1959), crossed cylinders; SKC, Staph *et al.* (1973), disks; VP, Vinogradov and Podolsky (1960), four ball; W, Watson (1957), disks. (From Dyson, 1975. Copyright IPC Science and Technology Press, Ltd. Reproduced by permission.)

information necessary for the calculations has been read from graphs, with varying degrees of precision. There is obviously much scatter, but the materials and the lubricants differ from author to author. Meng (1960) and also Carper and Ku (1975) preferred the total power of friction as a criterion for scuffing, but this is an extensive quantity rather than an intensive quantity, and it would be expected to vary with the size of the machine. Carper and Ku (1975) found that the correlation between the total power of friction and their parameter ξ, as defined previously, offered the best criterion, and that the friction power intensity often decreased significantly during a test before scuffing occurred.

Moorhouse (1973) reported that the scuffing of gears in service was best correlated with the friction power intensity divided by the sliding speed raised to the power 0.8. The mean tooth flank temperature gave the second-best correlation, followed by the total contact temperature of Blok (1939) and finally by the criterion of Drozdov and Gavrikov (1968). Matveevsky (1965) showed that the results quoted originally in terms of constant friction power intensity (Matveevsky, 1956; Remezova, 1959; Vinogradov and Podolsky, 1960) or of constant total friction power (Meng, 1960) satisfied equally well the criterion of constant total contact temperature. This illustrates the difficulty of the choice between different criteria.

In addition to the results shown in Fig. 4, the values of $\overline{p_H}V_s$ given by various authors may be converted into friction power intensity by multi-plication by an assumed value for the coefficient of friction. The values of 1.5×10^6 and 3.0×10^6 lb in.$^{-2}$ ft sec^{-1} quoted by Almen (1935) for spiral bevel gears and spur gears, respectively, both with mineral oils and with an assumed coefficient of friction of 0.06, become 180–360 W mm^{-2} approximately, while that of Yokoyama *et al.* (1972), 6.4×10^5 kg mm^{-1} sec^{-1}, becomes 300 W mm^{-2} for spur gears with the same value assumed for the coefficient of friction. Schnurmann and Pederson (1971) quoted a limit of $(1.2–1.6) \times 10^6$ kg cm^{-1} sec^{-1} for the four-ball machine. Here a rather higher coefficient of friction seems appropriate, and an assumed value of 0.09 gives 110–140 W mm^{-2} for the friction power intensity. This was the limit for the applicability of Amonton's law (friction proportional to normal force), which Schnurmann and Pederson identified as the limit for the avoidance of catastrophic wear. It is probably permissible to iden-tify this again with scuffing since this is the known mode of failure in most tests on the four-ball machine.

D. The Minimum Thickness of the Elastohydrodynamic Oil Film

1. HISTORICAL

Application of classical hydrodynamic theories to heavily loaded "counterformal" contacts (i.e., those between two convex surfaces) in practical machinery, such as gears, led to the prediction of minimum oil film thicknesses an order of magnitude or more less than the scale of the surface roughness. The persistence of the original grinding marks after years of operation, however, showed that lubrication could be very effec-tive, and this was ascribed to the boundary lubrication mechanism. At that time, this mechanism seemed to include everything that could not be explained by classical hydrodynamic theory.

2. ELASTOHYDRODYNAMIC LUBRICATION

It is now known that contacts of this type are lubricated hydrodynam-ically but in a regime quite different from that of classical hydrodynamics. In the new regime the elastic deflections of the surfaces play an impor-tant part and are usually much greater than the minimum film thickness. The regime is therefore known as that of elastohydrodynamic lubrication. The effect of pressure on the viscosity of the lubricant is also important for materials with a high elastic modulus, such as metals. Calculations of the minimum film thickness according to elastohydrodynamic theory resulted in values rather greater than, or of the same order of magnitude as, the scale of the surface roughnesses. But the theory did not offer any new criterion to help predict why or when failure should occur. The irony of the situation is that at one time it was not understood how these mechanisms could operate successfully without damage. Now we understand how they operate, but we cannot understand how they fail.

3. A POSSIBLE FAILURE CRITERION

The normal elastohydrodynamic theories assume ideally smooth surfaces and may therefore be expected to break down when the minimum film thickness is of the order of the surface roughness. Thus one criterion that suggested itself was that there is no danger of scuffing if the film thickness is greater than the surface roughness, but that there would be such a danger if the film thickness is less than the surface roughness.

This turned out to be true, but it is impossible to predict the likelihood of scuffing from the film thickness in the range in which it is less than the surface roughness, i.e., in which there is a danger of scuffing. Bell and Dyson (1972a) showed that the film thickness just before scuffing was independent of initial surface roughness over a range of more than 3:1 in roughness; Bell *et al.* (1975) reported results ranging from failure at a film thickness of 0.5 μm to no failure at 0.065 μm, both with initial roughness of 0.4 μm CLA for each surface. Bartz (1973) reached a similar conclusion. Bodensieck (1967), however, saw film thickness as a useful criterion for gear scoring.

4. CONCLUSION

The resolution of the conflicting evidence may be along the same lines as have been proposed for the contact temperature criterion. Bell *et al.* (1975) found that the minimum film thickness just before scuffing failure is approximately constant over a range of rolling and sliding speeds, and their results are shown in Fig. 5. Surprisingly, this range seems to be almost the same as that over which the total contact temperature was approximately constant, and again large variations were observed outside this range.

Staph *et al.* (1973) found that cross-ground disks gave a lower coefficient of friction and a higher scuffing load than did circumferentially ground disks of the same "composite" surface roughness. Since there are

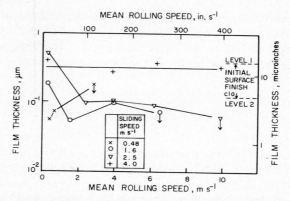

Fig. 5. Film thickness calculated for conditions just before failure. (From Bell *et al.*, 1975. Copyright American Society of Lubrication Engineers. Reproduced by permission.)

both theoretical (Johnson *et al.*, 1972) and experimental (Jackson, 1974) grounds for expecting thicker films between the cross-ground disks than between the circumferentially ground disks, this finding confirms the importance of elastohydrodynamic film thicknesses in scuffing.

E. Exposure Time

Borsoff and Godet (1963) proposed a "scoring factor for gears," defined as a time equal to the ratio of the width of the Hertzian contact zone divided by the maximum sliding speed. They claimed a linear correlation between the factor and the failure load in tests with any one lubricant in any one set of gears, but quite similar lubricants gave different correlations. Application for predictive purposes was rendered rather difficult by the fact that the failure load was itself contained in the "scoring factor." Furthermore, it is not easy to see the physical significance of a relation in which the failure load increased with increasing exposure time; a decrease would have been more plausible.

F. Conclusion

The friction power intensity, and particularly its variation with rolling speed, appears to the reviewer to be the most satisfactory way in which the results from various sources may be compared on the same basis.

Since both quantities are dimensional, some influential physical quantities must have been omitted, and the scatter in Fig. 4 makes it useless for design and prediction *ab initio*. Nevertheless, it is the best correlation of published results that the reviewer has been able to find.

VII. Screening Tests for Scuffing

Scuffing tests on full-scale machinery are often time consuming and expensive, and since, as we have seen, there is no simple and reliable empirical criterion of scuffing, many attempts have been made to develop simpler and cheaper screening tests both for lubricants and for metals. In general, the utility and reliability of such screening tests increase with the degree to which the devices used in the tests approach the conditions of the full-scale machinery. The correlation between the results of different rigs is often not very good. This may be due, at least partly, to metallurgical differences, and so far as possible the metallurgy of the test device should be identical with that of the intended service. Cam and tappet rigs and piston-ring and cylinder-liner rigs grow more and more complicated the closer it is desired to match conditions in the engine, and it frequently happens that the cheapest and simplest way of obtaining such conditions is by the use of an engine. The metallurgy of the test device should be identical with that of the practical machine.

VIII. The Mechanism of Scuffing

A. *The Importance of Surface Films*

Since scuffing necessarily involves welding, the interposition of some nonweldable substance between the two metallic surfaces would prevent scuffing. Clean metallic surfaces brought into contact in a hard vacuum adhere strongly (Bowden and Tabor, 1950, 1964, A); the coefficient of friction in sliding rises to very high values (2–5), and the surfaces may be regarded as having scuffed at very light loads.

Bjerk (1973) found that disks failed catastrophically in an inert atmosphere, even under the lowest load he applied; running in in the presence of oxygen generated a visible protective film that supported quite heavy loads in a subsequent test in an inert atmosphere. These results were obtained with a plain mineral oil. If a diester lubricant containing oxygen in the molecule was used, the disks could be run in in an inert atmosphere without distress. Protective films may therefore be formed from oxygen obtained either from the atmosphere or from the lubricant.

Baber *et al.* (1968) found that the scuffing load of gears at first decreased with increasing temperature because of the reduction in the viscosity of the lubricant. In an oxidizing environment further increases in temperature resulted in a minimum and finally an increase in the load-carrying capacity. This increase did not take place in an atmosphere of nitrogen and was attributed to the formation on the tooth surfaces of a visible carbonaceous film derived from oil oxidation products.

Protective films may be formed, or their formation may be accelerated, by other means, e.g., by fatty acids in solution in the oil or by EP additives. These latter normally contain one or more of the elements sulfur, chlorine, and phosphorus, but other elements are also effective, e.g., boron, iodine. The chemistry of the reaction between these substances and the surfaces is complicated and not well understood; the organic parts of the additive molecules play an important role, and mixed organic/inorganic compounds are found in the reaction products.

At one time it was thought that an essential feature of the action of EP additives was the production of soft films with a low melting point, but Kreuz *et al.* (1967) showed that borates also functioned as EP additives, although they produced films that were harder than the steel substrate and had a high melting point. Iodine is also effective as an EP agent and gives a very low friction coefficient: It is effective even on glass surfaces (Fein, 1966), so that the formation of a halide is not a necessary condition for the effectiveness of an EP additive. It is not even necessary for the formation of a protective film to have a substance present that is aggressive in a chemical sense, or even a metal that is recognized as liable to chemical action. A rather mysterious solid film known as "frictional polymer" is often seen on the surfaces of rubbing bodies, and protective qualities are attributed to it (Fein and Kreuz, 1965)—it may be formed on platinum surfaces (Chaikin, 1967). A similar film causing low friction has been reported by Macpherson (1959). Kreuz *et al.* (1973) have interpreted the breakdown of boundary lubrication with increase in temperature in terms of the solubilization of solid films formed on the surfaces. Furey (1973) has suggested antiscuffing and antiwear additives of a new type in which polymeric films are deposited on the rubbing surfaces at high temperatures.

In conclusion, scuffing is dominated by lack of oxygen at very low partial pressures with chemically inert lubricants, and the formation, wear, and renewal of protective films must play an important part in the mechanism of scuffing under more common circumstances. But very little progress has been made towards a quantitative assessment of the role of such films in scuffing under more "normal" conditions.

B. Experimental Results with a Bearing on the Mechanism of Scuffing

1. ENERGY CONSIDERATIONS

In addition to the experimental evidence discussed previously in relation to the empirical criteria of scuffing and to the importance of surface films, many experimental results with a bearing on the mechanism of scuffing have been reported. The importance of the rate of supply of energy was emphasized by Cameron (1954), who passed a heavy electrical current through the contact between two loaded disks. He showed that as the electrical power input to the contact increased, the mechanical power needed to scuff the disks decreased, the total electrical and mechanical power remaining constant. The ratio of electrical to total power reached approximately 45%.

The importance of the effectiveness of the means of removal of the power generated in the contact was emphasized by Crook and Shotter (1957), who ran a pair of disks with the outer rims thermally insulated from the hubs by annuli of resin board. These disks scuffed at a lower load than did a similar pair of disks without the thermal insulation. This importance of the production and removal of power led Barwell and Milne (1952) to propose a thermal instability mechanism of scuffing: At some critical combination of load and speed, an increase in temperature would lead to an increase in the area of contact and in the coefficient of friction sufficient to produce an unstable temperature condition.

Rozeanu (1973) developed a stability criterion involving the gradient of the viscosity in the superficial layers, and produced experimental results to support his model. He also stressed the importance of transient effects in provoking instability that may lead to scoring or scuffing (Rozeanu, 1976). Rozeanu's experimental results are very striking and do not seem to be explicable in terms of conventional ideas.

2. THE CUMULATIVE NATURE OF SCUFFING

Barwell and Milne (1957) noted that scuffing appeared to grow from small score marks distributed intermittently on the surfaces and increasing in size and frequency with increase in load. They suggested that these marks could act as centers for the further propagation of damage by repeated contact.

This idea of the cumulative nature of scuffing, in spite of its apparently sudden appearance, recurs in many experiments. Padmore and Rushton (1964–1965), for example, found that the scuffing load of gears decreased with increasing speed, passed through a minimum, and then increased rapidly with further increase in speed if the load was first applied at test speed. If the load was applied from rest, not only was the scuffing load lower, but the recovery after the minimum was very much less pronounced. Furthermore, the difference in the loading procedure affected the appearance of the scuffed surfaces at high speeds, beyond the minimum in the load-carrying capacity. If the load was first applied at test speed, there were indications of scoring, that is, fine grooves in the tooth surfaces in the direction of sliding, and also indentations on the flanks. These changes in the appearance of failed teeth at high speeds did not occur if the load was applied at rest. The most plausible explanation of these observations is that when the load was applied at rest the teeth suffered some form of preliminary damage (presumably during the period of running up to test speed and at a speed corresponding to that of the minimum in the load-carrying capacity) and that this preliminary damage resulted in the occurrence of scuffing at a later stage in the run. Furthermore, the nature of the preliminary damage dictated the appearance of the surfaces after the subsequent scuff.

The most direct proof of the cumulative nature of scuffing was obtained by Carper et al. (1973), who found that disks driven independently had considerably higher scuffing loads than did those coupled with gears when both were operated under the same conditions. The only difference between the two cases is that contacts occur repeatedly between the same pairs of points on the surfaces of disks coupled by gears, so that any damage that occurs is aggravated by repeated contacts. With disks driven by independent electric motors, however, the synchronization is not so precise, and in general repeated contacts do not occur. The coefficient of friction was also appreciably higher for the gear-coupled disks than for those driven independently. This demonstration of the importance of repeated contacts is related to the general experience that hunting-tooth gears, in which contacts between any two teeth recur with a relatively low frequency, give higher scuffing loads than do gears with speed ratios expressed as a simple fraction. The cumulative effect of repeated contacts is presumably responsible for the decrease of scuffing load with increasing running time, noted by Fowle and Hughes (1969–1970). Macpherson and Cameron (1973) report the same effect and suggest that it is caused by the formation of microcracks and micropits, which drain oil away from the contact.

3. RELATION BETWEEN WEAR AND SCUFFING

Borsoff (1959) made the interesting observation that gear wear rate at constant load passed through a maximum as the rotational speed increased and that the speed at which this maximum occurred was identical with the speed at which the scuffing load was a minimum.

4. HYDRODYNAMIC EFFECTS

The importance of hydrodynamic factors has been stressed by many authors. It has been pointed out before in this review that the decrease in scuffing load with increasing speed at constant slide/roll ratios is consistent with the postulate of a constant total contact temperature at failure, but that the subsequent increase in scuffing load with still further increase of speed is not consistent with the postulate. One interpretation of these results is that temperature is important at low speeds, while hydrodynamic factors dominate at the high speeds (Godet, 1963).

However, it can be shown that hydrodynamic effects are extremely important even in the low-speed regime, in which the scuffing load decreases with increasing speed at constant slide/roll ratio. Table II shows some results obtained by Bell et al. (1975) with a disk machine in which the load was increased in stages until scuffing was observed or until the load limit of the machine was reached. The only difference in the running conditions of the five tests recorded in the table was the mean rolling speed. In the first test, at a mean rolling speed of 0.38 m s^{-1}, the peripheral velocities of the two disks were in opposite senses, and lubrication is known to be difficult in these circumstances. But the remaining four tests, in which the peripheral velocities are in the same sense, show considerable differences at different rolling speeds. All of the obvious indications—load, temperatures, and film thickness—suggest that the conditions corresponding to the rolling speed of 10 m s^{-1} were more severe than were those for the

TABLE II

EFFECT OF ROLLING SPEED ON SCUFFING CONDITIONS[a]

| Mean rolling speed (m s^{-1}) | Load | | Temperatures, °C | | | Film thickness | | Remarks |
	(kN)	(lb)	Disk bulk	Flash	Total contact	(μm)	(μin.)	
0.38	0.067	15	40	13	53	0.48	19	Conditions
2.5	2.89	680	110	62	172	0.17	6.8	immediately
4.0	5.69	1280	162	67	229	0.10	4.0	before
6.25	8.36	1880	216	67	283	0.081	3.2	scuffing
10.0	17.7	3980	327	60	387	0.054	2.1	No failure

[a] Sliding speed: 2.5 m s^{-1}; initial surface finish: 0.4 μm CLA; lubricant: straight mineral oil.

other rolling speeds, yet the disks did not scuff within the load limit of the machine at 10 m s^{-1}, whereas they did scuff at the other speeds. The only favorable factor at 10 m s^{-1} is the rolling speed, which is a hydrodynamic factor. A similar conclusion was reached by Cocks and Tallian (1971).

5. MIXED FRICTION

In spite of the importance of hydrodynamic factors, the regime of lubrication is very mixed; that is, boundary lubrication is also important. Bell and Dyson (1972b) take the friction obtained with very smooth disks as the upper limit of elastohydrodynamic friction. They show that between one-half and two-thirds of the friction observed with rougher disks under conditions approaching scuffing is caused by some factor other than elastohydrodynamic friction, presumably boundary friction. Their estimate of the coefficient of boundary friction under these circumstances, about 0.5, is surprisingly high.

C. Breakdown of Boundary Lubrication

Hydrodynamic factors are invariably present in practical machinery, but it was seen in the previous discussion of the film thickness criterion that contacts between asperities of the surfaces through the elastohydrodynamic film are thought to be important in scuffing. Thus valuable insight into scuffing might be gained from a study of such contacts in the absence of disturbing hydrodynamic factors. This is the province of boundary lubrication, where experiments are usually conducted at such low speeds (typically less than 1 mm s^{-1}) that hydrodynamic effects are very small.

Bowden and Tabor (1950, 1964, B) found that many systems showed a well-defined rise in the coefficient of friction as the bulk temperature was increased. Because of the low sliding speed, the flash temperature was negligible and the total contact temperature was identical with the bulk temperature. With pure paraffins and alcohols as lubricants, the transition temperatures were identical with the bulk melting points, but with fatty acids the transition temperatures were much higher. They were often identified with the melting points of the metal soaps of the fatty acids. Similar results were obtained from fatty acids or other polar materials in solution in an inert solvent. These transitions were reversible. The friction decreased again as the temperature was lowered.

If the Blok postulate of the constancy of the total contact temperature at scoring is accepted, it is an obvious move to compare this temperature with that for the breakdown of boundary lubrication. If the two temperatures are identical, then first, the mechanism of scoring is explained, or at least related to another known phenomenon, and second, the probability of scuffing may be predicted from the very much easier experiments in boundary lubrication.

Some authors have made the comparison and have reported the identity of the two temperatures (O'Donoghue *et al.*, 1967–1968; Matveevsky, 1965). Sharma and Cameron (1973) referred to a common failure temperature of 150°C for steel surfaces lubricated by mineral oils, and Grew and Cameron (1972) conferred the accolade of thermodynamic respectability on this finding. They explained it in terms of the physical desorption from the surfaces of surface-active materials present at low concentration in an inert carrier.

Certain reservations must be made here. Lovell and Cameron (1967–1968), for instance, reported that they could not obtain repeatable results with mild or hardened steels and mineral oils and that they were unable to detect a critical temperature, with an abrupt rise in friction, with these materials. They attributed this to variations in the ratio of martensite to austenite in the steel. Mould (1971) found a transition with EN 31 steel and a mineral oil at high rates of increase of temperature, but not at low rates of increase. Grew and Cameron (1972) could detect transition temperatures with mild steel and fatty acids but not with mineral oils, and the combination of stainless steel with a solution of a fatty acid in white oil now seems to be commonly used in studies of this nature (Bailey and Cameron, 1973; Hirst and Stafford, 1972). There are difficulties even with this system, however. Hirst and Stafford (1972) found sharp transitions in some conditions but very ill-defined transitions in other conditions, and their transition temperatures covered a range of more than 200°C for the same materials and lubricant. The identification of scoring with the breakdown of boundary lubrication depends on both temperatures being constant, or at least correlated in a defined manner with experimental conditions. If neither temperature is constant or is so correlated, then there is no basis for a comparison.

Hirst and Hollander (1974) have provided some very interesting information on the relation between the characteristics of the surface topology and the breakdown of boundary lubrication. Both the vertical and the horizontal scales of roughness were found to be important, and the authors concluded from their results that failure occurs as a result of the plastic deformation of those features of the topography that they identify with the main structure of the surface; deformation of the fine structure, of much shorter wavelength, that is superimposed on the main surface is not important.

D. Breakdown of Hydrodynamic Lubrication

1. ABSENCE OF THE GEOMETRIC WEDGE

It has been seen that a condition of mixed lubrication, hydrodynamic and boundary, is present in situations leading to scuffing. The breakdown of boundary lubrication has been discussed as a mechanism of scoring and the possible failure of the hydrodynamic component must now be considered. The main hydrodynamic wedge mechanism fails when the peripheral velocities of the components relative to their conjunction are equal and opposite, i.e., $U_1 + U_2 = 0$. Hydrodynamic action then depends on the much weaker viscosity wedge mechanism, resulting from the difference in temperature between the surfaces approaching and leaving the contact (Cameron, 1951). There will also be some support by the squeeze film mechanism. This condition of equal and opposite peripheral velocities occurs twice during the operating cycle of a cam and tappet valve mechanism and Müller (1966) showed the importance of passing through this condition as rapidly as possible if scuffing is to be avoided. The entrainment velocity also falls to zero at the top and bottom dead-center positions of pistons, but the sliding speed is also zero at this position. Nevertheless, the momentary lack of hydrodynamic support, occurring as it does near the conditions of maximum temperature and pressure, may play an important part in ring and cylinder scuffing.

2. BREAKDOWN OF THE GEOMETRIC WEDGE

The breakdown of hydrodynamic lubrication under more normal conditions, in which the classical hydrodynamic wedge action operates, will now be considered. Petrusevitch (1971) finds that there are circumstances under which there is no solution of the simultaneous elastic and hydrodynamic equations at certain regions of the film. This does not necessarily mean that no lubrication film is present; support may still be obtained from the squeezing of the film established at earlier stations in the inlet region.

3. EFFECT OF SURFACE ROUGHNESS

a. *Introduction.* It has been seen in Section VI,D,3 that the minimum thickness of the elastohydrodynamic film calculated for conditions approaching scuffing is of the same order as the height of the roughness of the surfaces and that conventional smooth-surface theory is therefore inapplicable.

Some progress in understanding scuffing may therefore be possible if elastohydrodynamic theory could be extended to take into account the roughness of the surfaces. This can be done only for certain types of surfaces, and an approximate treatment for circumferentially ground disks has been worked out by Dyson (1976).

This treatment is based on the analysis of the unscuffed parts of a pair of disks that had been used in a scuffing experiment by Bell and Dyson (1972a). Since the band of scuffing spreads in an axial direction, the statistical properties of these run but unscuffed parts were taken as representative of those of the scuffed parts just before they scuffed. For the purposes of the hydrodynamic calculations, the roughness was assumed to take the form of ridges and valleys running in a circumferential direction.

b. Load-Carrying Mechanism. Preliminary calculations showed that a very large part of the load must be carried by asperity contacts under conditions approaching scuffing, and it was therefore assumed that all of the load was carried by asperity contacts at scuffing. This must inevitably cause severe wear and surface damage in sliding unless the asperity contacts are protected by a thin film of a very viscous fluid.

c. A Failure Criterion. The high viscosity could come only from the existence of a high hydrostatic pressure, which in turn could be generated by the main elastohydrodynamic system by the operation of the Reynolds equation, as modified for the effects of surface roughness. Now, in an elastohydrodynamic system with a fixed geometry there is a rather sharp transition from effective to ineffective lubrication, and it was estimated that this transition would occur at a disk bulk temperature of 150°C for the conditions of the test. The temperature recorded just before failure by thermocouples embedded in the disks was 180°C, and the agreement is quite encouraging for a calculation from first principles: No disposable constants have been introduced. If one of the constants in the calculated relation between the separation of the two surfaces and the mean pressure on them as a result of asperity deformation is regarded as disposable, exact agreement of the two temperatures can of course be obtained.

d. Extension to the Results of Other Runs. So far, only one pair of disks used in a single scuffing test has been considered. It is possible to extend the comparison to include runs on other pairs of disks produced by the same methods and to the same initial surface finish if it is assumed that the statistical properties of the surface of these disks just before failure were identical with those of the pair analyzed. Figure 6 shows a comparison between predicted and observed failure temperatures for the results of Bell and Dyson (1972a) and of Bell *et al.* (1975). The agreement is remarkably good, but the test of the theory is not very realistic since the failure load has been assumed to be known, and the comparison has been

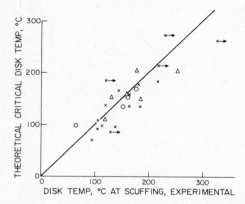

Fig. 6. Comparison of predicted and experimental scuffing temperatures. Arrows denote tests for which no failure was obtained within the load limit of the machine. ○ mineral oil, Bell and Dyson (1972); △ additive oil, Bell and Dyson (1972a); × mineral oil, Bell *et al.* 1975. [Reproduced by permission of the Council of the Institution of Mechanical Engineers from Dyson (1976). *Proc. Inst. Mech. Eng. London* **190**, 52176, 699–711.]

made on the basis of the failure temperatures. It is the object of a scuffing test to determine the failure load.

e. Prediction of Failure Loads. The failure load may be predicted if use is made of the observed relation between the disk bulk temperature and the power dissipated in friction. If a coefficient of friction is assumed, there is then a relation between the disk bulk temperature and the load for a given sliding speed. The relation between the critical disk bulk temperature and load for a given rolling speed is obtained from the rough-surface elastohydrodynamic theory just outlined. The intersection of the two temperature curves gives the failure conditions, including the failure load for that particular combination of rolling and sliding speeds. A comparison of these predicted failure loads with those observed by Bell *et al.* (1975) is shown in Fig. 7. There is a lot of scatter, but this may perhaps be acceptable in view of the difficulty of the task attempted; the theoretical prediction of failure loads from first principles.

f. Conclusion. The theory just described is rather narrowly based on the properties of a single pair of disks, and much more work will be required before its status may be judged. Furthermore, it is a physical theory, and seems to offer the possibility of accounting for the effects of physical factors, e.g., rolling and sliding speeds, viscosity–temperature relations, etc., on scuffing.

But scuffing is known to be very sensitive to chemical and metallurgical

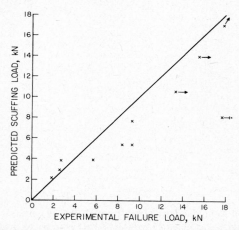

Fig. 7. Comparison of predicted and experimental scuffing loads, showing results of Bell *et al.* (1975). [Reproduced by permission of the Council of the Institution of Mechanical Engineers from Dyson (1976). *Proc. Inst. Mech. Eng. London* **190,** 52176, 699–711.]

factors, which are not accommodated in the theory. It is possible that such factors influence the coefficient of friction and also the statistical properties of the surfaces just before failure. An additive that filled in the bottoms of the valleys would improve the hydrodynamics and thereby increase the scuffing load.

E. A Possible Relation between Scoring and Scuffing and the Breakdown of Boundary and Elastohydrodynamic Lubrication

1. TRANSITIONS IN POINT CONTACT AND SIMPLE SLIDING

A notable advance in the understanding of failure at point contacts in simple sliding, e.g., in four-ball machines and pin-on-disk machines with spherically ended pins, has been made by workers of the International Research Group of the OECD (Begelinger and de Gee, 1974,1976a,b; Salomon, 1976; Czichos, 1977). Their results are sketched in Fig. 8 in the form of a graph of failure load versus sliding speed. As the load in a scuffing test was increased at constant sliding speed, three regimes of operation and three different types of transition were identified. At low sliding speeds the first transition was between regimes I and II, marked by a moderate increase in wear rate and a comparatively large increase in friction. Further loading produced a second transition, from regime II to regime III, marked this time by a large increase in wear rate and a modest

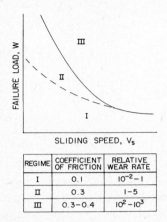

REGIME	COEFFICIENT OF FRICTION	RELATIVE WEAR RATE
I	0.1	$10^{-2}-1$
II	0.3	$1-5$
III	0.3-0.4	10^2-10^3

Fig. 8. Regimes of operation of a four-ball machine. – – – hydrodynamic film failure; ——— boundary film failure.

increase in friction. There was some evidence that this transition occurred at an approximately constant total contact temperature. At higher sliding speeds regime II was absent, and there was only one transition, that between regimes I and III, marked by large increases in both friction and wear rate. Regime I was identified as that of hydrodynamic lubrication and regime II as that of boundary lubrication, while regime III was described as one of complete failure of lubrication. Thus at low sliding speeds the first transition, between regimes I and II, was a failure of hydrodynamic lubrication, while the second transition, between regimes II and III, represented a further failure, that of boundary lubrication. At higher sliding speeds the hydrodynamic and boundary failures occurred simultaneously.

2. POSSIBLE EXTENSION TO OTHER CONDITIONS

The author now wishes to suggest a rather speculative extrapolation of this scheme to the conditions pertaining in gears and disks, i.e., line contact with sliding and rolling. The most obvious difference between this case and the previous one is that the hydrodynamics of the system are now much more favorable, and the line of hydrodynamic failure must therefore be moved upwards. The boundary film failure occurs at approximately constant total contact temperatures and will be little affected by the change in the hydrodynamics. The author's suggested extension of Fig. 8 to cover the present case is shown in Fig. 9. The line of hydrodynamic failure now cuts the line of boundary failure at a much larger angle,

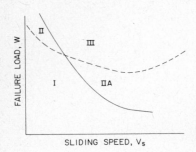

Fig. 9. Suggested extension of Fig. 8 to cover disks and gears. – – – hydrodynamic film failure; ——— boundary film failure.

and it seems logical to extrapolate it past the boundary failure line in the manner shown. A similar prolongation was shown in Fig. 7 of Begelinger and de Gee (1976a). There is now an additional regime II A characterized by failure of the boundary film in the presence of a still effective hydrodynamic film. The minimum in the hydrodynamic failure load is a consequence of the form of the viscosity–temperature relation of the lubricant, as explained in the Appendix.

It is now suggested that most work with disk machines and gears is carried out at sliding speeds greater than that corresponding to the interception of the two failure lines in Fig. 9, and that consequently the boundary film failure occurs before the hydrodynamic failure.

3. SCORING AND SCUFFING

The author has provided some evidence that the final scuffing stage is a hydrodynamic failure (Dyson, 1976), and it may therefore be tentatively identified with the hydrodynamic failure line in Fig. 9. What can be said of the boundary failure line?

In tests with parallel-sided disks, scuffing starts from an edge and works inward (Bell and Dyson, 1972; Bell et al, 1975). The scuffed surface is rough and heavily distorted, and the unscuffed part is lightly scored and contains a thin patchy layer of transformed material (Campany, 1977) identified as W_1 (Rogers, 1969, 1970). The other transformed material W_2 is present in the scuffed part (Campany, 1977).

It is suggested that production of the W_1 phase must have been the result of a partial breakdown in lubrication, occurring on a small scale at many discrete points before the final scuff took place. This small-scale breakdown will be called "scoring." The scuffing stage has been identified with the hydrodynamic failure, and it is therefore tempting to identify the scoring stage with boundary film failure.

The two forms of failure are obviously related intimately to one another. As the load is increased during a scuffing test, scuffing usually follows fairly closely on scoring, although there is some evidence that the load interval between them may vary with the conditions of the test. At some intermediate stage between scoring and scuffing, transverse cracks appear in the scored regions and material is transferred from one disk to the other. The transverse cracks appear mainly on the slower surface and are confined to the polished areas, which are covered with a thin layer of W_1, and the transfer of material is mainly from the slower to the faster surface (Campany, 1977).

It seems possible that scuffing is triggered by the deterioration in the hydrodynamics of the system caused by the cracks and the presence of the transferred material. Continued operation in the scoring regime, regime II A in Fig. 9, may promote further cracks and transfer by a fatigue mechanism (Macpherson and Cameron, 1973). It is commonly reported that the presence of EP additives and operation at very high speeds tend to promote scoring rather than scuffing as the mode of failure (Hughes and Waight, 1958).

4. FAILURE IN PRACTICAL MACHINERY

Which is the true failure point? The answer must depend on the conditions and on the design of the equipment. It seems quite feasible to run some machinery past the scoring point, although at the cost of some wear. Scoring is a gradual phenomenon, and it may be difficult to specify exactly the point of onset. A disk machine could certainly be run satisfactorily in the scoring condition; the behavior of any other machine would probably depend on its sensitivity to wear. If a moderate amount of wear could be tolerated, the machine could probably run up to the scuffing point; if not, then the scoring point is probably the limit of safe operation. No machine could be run beyond the scuffing point; this is an unmistakable "sudden death."

5. EFFECT OF HYDRODYNAMIC SITUATION AT SCORING

One important observation remains to be made. The phenomena observed at the scoring point will depend on the hydrodynamic situation at that time. If this is bad, then the failure of the boundary film will be evident immediately. If the hydrodynamics are so favorable that there is no contact between the surfaces, the scoring point will escape detection. It is probable that this is what is happening in both gears and disks at speeds beyond the minimum in the failure load. From the argument given in Section VI,A,3,c, this failure cannot take place at an approximate constant total contact temperature if the failure load increases with speed at constant slide/roll ratio, but nevertheless the failure is of the scoring type. The explanation suggested is that the critical total contact temperature is reached when the lubricant films are quite thick, but the subsequent boundary film failure is noticed only when the temperatures have increased to such an extent that the hydrodynamic film is thin enough to allow some interaction between the surfaces. It is shown in the Appendix that this can lead to a failure load increasing with speed at constant slide/roll ratio. Thus the scoring failure at high speeds will be a gradual process with a rather indeterminate starting point, as opposed to the "sudden death" type of scuffing failure characteristic of the hydrodynamic breakdown at lower speeds.

The foregoing model is highly speculative, and it will require a lot more work before it can be confirmed or rejected.

Acknowledgments

This chapter is based largely on a review paper that was published in *Tribology* (Dyson, 1975) and the author is most grateful to the publishers, IPC Science and Technology Press, Ltd., for permission to use material freely from that paper.

Appendix: The Hydrodynamic Minimum in the Curve of Failure Load Against Sliding Speed

It will be assumed that failure occurs at a constant value of the EHD film thickness calculated on conventional smooth-surface theory. If the minor effects of the load and pressure coefficient of viscosity on film thickness are neglected, then at constant slide/roll ratio the condition is

$$\eta V_s = C \tag{1}$$

where η is the viscosity of the lubricant at atmospheric pressure, V_s is the sliding speed, and C is a constant. The viscosity is related to the bulk temperature of the metal parts by a relation of the form

$$\eta = A \exp[B(\theta + \theta_0)^{-1}] \tag{2}$$

where A, B, and θ_0 are constants.

If the coefficient of friction is approximately constant, the bulk temperature θ will be linearly related to the product of the sliding speed and the load W,

$$\theta = \theta_1 + \theta_2 WV_s$$

where θ_1 and θ_2 are further constants. The three equations give a relation of the form

$$\theta_0 + \theta_1 + \theta_2 WV_s = -B[\ln(DV_s)]^{-1} \tag{3}$$

where D is another constant. This relation gives a minimum in the curve of W against V_s.

References

Abell, R. F. (1977). Reprint 770019 Society of Automotive Engineers, Warrendale, Pa.

Ali, M. S., and Thomas, M. P. (1964). *Proc. Symp. Gear Transm. Oils and Their Testing* Vol. V, p. 1. Institute of Petroleum, London.

Almen, J. O. (1935). *Automot. Ind.* **73**, 662–668.

Almen, J. O. (1942). *J. Soc. Automot. Eng.* **50**, 9, 373–380.

Almen, J. O. (1948). *In* "Mechanical Wear" (J. T. Burwell, ed.), Chapter 12, pp. 229–238. American Society for Metals, Metals Park, Ohio.

Appeldoorn, J. (1970). *Conf. Lub. Test Devices and Their Relation to Service Conditions.* Paisley College of Technol., Scotland.

Baber, B. B., Anderson, E. L., and Ku, P. M. (1968). *J. Lub. Technol.* **90**, 117–124.

Bailey, M. W., and Cameron, A. (1973). *ASLE Trans.* **10**, 121–131.

Bartz, W. J. (1973). *Ver. Deut. Ing. Ber.* **195**, 87–102.

Barwell, F. T., and Milne, A. A. (1952). *J. Inst. Pet. London* **38**, 624–632.

Barwell, F. T., and Milne, A. A. (1957). *Proc. Conf. Lub. Wear, London, England* pp. 735–741. Institution of Mechanical Engineers, London.

Begelinger, A., and de Gee, A. W. J. (1974). *Wear* **28**, 103–114.

Begelinger, A., and de Gee, A. W. J., (1976a). *J. Lub. Technol.* **98**, 575–579.

Begelinger, A., and de Gee, A. W. J. (1976b). *Wear* **36**, 7–11.

Bell, J. C., and Dyson, A. (1972a). *Proc. Symp. Elastohydrodynam. Lub.* pp. 61–67. Institution of Mechanical Engineers, London.

Bell, J. C., and Dyson, A. (1972b). *Proc. Symp. Elastohydrodynam. Lub.* pp. 68–76. Institution of Mechanical Engineers, London.

Bell, J. C., Dyson, A., and Hadley, J. W. (1975). *ASLE Trans.* **18**, 62–73.

Bjerk, R. O. (1973). *ASLE Trans.* **16**, 97–106.

Blok, H. (1937a). General discussion on lubrication and lubricants, *Proc. Inst. Mech. Eng. London* **2**, 222–235.

Blok, H. (1937b). *Proc. World Pet. Congr., 2nd, Paris, Sec. IV* **III**, p. 471.

Blok, H. (1939). *J. Soc. Automot. Eng.* **44**, 193–210, 220.

Blok, H. (1970). NASA Rep. No. SP-237, pp. 153–241, Washington, D.C.

Blok, H. (1972). *Proc. Symp. Elastohydrodynam. Lub.* p. 166 (contribution to discussion). Institution of Mechanical Engineers, London.

Bodensieck, E. J. (1967). *Power Transm. Des.* November, 53–56.

Borsoff, V. N. (1959). *J. Basic Eng.* **81**, 79–88.

Borsoff, V. N., and Godet, M. (1963) *ASLE Trans.* **6**, 147–153.

Bowden, F. P., and Tabor, D. (1950, 1964). "The Friction and Lubrication of Solids." Oxford Univ. Press (Clarendon), London and New York.
  A: Pt. I, Chapter VII, pp. 145–160; Pt. II, Chapter V. pp. 87–107.
  B: Pt. I, Chapter IX, pp. 181–183; Pt. II, Chapter XVIII, pp. 367–369.

British Technical Council of the Motor and Petroleum Industries (1972). Cams and Tappets, A Survey of Information, Rep. BTC/L2/72. Motor Industries Research Association, Nuneaton, Warwickshire, England.

Cameron, A. (1951). *J. Inst. Pet.* **37**, 471–486.

Cameron, A. (1954). *J. Inst. Pet.* **40**, 191–196.

Campany, R. G. (1977). Private communication.

Carper, H. J., and Ku, P. M. (1975) *ASLE Trans.* **18**, 39–47.

Carper, H. J., Ku, P. M., and Anderson, E. L. (1973). *Mech. Mach. Theory* **8**, 209–225.

Chaikin, S. W. (1967). *Wear* **10**, 49–60.

Christensen, H. (1966). *Acta Polytech. Scand. Mech. Eng. Ser. E.* No. 25.

Cocks, M. and Tallian, T. E. (1971). *ASLE Trans.* **14**, 31–40.

Crook, A. W., and Shotter, B. A. (1957). *Proc. Conf. Lub. Wear, London* pp. 205–209. Institution of Mechanical Engineers, London.

Czichos, H. (1977). *Wear* **41**, 1–14.

Czichos, H., and Kirschke, K. (1972). *Wear* **22**, 321–336.

De Gruchy, V. J., and Harrison, D. W. (1963). *Proc. Lub. Wear Convent., Bournemouth, England* pp. 160–180. Institution of Mechanical Engineers, London.

Drozdov, Yu. N. (1972). *Wear* **20**, 201–209.

Drozdov, Yu, N., and Gavrikov, Yu. A. (1968). *Wear* **11**, 291–302.

Dyson, A. (1975). *Tribology* **8**, 77–87, 117–121.

Dyson, A. (1976). *Proc. Inst. Mech. Eng. London* **190**, 52/76, 699–711.

Eyre, T. S., and Baxter, A. (1972). *Met. Mater.* **6**, 435–439.

Fein, R. S. (1960). *ASLE Trans.* **3**, 34–39.

Fein, R. S. (1965). *ASLE Trans.* **8**, 59–68.

Fein, R. S. (1966). *Wear* **9**, 367–387.

Fein, R. S. (1967). *ASLE Trans.* **10**, 373–385.

Fein, R. S., and Kreuz, K. L. (1965). *ASLE Trans.* **8**, 29–35.

Fleck, W. (1963). *Maschinenbautechnik* **12**, 209–216, 438–444.

Fowle, T. I., and Hughes, A. (1969–1970). *Proc. Inst. Mech. Eng. London* **184** (Pt. 3-O) 122–130.

Furey, M. J. (1973). *Wear* **26**, 369–392.

Genkin, M. D., Kuz'min, N. F., and Misharin, Yu. A. (1949). *In* "Izdatelstvo Akademiyi Nauk SSSR, Moscow," Chapter 5. Condensed English translation in National Research Council of Canada Tech. Transl. 1056, Ottawa, Canada (1963). Complete German Translation in *Maschinenbau Fertigungstech. USSR* **3**, 248–267 (1961).

Godet, M. (1963). *C. R. Acad. Sci. Paris* **257**, 48–51.

Goldmann, I. B. (1969). *Wear* **14**, 431–444.

Grew, W. J. S., and Cameron, A. (1972). *Proc. R. Soc. London Ser. A* **327**, 1568, 47–59.

Hirst, W. (1976). *Chart. Mech. Eng.* **21**, 4, 88–92.

Hirst, W., and Hollander, A. E. (1974). *Proc. R. Soc. London Ser. A* **337**, 379–394.

Hirst, W., and Stafford, J. V. (1972). *Proc. Inst. Mech. Eng. London* **186**, 15/72, 179–192.

Hughes, J. R., and Waight, F. H. (1958). *Proc. Int. Conf. Gearing* pp. 135–143. Institution of Mechanical Engineers, London.

Institution of Mechanical Engineers (1957). *Proc. Conf. Lub. Wear, London* p. 4.

Jackson, A. (1974). Optical Elastohydrodynamics of Rough Surfaces, Ph.D. Thesis, Imperial College, Univ. of London.

Johnson, K. L., Greenwood, J. A., and Poon, S. Y. (1972). *Wear* **19**, 91–108.

Kelley, B. W. (1953). *J. Soc. Automot. Eng.* **61**, 175–185.

Kelley, B. W., and Lemanski, A. J. (1967–1968). *Proc. Inst. Mech. En. London* **182**, 3A, 178–184.

Kreuz, K. L., Fein, R. S., and Dundy, M. (1967). *ASLE Trans.* **10**, 67–76.

Kreuz, K. L., Fein, R. S., and Rand, S. J. (1973). *Wear* **23**, 393–407.

Ku, P. M., and Baber, B. (1959). *ASLE Trans.* **2**, 184–194.

Ku, P. M., Staph, H. E., and Carper, H. J. (1975). USAAMRDL Tech. Rep. 75–33, AD AO 13527. Defense Documentation Center, Alexandria, Virginia.

Leach, E. F., and Kelley, B. W. (1965). *ASLE Trans.* **8**, 271–285.

Lemanski, A. J. (1965). Am. Gear Manuf. Assoc., Inf. Sheet No. 217.01.

Lovell, A. C., and Cameron, A. (1967–1968) *Proc. Inst. Mech. Eng. London* **182**, 3G, 122–124.

Macpherson, P. B. (1959). Discussion of Borsoff.

Macpherson, P. B., and Cameron, A. (1973). *ASLE Trans.* **16**, 68–72.

Matveevsky, R. M. (1956). "A Temperature Method for the Estimation of the Limiting Lubrication of Machine Oils." USSR Academy Press, Moscow (in Russian).

Matveevsky, R. M. (1965). *J. Basic. Eng.* **89**, 754–760.

Meng, V. V. (1960). *Frict. Wear Mach. USSR* **14** 202–217.

Michaelis, K. (1973). *Proc. FZG Colloq. Fressen Zahnrädern, München, Germany* pp. 31–60.

Miller, B. J., and Mackenzie, D. A. (1973). *ASLE Trans.* **16**, 245–251.

MIRA (1972). See Br. Tech. Council of the Motor and Pet. Ind.

Moorhouse, P. (1973). *Proc. FZG Colloq. Fressen Zahnrädern München, Germany* pp. 61–82.

Mould, R. W. (1971). *Nature (London)* **253**, 38, 62–63.

Müller, H. (1966). *Motor Tech. Z.* **27**, 58.

Neale, M. J. (1971). *Proc. Inst. Mech. Eng. London* **185**, 21–32.

Niemann, G., and Lechner, G. (1967). *Erdoel Kohle Erdgas Petrochem.* **20**, 2, 96–106.

Niemann, G., and Seitzinger, A. (1971). *Ver. Deut. Ing. Z.* **13**, 97–172.

O'Donoghue, J. P., Manton, S. M., and Askwith, T. C. (1967–1968) *Proc. Inst. Mech. Eng. London* **182**, 3N, 18–23.

Padmore, E. L., and Rushton, S. G. (1964–1965). *Proc. Inst. Mech. Eng. London* **179**, 3J, 278–289.

Petrusevitch, A. I. (1971). *Mashinovednie* No. 6, 72–76 (in Russian).

Pike, W. C., and Spillman, D. T. (1970). *ASLE Trans.* **13**, 127–133.

Remezova, N. E. (1959). *Vestn. Mashinostr.* **9**, 19 (*Engl. transl.: Russ. Eng. J.* **9**, 17).

Rogers, M. D. (1969). *Tribology* **2**, 123–127.

Rogers, M. D. (1970). *Wear* **15**, 105–116.

Rozeanu, L. (1973). *ASLE Trans.* **16**, 115–120.

Rozeanu, L. (1976). *ASLE Trans.* **19**, 257–266.

Sakmann, B. W. (1947). *J. Appl. Mech.* **14**, A-43–A-52.

Salomon, G. (1976). *Wear* **36**, 1–6.

Schnurmann, R., and Pederson, O. (1971). *Wear* **18**, 341–345.

Sharma, T. P., and Cameron, A. (1973). *ASLE Trans.* **16**, 258–266.

Shotter, B. A. (1977). Private communication.

Staph, H. E., and Ku, P. M. (1968). *Discussion of paper by D. L. Alexander, ASLE Trans.* **11**, 72–80.

Staph, H. E., Ku, P. M., and Carper, H. J. (1973). *Mech. Mach. Theory* **8,** 197–208.

Vinogradov, G. V., and Podolsky, I. I. (1960). Cited in Matveevski (1965). (In Russian.)

Watson, H. J. (1957). *Proc. Conf. Lub. and Wear* pp. 469–476. Institution of Mechanical Engineers, London.

Wilson, R. W. (1973). *Eur. Tribol. Conf., 1st, London,* Paper No. 49.

Wydler, R. (1973). *Proc. FZG Colloq. Fressen Zahnrädern, München, Germany* pp. 147–155.

Yokoyama, M., Ishikawa, J., and Hayashi, K. (1972) *Wear* **19,** 131–141.

Abrasive Wear

MARTIN A. MOORE†

National Institute of Agricultural Engineering
Wrest Park, Silsoe
Bedford, England

I. Introduction

The Research Group on Wear of Engineering Materials of the Organization for Economic Cooperation and Development has defined abrasive wear as wear by displacement of material from one of two surfaces in relative motion caused by the presence of hard protuberances on the second contact surface; or by the presence of hard particles either between the surfaces or embedded in one of them. Other, more specific, terminology has emerged from studies of abrasive wear and is used to describe the severity of wear, specific combinations of abrasive and wearing material, or the relative properties of the abrasive and wearing material. For example, in the United States abrasive wear is often classified as "gouging abrasion," "high-stress grinding abrasion," or "low-stress scratching abrasion," depending on severity of damage to the wearing material. The terms "two-body abrasive wear" and "three-body abrasive wear" are

† Present address: Fulmer Research Institute, Ltd., Stoke Poges, Slough, Bucks, England.

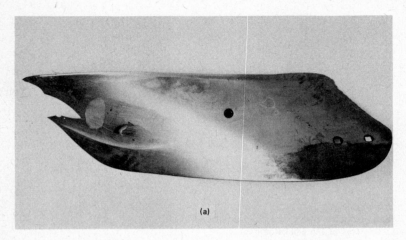

(a)

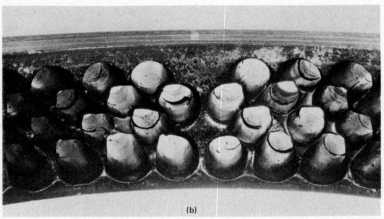

(b)

Fig. 1. Components failed through abrasive wear: (a) plow moldboard; (b) chicken feed cubing die. (Reproduced by permission of National Institute of Agricultural Engineering.)

often found in the literature—the first to describe wear caused by abrasive sliding on a material and in which a load is applied by gravity, the resistance of the abrasive medium to deformation, etc. and the second to describe wear caused by abrasive trapped between two moving surfaces. Richardson (1968) differentiated between wear caused by an abrasive of hardness equal to or less than that of the worn surface—a "soft abrasive"—and that caused by an abrasive harder than the worn surface—a "hard abrasive."

Although mankind uses abrasive wear to his advantage—for example, to shape and polish engineering components and to clean his teeth—its unwanted occurrence causes an enormous annual expenditure by industry and consumers. Most of this represents the cost of replacing or repairing equipment that has worn to the extent that it no longer performs a useful function. In some cases this may occur after only a very small percentage of the equipment's or component's total volume is worn away (Fig. 1). And for some industries, such as agriculture, as many as 40% of the components replaced on equipment have failed by abrasive wear (Richardson *et al.*, 1967). Other major sources of expenditure caused by abrasive wear are losses of production consequent to breakdown, the need to invest more frequently in capital equipment, and increased energy consumption as equipment wears.

Several groups have estimated the total cost of friction and wear. Jost (1975) stated that friction and wear in the United States accounted for an expenditure of $100 billion per annum; in the Federal Republic of Germany a committee of the Ministry for Research and Technology (1976) estimated friction and wear caused a national economic waste of 10 billion DM per annum, of which about half is due to abrasive wear alone; and Rabinowicz (1976a) has estimated that about 10% of all the energy generated by man is dissipated in friction processes. Ten years before these estimates were made a committee of the United Kingdom Department of Education and Science (1966) estimated that about 20% of the cost of friction, lubrication, and wear might be saved by technological improvements—a saving of about £500 million per annum at the time. Table I shows the percentage of savings attributed to various factors the committee assessed. These potential savings, plus the increasingly onerous conditions under which industry requires its machinery to function and consumers' demands for greater reliability, provide considerable incentive for both research and education in friction, lubrication, and wear.

TABLE I

Percentage Savings on the Cost of Friction, Lubrication, and Wear in the United Kingdom Attributed to Various Factors[a]

Estimated savings	Percentage
Savings in maintenance and replacement costs	45
Savings of losses consequent to breakdown	22
Savings in investment through increased life of machinery	19
Reduction in energy consumption through lower friction	6
Savings in investment due to higher utilization ratios and greater mechanical efficiency	4
Reduction in manpower	2
Savings in lubricant costs	2

[a] Data from Department of Education and Science (1966).

TABLE II

PERCENTAGE COSTS OF ABRASIVE WEAR TO VARIOUS INDUSTRIES IN THE
FEDERAL REPUBLIC OF GERMANY[a]

Industry	Percentage costs
Metallurgical extraction and ore mining	40
Coal mining, grading, and pulverizing	30
Agriculture, construction, stone, gravel, and sand quarrying and sundry comminution	20
Production engineering	10

[a] Data from Ministry for Research and Technology (1976).

An overwhelming problem in reducing the widespread occurrence of abrasive wear is that it is almost impossible to exclude material with abrasive properties from machinery. This is hardly surprising since the potentially abrasive materials silica, alumina, and iron oxides account for 81% of the earth's crust. With such an enormous reservoir of natural abrasive some will eventually manage to bypass seals and screens on machinery. However, the most severe problems occur in equipment that is in contact with soil, rock, and minerals as the worked material, i.e., in the metallurgical, mining, construction, and agricultural industries. Table II shows the cost, as a percentage of the total, of abrasive wear to various industries in the Federal Republic of Germany (Ministry for Research and Technology, 1976). In these industries some improvements have been brought about by design when this reduces the frequency of abrasive particle contact, contact loads, and sliding distances. But the most significant improvements have occurred by the application of existing knowledge on abrasive wear to the selection of materials to match the working environment. Unfortunately the most wear-resistant material does not always provide the most cost-effective solution.

The economics of materials use varies considerably from one operation to another and depends on such factors as relative cost of the replaceable components to cost of the equipment as a whole, the cost of having equipment idle while components are repaired or replaced, the effect of stoppages on the timeliness of an operation, and the basic cost effectiveness of different materials and of components manufactured from different materials. For example, if a sand and gravel dredger comes into port for repairs to its pumping equipment, the loss of production may amount to more than £1000. Thus the overall cost of abrasive wear for the operation may be reduced by using wear resistant materials which are not in themselves cost effective. At the other extreme, in the agricultural industry the loss of production and the labor cost in replacing worn components are often very small—about 10% of the cost of new components (National Institute of Agricultural Engineering, 1974)—so material selection is controlled largely by the basic material cost effectiveness.

We cannot start to make scientific material selection and design changes without first understanding something of the abrasive wear process, how material properties affect abrasive particle contact and material removal, and how component geometry affects abrasive particle contact, loading, and sliding. Thus I have attempted in this chapter to present a broad state-of-the-art review of abrasive wear and also to indicate where there are gaps in our knowledge.

II. Validity of a Simple Model of Abrasive Wear

Let us consider a simple model of abrasive particle contact (Archard, 1953; Mulhearn and Samuels, 1962). Figure 2 shows an abrasive particle in contact with a surface. If this contact results in material removal, the volume worn away is

$$\delta V = CAL \tag{1}$$

where C is the proportion of the groove volume removed, A is the cross-sectional area of the groove, and L is the sliding distance.

Now we make the following assumptions:

(i) The cross-sectional area of a groove

$$A \propto p^2 \tag{2}$$

where p is the depth of indentation of the abrasive particle, and the mean constant of proportionality does not vary for different sizes of abrasive particles.

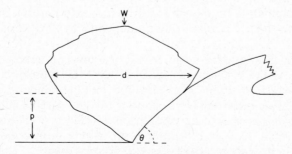

Fig. 2.   Idealized abrasive particle contact.

(ii)   The abrasive particle acts like a scratch indenter with a sharp point, and the projected area of contact of the particle, like A, is proportional to p^2. Then the hardness of the surface (O'Neill, 1967; Buttery and Archard, 1970/1971)

$$H \propto W/p^2 \tag{3}$$

where W is the load on the particle.

(iii)   The number of particles that contact the surface per unit area

$$N \propto d^{-2} \tag{4}$$

where d is the mean particle diameter, and the constant of proportionality does not vary for different loads per unit area.

(iv)   A proportion K of the abrasive particles that contact the surface result in material removal, and K is independent of d and the applied load per unit area σ.

Thus from Eq. (4) the mean load per particle that contacts the surface

$$W \propto \sigma d^2 \tag{5}$$

and from Eqs. (2), (3), and (5) the cross-sectional area of the groove

$$A \propto \sigma d^2/H \tag{6}$$

From Eqs. (1) and (6) the sum of δV for N contacts per unit area, the volume wear per unit area, is

$$V \propto KCNL\sigma d^2/H \tag{7}$$

which on substituting for N from Eq. (4) becomes

$$V \propto KCL\sigma/H \tag{8}$$

The constant of proportionality in Eq. (8) depends on the shapes of the abrasive particles and their dispersion. Thus this simple model of abrasive wear shows that the amount of material removed from the surface should be linearly dependent on the sliding distance, load per unit area, and the inverse of the surface hardness, but independent of the abrasive particle size.

In practice (Rabinowicz et al., 1961; Nathan and Jones, 1966; Richardson, 1967a) the amount of material removed from the surface increases linearly with sliding distance after initial nonlinear behavior (Fig. 3). The initial nonlinear behavior occurs when the surface deforms plastically and strain hardens to an equilibrium hardness H (Khrushchov and Babichev, 1960a). However, volume wear is only proportional to sliding distance when the other variables in Eq. (8) are constant. Thus Mulhearn

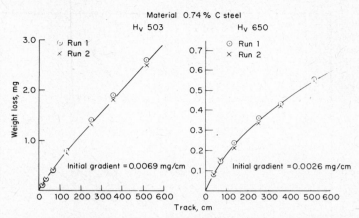

Fig. 3. Wear versus distance run for polished steel specimens on 500-μm glass paper, hardness H_V = 590 kg/mm². (From Richardson, 1967a.)

and Samuels (1962), Buttery and Archard (1970/1971), Billingham *et al.* (1974), and Date and Malkin (1976) have found that the wear rate decreases with increasing sliding distance (Fig. 4) when the abrasive is passed over many times so that it becomes clogged or deteriorates. Experimental data (Nathan and Jones, 1966; Larsen-Badse, 1968a; Lancaster, 1969; Wilshaw and Hartley, 1972) also show that the amount of

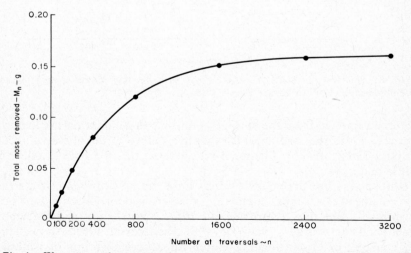

Fig. 4. Wear versus the number of traverses on the same track for steel on 70-μm silicon carbide paper. (From Mulhearn and Samuels, 1962.)

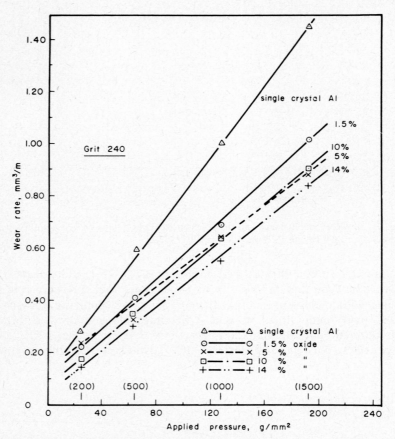

Fig. 5. Wear rate of some Al–Al$_2$O$_3$ alloys as a function of the applied pressure (load/abraded area). (From Larsen-Badse, 1968a.)

material removed from the surface is linearly dependent on the load per unit area up to the load at which massive failure of the abrasive occurs (Larsen-Badse, 1968b). Figures 5 and 6 show this for both ductile and brittle materials.

Khrushchov and Babichev (1956a) were the first to demonstrate that the volume wear of pure metals is inversely proportional to their bulk hardnesses. But they (Khrushchov, 1957; Khrushchov and Babichev, 1960b) and later workers (Selwood, 1961; Richardson, 1967b) showed that the constant of proportionality varied for different groups of materials such as pure metals, alloys, plastics, and ceramics. Data for pure metals and ceramics, plotted as the normalized inverse of volume wear—relative

wear resistance—in Fig. 7 show this. However, Khrushchov and Babichev (1960a), Avient *et al.* (1960), Alison and Wilman (1964), Larsen-
Badse (1966), Richardson (1967c), Lin and Wilman (1969), Moore *et al.*
(1972), and Moore and Douthwaite (1976) have shown that a surface can
undergo severe plastic deformation and strain harden during wear. In Eq.
(8) the hardness H is the surface hardness, i.e., the equilibrium hardness
of the surface after some wear has taken place. Thus Avient *et al.* (1960)
and Richardson (1967c) attempted to correlate relative wear resistance
with surface hardness after wear. While this correlation (Fig. 8) is better
than that between the relative wear resistance and hardness of the unworn
surface, clearly Eq. (8) does not completely account for the effect on volume wear of mechanical properties of the surface material.

Although Eq. (8) predicts that volume wear is independent of the abrasive particle size, in practice volume wear increases rapidly with particle
diameter up to a critical size after which it increases at a much reduced

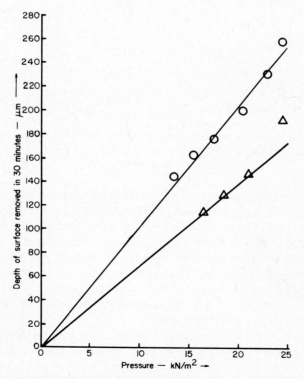

Fig. 6.   Wear rate of ○ soda-lime glass and △ silica on 30-μm SiC water slurry as a function of applied pressure. (After Wilshaw and Hartley, 1972.)

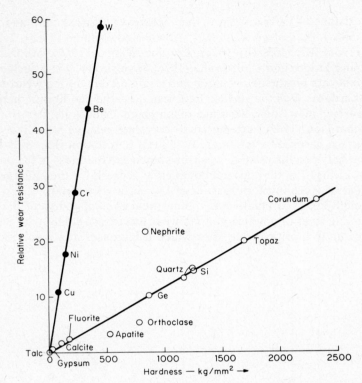

Fig. 7. Wear resistance versus hardness for some pure metals and minerals. (From Khrushchov and Babichev, 1960b.)

rate with increasing particle size (Fig. 9). This may occur (Avient *et al.*, 1960; Mulhearn and Samuels, 1962; Rabinowicz and Mutis, 1965; Richardson, 1968; Larsen-Badse, 1968b,c, 1972; Date and Malkin, 1976) if the geometry of the abrasive particles varies with size; if K in Eq. (8) is less for small abrasive particles because of their larger height distribution on commercial bonded abrasives, because wear debris interferes with abrasive particle contact, or because the small particles deteriorate more rapidly; and if C in Eq. (8) is less for small abrasive particles because they deteriorate more rapidly.

To summarize, a simple model of abrasive wear accounts well for the observed dependence of volume wear on sliding distance and applied load per unit area but not for the effect on volume wear of mechanical properties of the surface material and abrasive particle size. Clearly these latter variables influence abrasive particle contact in a subtle and complex way, and we shall consider them in greater detail in Sections III and IV.

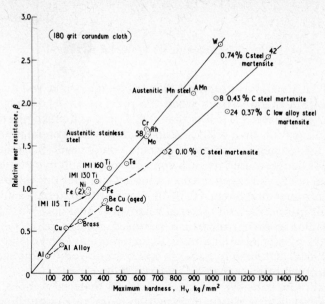

Fig. 8. Wear resistance on 84-μm corundum paper versus maximum hardness of the worn surface. (From Richardson, 1967c.)

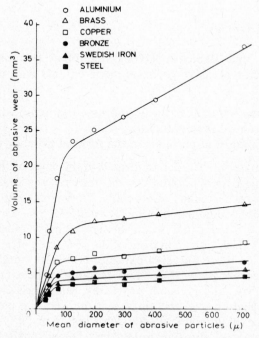

Fig. 9. Volume wear on SiC papers versus abrasive particle diameter. (From Nathan and Jones, 1966.)

III. Mechanisms of Material Removal

Kragelskii (1965) has proposed that a spherical slider will make elastic contact with a metal surface when the ratio of depth of indentation to diameter, p/d in Fig. 2., is less than 0.005, and will make plastic contact when p/d is greater than 0.005. Although this is an oversimplification, for a lightly loaded angular abrasive particle contact will be predominantly elastic. For ductile materials such contacts may result in material removal by surface molecular mechanisms (Rabinowicz, 1968) or removal of surface films (Aghan and Samuels, 1970). For brittle materials such contacts may result in fracture and material removal (Lawn and Wilshaw, 1975). As the load on an abrasive particle increases, contact on ductile materials will involve plastic deformation to a greater extent (Khrushchov and Babichev, 1960a). Ultimately material will be removed by fracture, but the rate-controlling process may be material flow.

We shall now consider the mechanisms of material removal in detail; first, those for which plastic deformation is the rate-controlling process; second, those for which fracture is rate controlling; and last, surface molecular and corrosion/mechanical mechanisms.

A. Plastic Deformation

Khrushchov and Babichev (1960a) identified two major processes taking place when abrasive particles made contact with the surface of a ductile material:

(i)   the formation of plastically impressed grooves which did not involve direct material removal; and

(ii)   the separation of particles in the form of primary microchips.

In both cases material was also deformed to the sides of the grooves, and Aghan and Samuels (1970) found that this material can sometimes become detached and form secondary microchips. They concluded, however, that primary microchips dominate in terms of material loss. Graham and Baul (1972) have also shown that when distinct primary microchips are not formed, material may pile up in front of the abrasive particle until fracture occurs. With fine loose abrasive particles (Blombery and Perrott, 1974) and fine bonded abrasives (Date and Malkin, 1976) some material loss may also occur by adhesive wear processes; this debris can be extremely small and be smeared over the surface of the abrasive particles.

Material removal involving plastic deformation occurs for a wide range of both material and abrasive properties. Dobson and Wilman (1963),

Stickler and Booker (1963), Cutter and McPherson (1973), and Swain (1975) have observed microchip formation and plowing during the wear of materials that have predominantly brittle properties. This may occur because temperature rises of about 800°C at the surface during wear increase the material's ductility. For loose abrasive particle contact (Rabinowicz *et al.*, 1961) particles spend about 90% of the time in rolling contact but for the remaining time form grooves in the wearing surface. Plastic deformation processes leading to material removal have also been observed for very fine abrasive polishing (Aghan and Samuels, 1970) and fine grinding at high speeds (Doyle and Aghan, 1975).

Thus there is strong experimental evidence for material removal mechanisms in which plastic deformation is the rate-controlling step. But we now need to consider how many of the contacting abrasive particles result in such material removal, K in Eq. (8), and what proportion of the groove volume is removed, C in Eq. (8).

1. NUMBER OF CONTACTS RESULTING IN MATERIAL REMOVAL

Goddard *et al.* (1959) observed that between 50 and 100% of bonded commercial abrasive particles in contact with a range of pure metals produced wear debris. But in a comprehensive study Mulhearn and Samuels (1962) estimated that although the number of contacts per unit area on a medium-carbon steel decreased as the mean abrasive particle diameter increased, the proportion of contacts producing microchips was approximately constant at about 12%. More recently, Larsen-Badse (1968c) estimated that 50–60% of the contacting abrasive particles on copper produce microchips. Larsen-Badse (1968c) also points out the experimental difficulties of measuring the number of microchip-forming contacts, which may lead to underestimates for contacts on material with brittle microchips.

Mulhearn and Samuels (1962) argued that microchips are only produced when the angle between the contacting face of the abrasive particle and material surface, the attack angle θ in Fig. 2., exceeds a critical value. They estimated this value to be 90° and thus showed (Fig. 10) that the majority of abrasive particles on commercial bonded papers had shapes or were oriented such that their attack angles were less than this critical value. Friedman *et al.* (1974) have also estimated, for commercial abrasives over a wide size range, that only 6–13% of possible particle contacts have active cutting points. Sedriks and Mulhearn (1963, 1964) carried out a series of experiments with pyramidal tools and found that the critical attack angle varied considerably from material to material. They also found good agreement (Fig. 11) between theoretical and observed groove

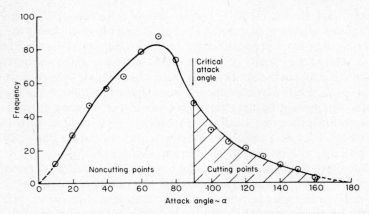

Fig. 10. Frequency distribution of the attack angle of the contacting abrasive particles on unused 70-μm SiC paper, with the critical angle assumed to be 90°. (From Mulhearn and Samuels, 1962.)

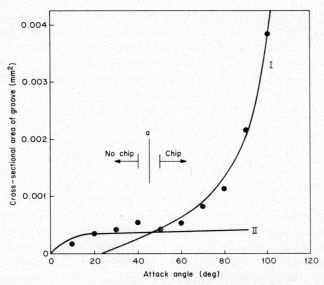

Fig. 11. Variation of the cross-sectional area of wear grooves with attack angle of the abrasive particle. Theoretical curves I for chip cutting and II for pure rubbing with experimental values for copper. (From Sedriks and Mulhearn, 1964.)

cross-sectional areas; they predicted that the critical attack angle should decrease as the coefficient of friction between the leading face of the abrasive particle and material increases, but has an upper limit of about 125°. Subsequently, Dean and Doyle (1975) found that a smooth diamond tool, with attack angle of 22°, produced no microchips, but did produce wear debris formed by plowing, whereas a damaged diamond tool, with attack angle of 30°, produced about equal quantities of microchips and plowing debris. This may be explained if the critical attack angle was lower for the damaged tool with a microuneven surface and thus higher friction. Malkin (1975) also points out that the critical attack angle in grinding operations becomes less at faster cutting speeds.

Khrushchov and Babichev (1960a), Kragelskii (1965), Aghan and Samuels (1970), Graham and Baul (1972), and Dean and Doyle (1975) have suggested that whether a particle contact results in material removal or not depends on the shape of the particle as well as the orientation of its leading face. For spherical particles Kragelskii (1965) has proposed that partial microchip formation takes place when the ratio p/d (Fig. 2) exceeds 0.05 and that plastic displacement with no material removal takes place when p/d is between 0.005 and 0.05. Graham and Baul (1972) and Basuray et al. (1977) confirm this. It is not possible to characterize the attack angle of spherical particles by a single value, but for a p/d of 0.05 it has a maximum value of about 25°. This is considerably less than the minimum critical attack angle for smooth angular particles, which supports the view that particle shape is important. Avient et al. (1960) measured the abrasive wear rate of copper and silver on 125-μm diameter glass spheres and angular glass abrasive of about 125-μm mean particle diameter. Both metals had zero wear on the glass spheres compared with wear of about 50 μg/mm^2 m MNm^{-2} (weight loss per unit area of contact per unit distance run per unit applied load) on the angular glass abrasive.

2. PROPORTION OF THE GROOVE VOLUME REMOVED

When a contacting abrasive particle forms a microchip or plows up material ahead of it some material from the groove is deformed to its sides and only a proportion may be removed. Goddard et al. (1959), Avient et al. (1960), and Stroud and Wilman (1962) estimated that a mean of between 10 and 64% of the volume of the groove below the original surface is removed. They made their measurements on ductile metals worn by commercial bonded abrasives with mean particle diameters from 5 to 150 μm, and suggested that particle size might affect the proportion of the groove volume removed because of shape and orientation variation.

TABLE III

PROPORTION BY VOLUME REMOVED FOR GROOVES FORMED BY A
VICKERS HARDNESS INDENTER[a]

Material	Vickers hardness H_V (MN/m^2)	Proportion of groove volume removed
0.9% C steel	8750	0.74
	6950	0.61
	5850	0.44
	4350	0.33
	3100	0.30
	2600	0.27
0.2% C steel		
Cold worked	2200	0.40
Annealed	1150	0.13
Copper		
Cold worked	1000	0.01
Annealed	490	0.20
Aluminum		
Cold worked	390	<0.01
Annealed	200	<0.01
Stellite 100	8100	0.69

[a] Data from Buttery and Archard (1970/1971).

Larsen-Badse (1968b) estimated that 15% of the groove volume is removed when copper is worn on commercial bonded abrasives.

Hata and Muro (1969) and Childs (1970) measured the geometry of grooves formed by cones and pyramids. Childs' (1970) results indicate that about 85% of the groove volume for annealed silver may be removed and that the ratio of ridge height to groove depth varies little with cone angle. Hata and Muro (1969) also found that this ratio varies little with depth of indentation and orientation of a pyramid on steel. However, Buttery and Archard (1970/1971) found that for grooves made with Vickers and Rockwell hardness indenters the proportion of the groove volume removed is strongly dependent on the material (Table III). This effect is similar to that in static hardness indentation when material piles up around the edge of the indentation. Buttery and Archard (1970/1971) also found that repeated transversals on the same track caused further material removal, as primary microchips and as secondary microchips from material at the sides of grooves.

B. Fracture

Although plastic deformation occurs during abrasive wear of brittle materials, fracture is often the rate-controlling mechanism (Swain, 1975).

Even during the wear of ductile materials fracture may occur (Zum Gahr, 1976) and aid the formation of microchips or other wear debris. For ductile materials fracture is most likely to happen just behind a contacting abrasive particle since this region is subject to a tensile stress (Ramalingam and Lehn, 1971). In addition, sliding motion significantly lowers the normal load required to produce cracking, by up to 95% for polystyrene (Bethune, 1976). For polymers Lancaster (1969) has postulated that the mechanism of material removal during abrasive wear changes from plastic deformation/microchip formation to tearing and fatigue as the elastic modulus decreases. And for brittle materials worn at high speeds thermal shock may cause fracture and high material removal rates (Miller, 1962).

The mode of fracture as a result of indentation in brittle materials (Lawn and Wilshaw, 1975) depends on the applied load on the indenter, the shape of the indenter, whether the indenter is sliding or static, and the environment. If the indenter is spherical, fracture often occurs under elastic contact—Hertzian fracture—and the cracks are conical, extending into the material. However, at small radii of curvature elastic–plastic contact occurs because the radius of curvature has a larger effect on the plastic indentation load than on the Hertzian fracture load. The critical radius for the transition from elastic–plastic indentation to Hertzian fracture increases as the hardness of the material decreases and as its fracture toughness K_c increases (Evans and Wilshaw, 1976). With sharp indenters, pyramids or cones, indentation results in plastic deformation, and fracture does not occur until the indentation reaches a critical size (Lawn et al., 1976). This critical indentation size increases as the hardness of the material decreases and as its fracture toughness increases, and is higher for blunt than for sharp indenters. These static indentation phenomena also apply qualitatively to sliding indentation but with reduced loads required to produce fracture (Lawn, 1967; Bethune, 1976). In both cases fracture may be enhanced in certain environments; for example, by the presence of water (Wiederhorn and Roberts, 1975) or acidic solutions (Lawn and Wilshaw, 1975) on glass.

Thus material removal in brittle materials is likely to be controlled by fracture rather than by plastic deformation except during wear by very lightly loaded blunt abrasive particles. Even under these conditions subsurface fracture of polycrystalline brittle materials occurs. Swain (1975) has proposed that dislocations and twins in alumina will propagate through the surface grains during wear and pile up at subsurface grain boundaries. Cracks will nucleate because stress relief by slip in adjacent grains is unlikely and the grain boundaries are weak. This accounts for the delamination mechanism of material removal Swain (1975) observed.

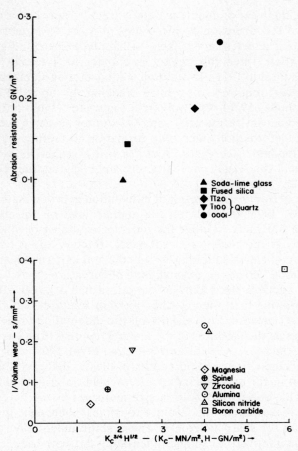

Fig. 12. Correlation between the reciprocal of volume wear (wear resistance) and the product $K_c^{3/4}H^{1/2}$. Upper data for wear on 30-μm SiC water slurry (after Wilshaw and Hartley, 1972) and lower data for diamond sawing (from Evans and Wilshaw, 1976).

When the critical indentation size is exceeded, several modes of cracking occur (Lawn and Wilshaw, 1975). During loading, radial and median vent cracks initiate at the plastic boundary around the indenter and grow. During unloading, lateral vent cracks initiate and grow, even after complete removal of the load. The crack lengths approximately scale with the size of the indentation and, for the same indentation size, increase with decreasing fracture toughness of the material (Evans and Wilshaw, 1976). If we apply these observations to wear by sharp highly loaded abrasive particles, we see that material is removed when the lateral cracks intersect each other or the surface (Lawn and Wilshaw, 1975;

Evans and Wilshaw, 1976). Evans and Wilshaw (1976) have predicted that for wear by sliding abrasive particles the volume of material removed is approximately proportional to the applied load per unit area and inversely proportional to the function $K_c^{3/4}H^{1/2}$, where K_c is fracture toughness and H is hardness of the material. Figures 6 and 12 show that there is good agreement between these predictions and experimental data.

During the wear of heterogeneous materials containing both ductile and brittle phases, fracture may also occur. The predominant material removal mechanism will depend on the properties of the individual phases and on their volume fractions. In materials such as sintered tungsten carbide–cobalt composites containing low volume fractions of the ductile cobalt phase, indentation fracture mechanisms are predominant (Perrott and Robinson, 1974). However, the mode of fracture is different from that observed for homogeneous brittle materials and depends on the cobalt content. Cracks occur close to the surface in the plastic zone and grow during both loading and unloading. Crack growth on unloading tends to occur for low-cobalt material and on loading for high-cobalt material.

Ogilvy et al. (1977) point out that there is unlikely to be a correlation between the fracture toughness of such materials in bulk form and their effective fracture toughness during indentation and wear. Table IV shows

TABLE IV

WEAR RESISTANCE OF SINTERED TUNGSTEN CARBIDE–COBALT COMPOSITES RELATIVE TO A 4% Co, 1–2-μm GRAIN SIZE COMPOSITE[a]

				Relative wear resistance		
Cobalt content (wt %)	Grain size (μm)	Vickers hardness (GN/m²)	Fracture toughness K_c (MN/m$^{3/2}$)	Laboratory 0.2–0.5-mm Al_2O_3/H_2O slurry	Percussive drilling granite	Rotary drilling sandstone
4	1–2	~18	~7	1.0		1.0
4	2–3	~16	~10			0.4
4	—	—	—		1.0	
6	1–2	~16	~10	0.5		0.6
6	2–3	—	—	0.4		0.2
6	—	—	—		~0.6	
8	—	—	—		0.2	
9	1–2	~15	~9	0.2		0.3
9	2–3	~13	~13	0.1		0.15
9	—	—	—		0.25	
11	—	—	—		0.1	
15	1–2	~11	~18			0.1
15	2–3	~10.5	~19	0.1		0.04

[a] Data from Larsen-Badse (1973) and Feld and Walter (1976).

that the volume wear of tungsten carbide–cobalt composites increases rapidly as the cobalt content increases, in spite of the increasing bulk fracture toughness of the material. (The wear data in Table IV are presented as relative wear resistances, defined as volume wear of a reference material divided by volume wear of the test material.) There is strong experimental evidence (Blombery et al., 1974; Kenny and Wright, 1974) that for the higher cobalt content grades the cobalt is more easily worn away by mechanisms involving plastic deformation. This exposes the carbide network to fracture damage. Thus we could argue that plastic deformation is the rate-controlling step although fracture predominates in terms of volume loss of material.

C. Surface Molecular and Corrosion/Mechanical Mechanisms

Rabinowicz (1971) has argued that mechanical mechanisms of material removal are energetically unfavorable for very small abrasive particle indentation depths. This is because the energy to create the new surfaces exceeds the plastic strain energy for formation of a wear groove. Rabinowicz calculated that the lower limit for groove widths in copper would be about 0.25 μm, but recent evidence (Moore and Douthwaite, 1976) suggests that the plastic strain energy is much higher than that estimated by Rabinowicz, and thus the limiting groove width in copper may be as low as 1 nm. It is very likely that the contact would change from elastic–plastic to purely elastic at indentations larger than this limiting value.

Andarelli et al. (1973) and Gane and Skinner (1973) have experimented with sliding contacts on ductile metal foils for which no visible plastic deformation occurred. Both groups agree that the stored elastic energy can only account for less than 1% of the friction energy. Gane and Skinner argue that the most likely other energy dissipation process would be some form of molecular mechanism. Furthermore, they point out that plastic deformation does not play a significant part in surface contacts on diamond and graphite and thus friction may be due to a surface molecular mechanism.

Rabinowicz (1968) has proposed surface molecular mechanisms to account for the removal of material in fine polishing. In support of this he shows that polishing rates increase as the latent heat of evaporation of the material—a measure of the molecule–molecule bond strength—decreases. But even for wear on large and highly loaded abrasive particles the wear rate decreases with increasing intermolecular cohesive energy (Giltrow, 1970; Vijh, 1975). Thus there is only indirect evidence to support surface molecular material removal mechanisms.

For very lightly loaded abrasive particle contact it is possible that indentation is limited to surface films, where these exist. Thus Aghan and Samuels (1970) have proposed a mechanism of material removal involving continuous removal and formation of surface films. They argue that this mechanism is supported by observation that some fine polishing processes are sensitive to small changes in the chemical activity of the liquid in which the abrasive particles are suspended. In air we can estimate the thickness of oxide film that would grow between particle contacts. For copper, assuming each point on the wearing surface is contacted about once every $10^{-2}-10^{-4}$ sec and that the surface temperature is about 1000°C, a 50-nm oxide layer would grow. A layer this thick would only have a significant effect on particle contact in very fine polishing processes. McKenzie and Hillis (1965), however, have shown that combined corrosion/mechanical wear mechanisms occur during wood cutting, for which the active liquid environment has a pH of between 2.6 and 4.

IV. Variables Influencing Abrasive Particle Contact

We have seen in Section II that a simple model of abrasive wear does not account for the dependence of volume wear on the abrasive particle size and on the mechanical properties of the surface material. Abrasive particle size may influence particle contact if particle geometry and deterioration rates vary with size. These properties may also vary for different abrasive types. In Section III we have seen that for both ductile and brittle materials plastic deformation occurs during material removal. Thus both low-strain mechanical properties and strain hardening properties of the material must influence particle contact. In this section we shall consider in more detail the ways in which properties of the abrasive and the wearing material influence particle contact.

A. *Properties of the Abrasive*

1. GEOMETRIC PROPERTIES

Although abrasive particle contact is dependent, in a complex way, on the mechanical properties of the surface material, if we assume that the surface hardness is an important factor, we can predict the effect of particle shape on contact. Regardless of the shape of the particle the projected area of the contact will remain constant provided the load on the particle and surface hardness of the material also remain constant. How-

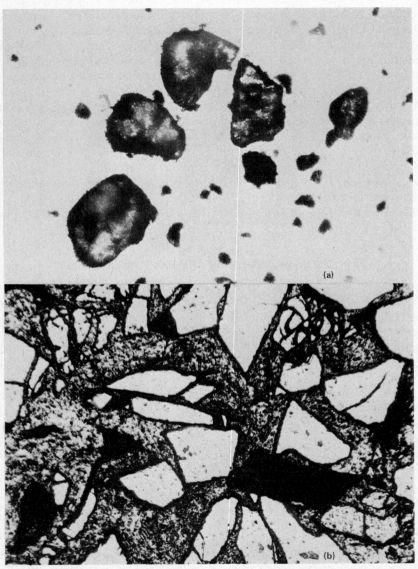

Fig. 13. The shape of abrasive particles: (a) 500-μm-diameter sandy loam soil particles; (b) 384-μm-diameter commercial flint abrasive. (Reproduced by permission of National Institute of Agricultural Engineering.)

ever, the cross-sectional area of a groove made by the contact will be dependent on the particle shape. And we might expect a high wear rate when the ratio of the cross-sectional area to the projected area of contact is high. Simple calculation for pyramidal, conical, and spherical contacts shows that this ratio increases as the included angle for pyramids and as the cone angle or radius for spheres decrease.

Because of the range of particle shapes found in any one sample of abrasive, it is very difficult to detect any differences in the mean particle shape. Most methods of describing particle shape account for the whole particle, whereas abrasive particle contact often involves only 0.1–0.3 times the particle perimeter. McAdams (1963) used a polyhedron model of an abrasive particle to calculate the statistical distributions of radii measured from the center of a particle to its surface and of projected areas of the particle. While this model provides a means of incorporating particle size, shape, and orientation into a single concept, McAdams (1963) points out that it does not provide information pertaining to the fine structure of the particle surface. However, Davis and Dexter (1972) developed a method for quantitative description of soil particle shape which compares cord lengths with arc lengths around particle outlines. Results from this analysis for soil particles and for commercial abrasive particles (Fig. 13) show (Fig. 14) that the method can detect mean differences in shape of small fractions of the particle perimeters.

In spite of the probable importance of particle shape in abrasive wear there is very little quantitative information in the literature. Mulhearn and

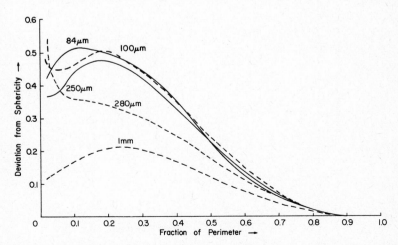

Fig. 14. Asphericities of various size abrasive particles from a sandy soil (broken lines) and from commercial SiC papers (solid lines). (After Davis and Dexter, 1972.)

Samuels (1962) studied the shape and arrangement of abrasive particles on commerical bonded silicon carbide papers. They found that the particles were mainly equiaxed for coarse particle size papers, whereas there was a much higher proportion of acicular particles for fine particle size papers. The contacting points were generally angular although the apex angle varied considerably. Zolotar (1973) characterized abrasive particle shape by a complex function which increases as the average number of peaks per particle and their apex angles increase and as the radius of curvature of the peaks decreases. His results show that the rate of wear by loose sand particles increases as this particle form factor increases.

The observations of Mulhearn and Samuels (1962) and the results of Fig. 14 seem to indicate that fine abrasive particles would cause more wear than coarse particles. However, the opposite is usually found, Fig. 9, so abrasive particle-size-dependent variables other than shape must have a greater effect on particle contact. One possibility is the height and size distribution of contacting particles. Commercial abrasives have a range of particle sizes in samples characterized by nominal sieve sizes, Fig. 15, and the finer sizes of commercial bonded abrasives tend to have

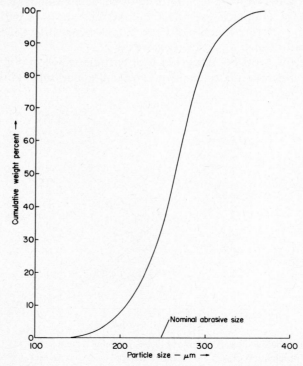

Fig. 15.   Particle size distribution for a standard 60-grit sand (British Standard 872:1965).

several layers of abrasive and a wide range of height distribution (Mulhearn and Samuels, 1962; Larsen-Badse, 1968b). Larsen-Badse (1968c) also found that for commercial bonded abrasives the number of contacting abrasive particles is a linear function of applied load for a wide range of particle sizes. But the average individual particle contact area does not vary with applied load for the fine particle size abrasives. Thus Larsen-Badse (1968b,c) and Date and Malkin (1976) concluded that for fine particle abrasives a proportion of the applied load is carried elastically by particles which are prevented by height and size distributions from making plastic contact with the surface material. This would account for the lower wear rates on small particle abrasives.

The height and size distributions of particles may also cause two further effects. The first is that loose wear debris formed by the most deeply indenting contacts will hinder the contact of smaller or less deeply indenting abrasive particles. Avient et al. (1960), Rabinowicz et al. (1961), Rabinowicz and Mutis (1965), and Date and Malkin (1976) have proposed this mechanism to account for the lower wear rates and the fall in efficiency with use of fine compared with coarse particle abrasives. Avient et al. (1960) also suggested that fine particles are bonded less firmly than coarse particles to the backing material on commercial abrasives. Thus fine particles can become embedded more easily in the wearing material. Johnson (1970) has confirmed that the pickup of particles from commercial bonded silicon carbide abrasive increases rapidly with decrease in particle size and with increasing sliding distance and applied load. Billingham et al. (1974) have shown that pickup of abrasive is almost negligible during wear on coarse particle commercial bonded aluminum oxide.

2. Mechanical and Chemical Properties

Khrushchov and Babichev (1956b) realized that mechanical properties of the abrasive are important in abrasive wear. They concluded that when the hardness of the abrasive is very much greater than that of the material, wear is independent of the abrasive hardness. But as the hardness of the wearing material approaches that of the abrasive, wear decreases, and when the hardness of the abrasive is less than that of the wearing material, wear decreases rapidly. Subsequently, Nathan and Jones (1966/1967) and Richardson (1967b,1968) showed that even when the abrasive is very much harder than the wearing material volume, wear still depends on abrasive hardness (Fig. 16).

Richardson (1968) found that scratching ceases for coarse abrasives when the yield stresses of the abrasive and material are equal, Fig. 17a. This occurs when the ratio of hardness of the material to that of the abrasive, H/H_a, is between 1 and 1.5. The exact value of this ratio depends on

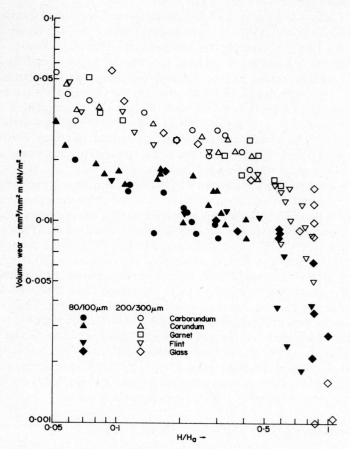

Fig. 16. Volume wear versus the ratio hardness of the material to hardness of the abrasive, H/H_a. (After Nathan and Jones, 1966/1967; Richardson, 1968.)

the ratio of yield stress to elastic modulus, Poisson's ratio, and strain hardening properties of the two materials (Attwood, 1967/68; Studman *et al.*, 1977). In contrast, on fine abrasives Richardson (1968) observed scratching when the ratio H/H_a was as high as 1.2. However, the scratches deteriorate rapidly, Fig. 17b, and material is still removed by the coarse grit even though continuous scratches are not formed. Even when the abrasive is very much harder than the wearing material the rate of abrasive deterioration, indicated by a decrease in the rate of wear, is much higher for fine than for coarse abrasive (Date and Malkin, 1976). Larsen-Badse (1972) has observed enhanced wear during initial contact and a rapid fall in the rate of wear as the length of contact increases.

Fig. 17. 0.8% C steel of hardness about 8000 MN/m^2 after sliding between 4 and 5 m on (a) 500-μm glass and (b) 75-μm glass papers of hardness about 6000 MN/m^2. (From Richardson, 1968.) Direction of abrasion from right to left.

These phenomena are partly due to conditioning of the abrasive during contact by plastic flow, fracture, and chemical degradation. Richardson (1968) proposed [after Bowden and Tabor (1954) and likening abrasive particles to cones] that plastic flow can occur in particles with small cone angle when the yield stress of the abrasive is very much greater than that of the wearing surface. For particles of large cone angle plastic flow occurs when the hardness of the wearing material exceeds that of the abrasive. Since many abrasive particles have large cone angles (Friedman *et al.*, 1974), blunting of the abrasive by plastic flow would account for the rapid deterioration in wear when H/H_a exceeds ~ 1.

Richardson (1968) then considered fracture of contacting abrasive particles. He applied to abrasives the criterion for fracture of a critical size of plastic zone which Bowden and Tabor (1964) and Marsh (1964) suggested and Lawn *et al.* (1976) recently confirmed for brittle materials. Patterson and Mulhearn (1969) and Komanduri and Shaw (1974,1976a) have pointed out that abrasive particles contain inherent cracks and defects. The effective tensile strength of abrasive particles decreases rapidly as their size, and thus the probability of the particle having a suitable oriented defect increases. For both commercial (Komanduri and Shaw, 1974,1976a) and natural (Hata and Muro, 1969) abrasives the tensile strength of the abrasive is approximately proportional to the inverse square root of the particle diameter. Thus coarse or highly loaded abrasive particles are more likely to fracture than fine or lightly loaded abrasive particles.

Since fracture may well regenerate cutting facets and produce loose abrasive fragments (Johnson, 1968; Patterson and Mulhearn, 1969), the wear rate is affected less by fracture than by plastic deformation alone of the abrasive. This accounts for the difference with different grit size in the reduction of wear when H/H_a exceeds ~ 1 (Fig. 16) and for the more rapid deterioration of fine abrasives with increasing sliding distance (Date and Malkin, 1976). It may also account for differences in the wear rate of a material on abrasives of similar hardness if their fracture strengths are different. For example, Richardson (1967d) found that the weaker abrasive particles in ironstone soils may be more effective in wear than the stronger abrasive particles in flint soils.

Chemical reactions between the abrasive and wearing material or environment can occur during wear and contribute to deterioration of the abrasive. Billingham *et al.* (1974) have identified chemical transfer from alumina abrasives to nickel alloy during wear, characterized by micropitting but macrosmoothing of the abrasive particle surface. They estimated that complex spinel compounds would form at the interface if the temperature exceeded 1100°C, and thus this type of deterioration might depend very much on the material being worn. Komanduri and Shaw (1976a,b) have observed deterioration of silicon carbide abrasive, characterized by smoothed surfaces, during wear of cobalt alloys under light loads and high speeds. They attribute this to oxidation of the abrasive at the high temperatures generated at the interface—a process resembling chemical etching. In contrast, Larsen-Badse (1975a,b) has suggested that the increase in wear rates with increase in atmospheric humidity between 0 and 65% occurs because enhanced fracture of the abrasive increases the number of cutting facets.

B.   Mechanical Properties of the Wearing Material

During wear, a surface is subject to applied stresses and temperature rises. These may cause plastic deformation, recrystallization, and phase changes, all of which affect the mechanical properties of the surface material and, consequently, abrasive particle contact.

Material removal during abrasive wear normally involves plastic deformation and because of strain hardening an extensive region below a worn surface is also deformed plastically (Fig. 18). It seems likely that as in machining (Haslam and Rubenstein, 1970; Ramalingam and Lehn, 1971) and in static hardness indentation (Tabor, 1970) a hydrostatic stress system is developed in front of an abrasive particle contact. Such a stress system will oppose void formation and growth, thus raising the ductile fracture stress and strain. However, the stress behind the contact is more likely to be tensile (Ramalingam and Lehn, 1971). The stress systems from adjacent particle contact may also produce cyclic deformation of the material between the contacts. This may lead to subsequent strain softening, particularly in hardened and tempered steels (Richardson, 1967c; Moore, 1974a). Extreme-surface softening during wear may also be caused by compatibility effects during deformation (Hirth and Rigney, 1976).

In discussing the extent of strain hardening at worn surfaces Moore et al. (1972) suggest that the limiting process is fracture. Thus the hardness of worn surfaces of, for example, aluminum and copper alloys in the solution-treated and precipitation-hardened states are the same, whereas the low-strain hardnesses are quite different (Lin and Wilman, 1969; Moore et al., 1972). However, the volume wear of the precipitation-treated aluminum and copper alloys is lower than that of the same alloys in their solution-treated states (Richardson, 1967b,c; Lin and Wilman, 1969). Tungsten, which strain hardens during wear by a factor of about 7, has a much lower volume wear than a medium-carbon steel, which strain hardens by a factor of about 1.5, even though their surface hardnesses after wear are the same. Thus the mechanical properties of the material

Fig. 18. Cross sections of copper–silver solder laminate (a) after trepanning with a blunted end-mill and (b) after abrasion by 250-μm SiC paper, showing extent of the surface plastic deformation. (Reproduced by permission of National Institute of Agricultural Engineering.)

influencing sliding indentation and/or material removal from the surface must be somewhat different from those influencing static indentation.

Moore (1974a) and Moore and Douthwaite (1976) investigated the plastic deformation below surfaces. The flow stress and strain distributions depend on the material's elastic properties, low-strain flow stress, strain hardening properties, and fracture properties, and on the abrasive particle size and applied load. The elastic properties and low-strain flow stress influence initial particle contact as in static hardness indentation (Studman et al., 1977), and the depth of the plastic zone scales with the depth of indentation. Subsequent particle contact depends on the properties of the deformed surface and thus on the strain hardening properties of the material. The material continues to deform plastically until the fracture strain is exceeded at the surface. At this state of equilibrium there is a strain gradient from the surface into the material. From measurements of this strain gradient Moore and Douthwaite (1976) estimated that for ductile materials the work done during wear is almost entirely balanced by the energy expended in plastic deformation at and below the worn surface. Thus abrasive particle contact on a worn surface is determined by the depth of the deformed zone, the strain history and strain hardening properties of the material, and the strain at the surface.

Because of the hydrostatic stress system, strains at worn surfaces can be very much higher than in conventional deformation processes. Moore and Douthwaite (1976) have estimated strains of 8 at a depth of 10 μm after trepanning—rubbing with a blunted end-mill—and 2.5 at 2.5 μm below an abraded surface of a copper–silver solder laminate. Larsen-Badse and Mathew (1969) have estimated a strain of about 5 for the removal of copper by abrasion, and Turley (1968, 1971) has estimated a strain of about 6.5 at the surface of brass after machining. However, for one material the strain as a function of depth below the surface is proportional to the nominal abrasive particle size and to the square root of the applied load per unit area (Fig. 19), i.e., proportional to the depth of indentation of the abrasive particles (Jacquet, 1950; Samuels, 1956/1957; Haslam and Rubenstein, 1970; Moore and Douthwaite, 1976). But at the extreme surface the strain is probably independent of the depth of indentation of the abrasive particles since the stress system and thus the fracture strain is the same here. This implies that the strain rate close to the surface increases as the applied load and abrasive particle size decrease, and as the speed of abrasion increases. Thus the speed of abrasion has little effect on the volume wear of a strain rate insensitive material such as copper; but for strain-rate-sensitive materials, such as Armco iron and steels, volume wear increases by up to 80% over the speed range 0.25–5 m/sec (Moore and McLees, 1975, 1977).

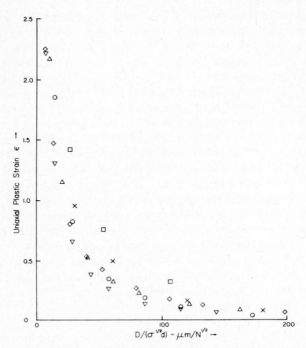

Fig. 19. Strain distribution below worn surfaces of copper–silver solder laminate as a function of $D/(\sigma^{1/2}d)$; □48 μm SiC, 0.97 MN/m²; ×84-μm SiC, 0.97 MN/m²; ○250-μm SiC, 0.48 MN/m²; △250-μm SiC, 0.97 MN/m²; ▽250-μm SiC, 1.94 MN/m²; ◇384 μm SiC 0.97 MN/m². (From Moore and Douthwaite, 1976. Copyright 1976 by The American Society for Metals.)

At least 90% of the energy expended in plastic deformation during wear must be dissipated as heat (Nabarro, 1967). For abraded copper (Moore and Douthwaite, 1976) the heat generated when an abrasive particle forms a wear groove is about 3.5 J/mm³. If this heat remains within the material from the groove, the adiabatic temperature rise would be about 1000°C. Direct and indirect evidence (Agarwala and Wilman, 1955; Moore, 1971; Cutter and McPherson, 1973; McPherson, 1973; Billingham *et al.,* 1974) suggests transient temperature rises during abrasion of this order and of about 10^{-2} sec duration. About 50% more heat would be needed to melt the volume of material from the groove and thus, as Nabarro (1967) states, extensive melting as a result of heat generated by plastic work is extremely unlikely. But limited melting is possible if the energy is dissipated on selected dislocation glide planes and over a very short period of time. This process of adiabatic shearing may well occur during abrasion since the strain rate for material removal is between 10 and 10^6 sec^{-1} (Stevenson and Oxley, 1969/70; Cutter and McPherson, 1973; Ramalingam and Black, 1973; Moore, 1974a).

The severe plastic deformation at worn surfaces can also lead to structural changes. Transient temperature rises of about 1000°C, having durations of about 10^{-2} sec (Moore, 1971; McPherson, 1973), and in small volumes under high pressure account for $\alpha-\gamma$ transformation close to the surface of abraded iron crystals (Agarwala and Wilman, 1955) and for the formation of a spinel compound on alumina abrasive (Billingham et al., 1974). The fraction of the plastic work stored in the form of strain energy of dislocations can be sufficient to provide the driving force for structural changes. This fraction depends on the material, amount of deformation, strain rate, stress system, and temperature (Bever et al., 1973). For copper the stored energy in material deformed from a wear groove is probably in the range 0.035–0.18 J/mm^3. This is sufficient (Friedel, 1964) to provide the driving force for very small grain nuclei, in the range 14–68 nm for copper. Furthermore, since there is a high probability of obtaining such nuclei by thermal fluctuations (Friedel, 1964), homogeneous nucleation is possible, at least in limited volumes. Doyle and Aghan (1975) have observed a very fine grain size, about 10 nm in microchips formed by fine grinding carburized mild steel, and Dance and Norris (1948) attribute surface softening to recrystallization of the heavily deformed material. This, in particular, demonstrates the complex and often unique phenomena associated with abrasive particle contact since Christian (1965) has argued that the probability of homogeneous nucleation is almost negligible during conventional deformation processes.

Both adiabatic shear and recrystallization to a fine grain size, as a result of high strain and strain-rate plastic deformation, may lead to local negative dynamic stress–strain behavior (Zener and Holloman, 1944; Ramachandron and Abrahamson, 1972). Thus deformation is confined to local regions while the surrounding material undergoes limited further strain, causing plastic instability. Plastic instability accounts for the lamellar structure of chips formed both in machining (Ramalingam and Black, 1973) and during abrasion (Doyle and Aghan, 1975). The interlamellar spacing depends on the depth of cut, or depth of indentation of an abrasive particle, and on microstructural features. Dispersed "hard" particles and large grain sizes increase the interlamellar spacing and so increase the chip formation energy per unit volume. During abrasion this will cause the depth of indentation of the abrasive and/or the microchip volume to decrease. Popov et al. (1972) found that "soft" inclusions such as MnS and FeS had an adverse effect on the wear of an austenitic steel, while "hard" inclusions such as Al_2O_3 and MgO reduced volume wear.

C. *Properties of Discrete Microstructural Phases*

We have seen that structural properties of heterogeneous materials are important during abrasive particle contact. Parameters such as volume fraction and distribution of a dispersed phase, its coherency, hardness, and other properties affect abrasive particle indentation, strain hardening, strain distribution, fracture, and recovery processes. Thus, for example, the volume wear of steels decreases as the carbide content increases (Khrushchov and Babichev, 1958), and the relationship between the wear resistance of steels and structure is similar to that between flow stress and structure (Larsen-Badse, 1966; Moore, 1974b).

However, Richardson (1968) noticed that the volume wear of materials containing relatively massive carbides was very sensitive to abrasive particle size (and load) (see Table V). He concluded that the dispersed phase becomes effective as a discrete component when its size is about the same as or greater than the depth of indentation of the contacting abrasive particles. The abrasive particles may blunt or fracture, by-pass the hard phase, ride over it, push it into the softer matrix, fracture it, or loosen it from the matrix.

As for particle contact on homogeneous materials, the relative hardnesses of the discrete phase and abrasive are critical. Thus, as Table V shows, volume wear decreases rapidly when the dispersed phase and abrasive have about the same hardness because of deterioration of the abrasive (see Section IV,A,2). Popov *et al.* (1969a,b, 1972) studied the fracture of carbides in alloy steels during abrasive wear and found a correlation between their fracture behavior and volume wear. Steels containing carbides that were not cut by the abrasive particles and that were firmly held by the alloy base—e.g., VC and NbC—had lower wear than those containing carbides that completely fractured on contact, were easily cut by the abrasive, or were not firmly held by the alloy base—e.g., Cr_7C_3 and WC.

If the hard phase resists fracture during contact it may be pushed into the softer matrix material. Blombery and Perrott (1974) identified this mechanism, occurring during the wear of cobalt alloys. In this process the matrix is extruded around the hard phase and, subsequently, removed by the abrasive or smeared over the surface (Kenny and Wright, 1974). Thus the relative mechanical properties of hard phase and matrix have a large influence on the wear of heterogeneous materials.

In practical abrasive wear environments heterogeneous materials are often used, but the abrasive particle size and applied load often cannot be characterized by single values. Thus it is only possible to select such materials for optimum wear performance, and both the relative wear

TABLE V

Dependence on Abrasive Size of the Volume Wear of Some Carbide-Containing Materials[a]

Material	Vickers hardness of carbides (MN/m²)	Vickers hardness of matrix (MN/m²)	Volume wear (mm³/mm² m MNm⁻²)			
			SiC		Flint	
Abrasive Vickers hardness (MN/m²)			30,000		10,600	
Size (μm)			84	250	84	384
2% C, 14% Cr die steel	12800	8600	0.0083	0.0132	0.0003	0.0051
3.6 % C white iron	11400	4800	0.0099	0.0168	0.0024	0.0112
3% C, 1.7% Cr, 3% Ni martensitic iron	8700	4100–5800	0.0101		0.0018	0.0119
3% C, 30% Cr cast iron	20600	4800	0.0086	0.0159	0.0001	0.0079
2.5% C, 33% Cr, 13% W Co alloy	21300	7000	0.0114	0.0209	0.0004	0.0078

[a] Data from Richardson (1967b, 1968).

TABLE VI

WEAR RESISTANCE OF SOME HETEROGENEOUS MATERIALS, RELATIVE TO A LOW ALLOY STEEL, AS INFLUENCED BY THE WEAR ENVIRONMENT[a]

Abrasive wear environment	Relative wear resistances					
	0.4% C, 1.5% Ni/Cr/Mo steel H_V ~5000	2% C, 12–14% Cr die steel H_V ~7000	~3% C chilled white iron H_V ~6000	3% C, 30% Cr white iron H_V ~7000	15/3 Cr/Mo white iron H_V ~9000	3% C, 2% Cr, white iron H_V ~7000
1. Laboratory bonded abrasives						
84-μm corundum	1.0	1.75	1.53	2.12		1.52
84-μm flint	1.0	11.7	4.32	129		5.95
384-μm flint	1.0	1.78	1.59	2.25		1.50
84-μm glass	1.0		~∞			
2. Agricultural soils						
Pumice	1.0			16.7		
Stone-free sand	1.0			9.6		
Ironstone loam/sand	1.0	1.94	2.32	2.28		1.71
Flint sand/loam	1.0	2.07	3.81	3.32		2.50
3. Iron ore comminution and steel works						
Quartz Fe ore, ball mill } sliding	1.0		~0.6	~1.1		~1.0
Blast furnace sinter } impact	1.0		1.35	8.2		15.0
Coke, sliding	1.0		0.84	3.2		3.2
impact	1.0		0.81	2.0		1.9
4. Quartz/feldspar Mo ore						
75-mm ball mill media	1.0	1.1	0.6		1.2	
125-mm ball mill liner	1.0				1.3	
Laboratory jaw crusher	1.0			1.6–2.1	3.1–4.8	1.0

[a] Data from Richardson (1968, 1969), Norman and Hall (1969), Borik and Scholz (1971), Fairhurst and Rohrig (1974), and Hocke (1972 and 1975).

resistance—volume wear of a reference material divided by volume wear of the test material—and the ranking order of these materials may vary considerably from one operation to another. Table VI shows this for a range of wear environments.

V. Concluding Remarks

The preceding sections have shown that abrasive wear is a complex problem, and one that continues to challenge the investigator. Very often we find that abrasive wear processes pose unique problems, for example, the high strain, high-strain-rate plastic deformation occurring during particle contact, the changing stress system at the surface, and the plastic deformation and fracture of microvolumes of material. But because it is difficult to devise experiments to measure the response of materials to such environments, and thus to understand the relative importance of material properties, we have in the past tended to rely on bulk material properties as a guide to abrasive wear resistance. This is largely unsatisfactory, particularly if only one property, such as hardness, is used.

In spite of this and the importance of abrasive wear in technology, Rabinowicz (1976b) estimates that at any one time there have been less than 50 people engaged in research on wear in general—as opposed to wear of specific components or materials. For abrasive wear alone the number would be very much less, perhaps as low as 10. Furthermore, wear did not exist as a technical subject *per se* until about 1950. Since that time the advances in our understanding have been enormous. Unfortunately, as highlighted by the Department of Education and Science (1966) in the United Kingdom, the available information has not always been given wide publicity nor put into practice. The technological and economical importance of wear provides an overwhelming case for increasing education, research, and development.

There still remain large gaps in our knowledge of abrasive wear processes. Some of these have been pointed out in the preceding sections; but such problems as the influence of geometry of components on their wear lives and macroscopic wear patterns have not been dealt with in this chapter. While an understanding of wear mechanisms is fundamental for material selection and development, an understanding of these geometric factors is essential for the development of wear-resistant designs.

References

Agarwala, R. P., and Wilman, H. (1955). *J. Iron Steel Inst. London* **179,** 124–131.
Aghan, R. L., and Samuels, L. E. (1970). *Wear* **16,** 293–301.

Alison, P. J., and Wilman, H. (1964). *Br. J. Appl. Phys.* **15**, 281–289.
Andarelli, G., Maugis, D., and Courtel, R. (1973). *Wear* **23**, 21–31.
Archard, J. F. (1953). *J. Appl. Phys.* **24**, 981–988.
Attwood, D. G. (1967/1968). *Proc. Inst. Mech. Eng.* **182**(3A), 369–373.
Avient, B. W. E., Goddard, J., and Wilman, H. (1960). *Proc. R. Soc. London Ser. A* **258**, 159–180.
Basuray, P. K., Misra, B. K., and Lal, G. K. (1977). *Wear* **43**, 341–349.
Bethune, B. (1976). *J. Mater. Sci.* **11**, 199–205.
Bever, M. B., Holt, D. L., and Titchener, A. L. (1973). *Progr. Mater. Sci.* **17**, 23–88.
Billingham, J., Lauridsen, J., and Bryon, J. F. (1974). *Wear* **28**, 331–343.
Blombery, R. I., and Perrott, C. M. (1974). *J. Aust. Inst. Met.* **19**, 255–258.
Blombery, R. I., Perrott, C. M., and Robinson, P. (1974). *Wear* **27**, 383–390.
Borik, F., and Scholz, W. G. (1971). *J. Mat.* **6**, 590–605.
Bowden, F. P., and Tabor, D. (1954). "The Friction and Lubrication of Solids." Oxford Univ. Press (Clarendon), London and New York.
Bowden, F. P., and Tabor, D. (1964). "The Friction and Lubrication of Solids," Part 2. Oxford Univ. Press (Clarendon), London and New York.
Buttery, T. C., and Archard, J. F. (1970/1971). *Proc. Inst. Mech. Eng.* **185**, 538–551.
Childs, T. H. C. (1970). *Int. J. Mech. Sci.* **12**, 393–403.
Christian, J. W. (1965). "The Theory of Transformations in Metals and Alloys," p. 720. Pergamon, Oxford.
Cutter, I. A., and McPherson, R. (1973). *J. Am. Ceram. Soc.* **56**, 266–269.
Dance, J. B., and Norris, D. J. (1948). *Nature London* **162**(4106), 71–72.
Date, S. W., and Malkin, S. (1976). *Wear* **40**, 223–235.
Davis, P. F., and Dexter, A. R. (1972). *J. Soil Sci.* **23**, 448–455.
Dean, S. K., and Doyle, E. D. (1975). *Wear* **35**, 123–129.
Department of Education and Science (1966). "Lubrication (Tribology), Education and Research—A Report on the Present Position and Industry's Needs." Her Majesty's Stationery Office, London.
Dobson, P. S., and Wilman, H. (1963). *Br. J. Appl. Phys.* **14**, 132–136.
Doyle, E. D., and Aghan, R. L. (1975). *Metall. Trans.* **6B**, 143–147.
Evans, A. G., and Wilshaw, T. R. (1976). *Acta Metall.* **24**, 939–956.
Fairhurst, W., and Rohrig, K. (1974). *Foundry Trade J.*
Feld, H., and Walter, P. (1976). *Z. Werkstofftech.* **7**, 300–303.
Friedel, J. (1964). "Dislocations," pp. 299–301. Pergamon, Oxford.
Friedman, M. V., Wu, S. M., and Suratkar, P. T. (1974). *J. Eng. Ind.* **96**, 1239–1244.
Gane, N., and Skinner, J. (1973). *Wear* **25**, 381–384.
Giltrow, J. P. (1970). *Wear* **15**, 71–78.
Goddard, J., Harker, H. J., and Wilman, H. (1959). *Nature London* **184**, 333–335.
Graham, D., and Baul, R. M. (1972). *Wear* **19**, 301–314.
Haslam, D., and Rubenstein, C. (1970). *Ann. CIRP* **18**, 369–381.
Hata, S., and Muro, T. (1969). *Mem. Fac. Eng. Kyoto Univ.* **31**, 456–489.
Hirth, J. P., and Rigney, D. A. (1976). *Wear* **39**, 133–141.
Hocke, H. (1972). Corporate Lab. Rep. No. PE/B/5/72, unpublished. British Steel Corp., Middlesborough.
Hocke, H. (1975). Private communication.
Jacquet, P. A. (1950). *Rev. Metall.* **167**, 355–364.
Johnson, R. W. (1968). *Wear* **12**, 213–216.
Johnson, R. W. (1970). *Wear* **16**, 351–358.

Jost, H. P. (1975). *In* "Principles of Tribology" (J. Halling, ed.), p. xii. Macmillan, New York.

Kenny, P., and Wright, A. C. (1974). *Wear* **30,** 377–383.

Khrushchov, M. M. (1957). *Proc. Conf. Lub. Wear* pp. 655–659. Institution of Mechanical Engineers, London.

Khrushchov, M. M., and Babichev, M. A. (1956a). *Frict. Wear Mach.* **11,** 5–18. N.E.L. Transl. No. 831, National Engineering Laboratory, Glasgow.

Khrushchov, M. M., and Babichev, M. A. (1956b). *Frict. Wear Mach.* **11,** 19–26. N.E.L. Transl. No. 830, National Engineering Laboratory, Glasgow.

Khrushchov, M. M., and Babichev, M. A. (1958). *Frict. Wear Mach.* **12,** 15–26. N. E. L. Transl. No. 828, National Engineering Laboratory, Glasgow.

Khrushchov, M. M., and Babichev, M. A. (1960a). Research on the Wear of Metals, Chapter 18. N. E. L. Transl. No. 893, National Engineering Laboratory, Glasgow.

Khrushchov, M. M., and Babichev, M. A. (1960b). Research on the Wear of Metals, Chapter 13. N. E. L. Transl. No. 891, National Engineering Laboratory, Glasgow.

Komanduri, R., and Shaw, M. C. (1974). *J. Eng. Mater. Technol.* **96,** 145–156.

Komanduri, R., and Shaw, M. C. (1976a). *J. Eng. Ind.* **98,** 1125–1134.

Komanduri, R., and Shaw, M. C. (1976b). *Wear* **36,** 363–371.

Kragelskii, I. V. (1965). "Friction and Wear," pp. 20–26, 80–110. Butterworths, London.

Lancaster, J. K. (1969). *Wear* **14,** 223–239.

Larsen-Badse, J. (1966). *Trans. AIME* **236,** 1461–1466.

Larsen-Badse, J. (1968a). *Wear* **12,** 357–368.

Larsen-Badse, J. (1968b). *Wear* **12,** 35–53.

Larsen-Badse, J. (1968c). *Wear* **11,** 213–222.

Larsen-Badse, J. (1972). *Wear* **19,** 27–35.

Larsen-Badse, J. (1973). *Powder Metall.* **16**(31), 1–32.

Larsen-Badse, J. (1975a). *Wear* **31,** 373–379.

Larsen-Badse, J. (1975b). *Wear* **32,** 9–14.

Larsen-Badse, J., and Mathew, K. G. (1969). *Wear* **14,** 199–206.

Lawn, B. R. (1967). *Proc. R. Soc. London Ser. A* **299,** 307–316.

Lawn, B. R., and Wilshaw, T. R. (1975). *J. Mater. Sci.* **10,** 1049–1081.

Lawn, B. R., Jensen, T., and Arora, A. (1976). *J. Mater. Sci.* **11,** 573–575.

Lin, D. S., and Wilman, H. (1969). *Wear* **14,** 323–335.

Malkin, S. (1975). *Wear* **32,** 15–32.

Marsh, D. M. (1964). *Proc. R. Soc. London Ser. A* **282,** 33–43.

McAdams, H. T. (1963). *J. Eng. Ind.* **85,** 388–393.

McKenzie, W. M., and Hillis, W. C. (1965). *Wear* **8,** 238–243.

McPherson, R. (1973). *Wear* **23,** 83–86.

Miller, D. R. (1962). *Proc. R. Soc. London Ser. A* **269,** 368–384.

Ministry for Research and Technology (1976). Tribologie, Res. Rep. T 76–38. Zentralstelle fur Luft-und Raumfahrtdokumentation und information, Munich.

Moore, M. A. (1971). *Wear* **17,** 51–58.

Moore, M. A. (1974a). The Strain Hardening and Fracture of Ferritic Materials under Abrasive Wear. Ph.D. thesis, Univ. of Newcastle upon Tyne.

Moore, M. A. (1974b). *Wear* **28,** 59–68.

Moore, M. A., and Douthwaite, R. M. (1976). *Metall. Trans.* **7A,** 1833–1839.

Moore, M. A., and McLees, V. A. (1975). NIAE Dept. Note DN/ER/584/1160, unpublished. National Institute of Agricultural Engineering, Bedford.

Moore, M. A., and McLees, V. A. (1977). Unpublished results.

Moore, M. A., Richardson, R. C. D., and Attwood, D. G. (1972). *Metall. Trans.* **3,** 2485–2491.

Mulhearn, T. O., and Samuels, L. E. (1962). *Wear* **5,** 478–498.

Nabarro, F. R. N. (1967). "Theory of Crystal Dislocations," pp. 696–698. Oxford Univ. Press (Clarendon), London and New York.

Nathan, G. K., and Jones, W. J. D. (1966). *Wear* **9,** 300–309.

Nathan, G. K., and Jones, W. J. D. (1966/1967). *Proc. Inst. Mech. Eng.* **181**(30), 215–275.

National Institute of Agricultural Engineering (1974). Wear, Project Rev. PR/ER/74/1160, unpublished. National Institute of Agricultural Engineering, Bedford.

Norman, T. E., and Hall, E. R. (1969). *In* "Evaluation of Wear Testing," ASTM STP 446, pp. 91–114. American Society for Testing and Materials, Philadelphia, Pennsylvania.

Ogilvy, I. M., Perrott, C. M., and Suiter, J. W. (1977). *Wear* **43,** 239–252.

O'Neill, H. (1967). "Hardness Measurements of Metals and Alloys," p. 75. Chapman and Hall, London.

Patterson, G. W., and Mulhearn, T. O. (1969). *Wear* **13,** 175–182.

Perrott, C. M., and Robinson, P. M. (1974). *J. Aust. Inst. Met.* **19,** 229–240.

Popov, V. S., *et al.* (1969a). *Russ. Eng. J.* **49**(7), 78–81.

Popov, V. S., Nagorny, P. L., and Garbuzov, A. S. (1969b). *Phys. Met. Metall.* **28**(2), 150–154.

Popov, V. S., Shumikin, A. B., Garbuzov, A. S., and Pirozhkova, V. P. (1972). *Russ. Metall.* (1), 164–167.

Rabinowicz, E. (1968). *Sci. Am.* **218**(6), 91–99.

Rabinowicz, E. (1971). *Wear* **18,** 169–170.

Rabinowicz, E. (1976a). *In* "Materials Technology—1976 (APS New York Meeting)" (A. G. Chynoweth and W. M. Walsh, eds.), pp. 165–174. *Am. Inst. Phys. Conf. Proc.* No. 32, New York.

Rabinowicz, E. (1976b). *Mater. Sci. Technol.* **25,** 23–28.

Rabinowicz, E., and Mutis, A. (1965). *Wear* **8,** 381–390.

Rabinowicz, E., Dunn, L. A., and Russel, P. G. (1961). *Wear* **4,** 345–355.

Ramachandron, V., and Abrahamson, E. P. (1972). *Scripta Metall.* **6,** 287.

Ramalingam, S., and Black, J. T. (1973). *Metall. Trans.* **4,** 1103–1112.

Ramalingam, S., and Lehn, L. L. (1971). *J. Eng. Ind.* **93,** 527–544.

Richardson, R. C. D. (1967a). NIAE Note No. 12, unpublished. National Institute of Agricultural Engineering, Bedford.

Richardson, R. C. D. (1967b). *Wear* **10,** 291–309.

Richardson, R. C. D. (1967c). *Wear* **10,** 353–382.

Richardson, R. C. D. (1967d). *J. Ag. Eng. Res.* **12,** 22–39.

Richardson, R. C. D. (1968). *Wear* **11,** 245–275.

Richardson, R. C. D. (1969). The Wear of Metal Shares in Agricultural Soil. Ph.D. thesis, Univ. of London.

Richardson, R. C. D., Jones, M. P., and Attwood, D. G. (1967). *Proc. Agr. Eng. Symp.* Div. 2, Paper 26. Institute of Agricultural Engineers, London.

Samuels, L. E. (1956/1957). *J. Inst. Met.* **85,** 51–62.

Sedriks, A. J., and Mulhearn, T. O. (1963). *Wear* **6,** 457–466.

Sedriks, A. J., and Mulhearn, T. O. (1964). *Wear* **7,** 451–459.

Selwood, A. (1961). *Wear* **4,** 311–318.

Stevenson, M. G., and Oxley, P. L. E. (1969/1970). *Proc. Inst. Mech. Eng.* **184,** 561–574.

Stickler, R., and Booker, G. R. (1963). *Phil. Mag.* **8,** 859–876.

Stroud, M. F., and Wilman, H. (1962). *Brit. J. Appl. Phys.* **13,** 173–178.

Studman, C. J., Moore, M. A., and Jones, S. E. (1977). *J. Phys. D: Appl. P.* **10,** 949–956.

Swain, M. V. (1975). *Wear* **35**, 185–189.
Tabor, D. (1970). *Rev. Phys. Technol.* **1**, 145–179.
Turley, D. M. (1968). *J. Inst. Metall.* **96**, 82–86.
Turley, D. M. (1971). *J. Inst. Metall.* **99**, 271–276.
Vijh, A. K. (1975). *Wear* **35**, 205–209.
Wiederhorn, S. M., and Roberts, D. E. (1975). *Wear* **32**, 51–72.
Wilshaw, T. R., and Hartley, N. E. W. (1972). *Proc. Eur. Symp. Comminut., 3rd, Cannes 1971* pp. 33–50. Verlag Chemie GmbH, Weinheim.
Zener, R., and Holloman, J. H. (1944). *J. Appl. Phys.* **15**, 22–32.
Zolotar, A. I. (1973). *Ind. Lab.* **39**(7), 1134–1136.
Zum Gahr, K-H. (1976). *Z. Metallk.* **67**, 678–682.

Fretting

R. B. WATERHOUSE

Department of Metallurgy and Materials Science
University of Nottingham
Nottingham, England

I. Introduction

Fretting was first recorded by Eden *et al.* (1911), who experienced problems with the formation of a red rust in the grips of fatigue machines in which they were testing steel specimens. The rust made removal of the specimens difficult and its formation was attributed to the varying stress between specimen and holder. Tomlinson (1927) first investigated this phenomenon experimentally and coined the term "fretting corrosion," by which name it is generally known today. Reports of its occurrence in practical applications have grown over the years, in part because of the increasingly close tolerances to which engineers are called to work and also because of the increased demands placed on materials by virtue of the harsher environmental conditions in which they are required to operate. The growing interest in the subject is indicated by the appearance of two monographs, one by the author (Waterhouse, 1972) and the other by Golego *et al.* (1974), which is currently being translated into English.

A. *Occurrence*

Fretting damage results when two loaded surfaces in contact undergo relative oscillatory tangential movement, known colloquially as "slip," as a result of vibration or cyclic stressing. The amplitude of the relative motion is small and is often difficult to measure. One of the complicating features is that in most practical cases slip occurs only over part of the surface areas in contact, as predicted by Mindlin's analysis of vibrating contacts (Mindlin, 1949) and verified experimentally by Johnson (1955). As fretting proceeds, the area over which slip is occurring usually increases due to the incursion of debris, with a consequent lowering of the coefficient of friction, into neighboring areas (Rollins and Sandorff, 1972; Waterhouse and Taylor, 1971a). In many practical instances of fretting the amplitude of slip appears to be within the range 2–20 μm. It must be remembered that in practice the movement is fortuitous, whereas in many experimental investigations the vibrations are forced and amplitudes have been considerably greater. As will be seen in Section II,A, where the effect of amplitude is considered, the type and volume of damage change at amplitudes above 75 μm, becoming characteristic of reciprocating wear. One of the features of fretting is that the surfaces are always in close contact, resulting in entrapment of debris and restriction of access of the environment. With increasing amplitudes these features are lost. Unfortunately the small amount of debris produced in short-term tests has often encouraged investigators to work with uncharacteristically high amplitudes.

One of the particularly important features of fretting, especially where the movement is caused by cyclic stressing of one of the contacting surfaces, is the initiation of fatigue cracks, but as the subject of this chapter is fretting wear, it will not be discussed here. It should be said, however, that in cases in which large amounts of fretting debris are produced, the effect on fatigue strength is minimal (Field and Waters, 1967). This is thought to be due to the lubricant effect of the debris and to the removal of surface material in which fatigue cracks are being initiated before the cracks have had time to propagate away from the region influenced by the fretting action.

The possible situations in which fretting can arise in practice are legion, but some of the more common ones are the wheel seat of a hub-on-axle assembly or a flywheel on shaft, marine propeller shafts that carry a bronze liner, and all types of flexible coupling of the toothed or splined variety, together with keys and keyways. A further group of situations concerns static joints subjected to vibration, e.g., riveted air frame structures and bolted joints in vehicle chassis. Akin to this is fretting which can occur between the undersides of screw heads and the countersink of the holes in bone plates used in orthopedic surgery to fix difficult fractures of limbs. Other common sites are between the strands in steel ropes and stranded conductors, and in electrical contacts and knife edge contacts in scientific equipment subject to vibration. Finally, the transport of goods by road, rail, or sea has been a frequent source of trouble including the classic case of the damage to ball and roller bearings in the road transport of automobiles from Detroit to the west coast.

B. Detrimental Effects

The debris produced by fretting is largely the oxide of the metals involved and therefore occupies a greater volume than the volume of metal destroyed. In a confined space this can lead to the buildup of pressure or to the seizure of the contact, which can be serious when the two surfaces are called upon to move relative to each other at certain intervals. A particular instance is a machine governor working on the familiar principle of increasing the angular momentum by the movement of a sliding collar on a rotating shaft. Fretting between the collar and shaft can lead to seizure, rendering the governor inoperative. Another area in which entrapment of debris leads to damage is in locked coil steel wire ropes. These ropes have specially shaped strands which fit together closely and are used in overhead cable ropeways. If fretting debris forms inside the rope it forces the strands apart and allows the atmosphere to get in and promote rapid corrosion damage. In situations in which the debris can escape, loss of fit will result, which is serious in itself, but the escaping debris, being an abrasive oxide, may cause further damage by falling onto moving parts. The presence of oxide debris in electrical contacts leads to distortion of signals and high resistance. Foodstuffs transported in stacked aluminum trays have become contaminated with black debris. This black debris formed on stacked ammunition boxes and, accumulating in the holds of ships, was found to be pyrophoric and a potential hazard. Fretting in surgical implants leads to the liberation of base metal ions into the body fluids which may have a toxic effect. Recent work suggests that fretting may encourage break-away corrosion in stainless steel operating in carbon dioxide

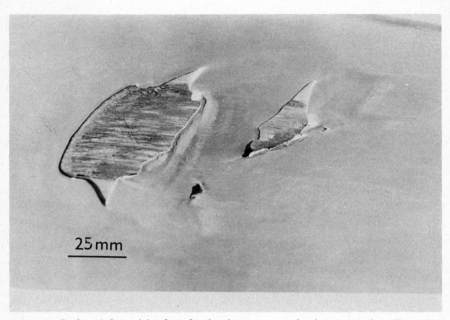

25mm

Fig. 1. Surface defect arising from fretting damage on an aluminum extrusion. (From Waterhouse, 1977a.)

atmospheres at temperatures around 600°C, conditions to be found in some nuclear reactors. Fretting debris on the surface of aluminum ingots transported by rail resulted in serious surface defects in subsequent extrusion operations (see Fig. 1) (Waterhouse, 1977a).

C. *Nature of Debris*

On steel the debris when it escapes from the contact area is reddish in color and is mainly the hexagonal α-Fe_2O_3, which is nonmagnetic but may contain a small proportion of metallic iron. If it is produced under pressure, where it cannot escape, it is usually flaky with a black irridescent appearance, but is still α-Fe_2O_3. The amount of metallic iron in the debris is a function of the mechanical conditions prevailing in its production: the lower the frequency, the larger the amplitude, and the heavier the applied load, the greater the amount of metallic iron. It also depends to some extent on the hardness of the surface, softer surfaces giving rise to more metallic particles (Toth, 1972). Under the converse of these conditions the debris is more likely to be the fine red powder. On aluminum and its alloys the debris is black and contains about 23% metallic aluminum, the remainder being alumina (Andrew *et al.*, 1968). The oxide was thought to be amorphous since only lines of metallic aluminum were obtained in x-ray diffraction patterns, but recent electron diffraction analysis has shown that the platelike particles are cubic γ-alumina, oriented with the (211) planes parallel to the surface (Markworth, 1976). The author has found the platelike debris on titanium alloys to have a high degree of preferred orientation, and its composition is thought to be the cubic TiO.

There is increasing evidence to suggest that fretting debris on a number of metals and alloys is platelike in character, ranging in size on steel from 0.01 to 0.1 μm and on aluminum up to 0.15 μm. Some recent work on fretting steel against a number of polymers indicates that the debris is largely α-Fe$_2$O$_3$ and platelike in form (Fig. 2) and the particle size is independent of load, frequency, amplitude, and time of sliding, but depends on the particular polymer, Table I (Higham *et al.*, 1978a). Also included in the table are values of the critical surface tension, which is the maximum surface tension of a liquid that completely wets the polymer, and is related to the surface energy of the polymer; there appears to be some correlation, as one would expect if it is assumed that particle size is directly proportional to the work of adhesion between the two surfaces (Rabinowicz, 1961). A platelike wear particle is predicted by the delamination theory of wear, but subsequent grinding between the surfaces may roll it up if plastic or break it down further if brittle. Spherically shaped particles have been observed by Stowers and Rabinowicz (1972) on fretted silver surfaces and

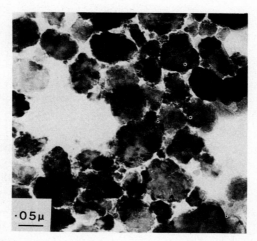

Fig. 2. Transmission electron microscope picture of wear debris transferred from steel surface to nylon 66 surface during fretting (amplitude 6.3 μm, load 330 g, frequency 60 Hz, 3.45×10^6 cycles). (From Higham *et al.*, 1978a, by courtesy Pergamon Press.)

TABLE I

Oxide Particle Sizes When Mild Steel Is Fretted
against Various Polymers

Polymer	Critical surface tension (N/m)	Particle size (μm)
Polyethylene	0.031	0.008
Polyvinylchloride	0.039	0.020
Polymethylmethacrylate	0.040	0.045
Polysulphone	0.041	0.040
Polycarbonate	0.042	0.045
Nylon 66	0.046	0.070

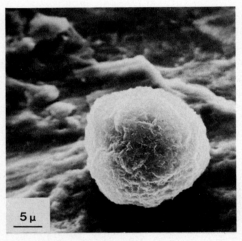

Fig. 3. Spherical particle produced by fretting mild steel against a wear-resistant surface in an oxidizing atmosphere at 280°C (amplitude 50 μm, frequency 100 Hz, 21.6×10^6 cycles). (From Taylor, 1977.)

also on steel surfaces fretted in argon at temperatures between room temperature and 500°C (Hurricks, 1974). These particles have a layered structure and appear to be formed by the rolling up of a lamellar particle. Figure 3 shows such a particle formed by fretting mild steel against a wear-resistant surface in an oxidizing atmosphere at 280°C; the amplitude was 50 μm and the frequency 100 Hz. Analysis showed that the particle derived from the steel surface.

II.   Effect of Variables

A.   *Mechanical Variables*

1.   NUMBER OF CYCLES

Figure 4 is a schematic representation of the type of wear curve seen in fretting investigations. There is an initial rapid removal of material which decreases with time. It is in this period that intermetallic contact indicated by low contact resistance is occurring. The coefficient of friction is usually high, which is characteristic of metal-to-metal contact. As oxidized debris is formed the rate becomes steady or continues to fall as shown in curves B and D. This is typical of mild steel. Higher amplitudes of slip result in curve B and lower amplitudes in curve D. Some materials show an accelerating damage rate as in curve A or C. This behavior appears typical of materials with a low hardness but with an oxide of high hardness such as aluminum, where abrasion is a contributory factor. Lewis and Didsbury (1977) have made a comparison among results obtained by various investigators in fretting mild steel in terms of the specific wear rate (Fig. 5). The amplitudes of slip ranged from 2.5 to 200 μm, but the results show that the specific wear rate is falling markedly with increasing number of cycles. This is referred to in Section IV.

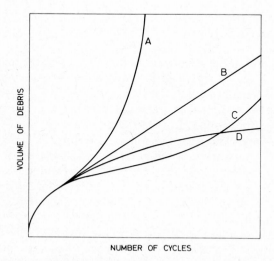

Fig. 4. Schematic representation of the production of wear debris with increasing number of cycles.

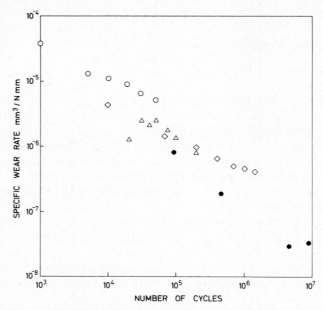

Fig. 5. Specific wear rate for fretting mild steel surfaces as a function of number of cycles. ● Lewis and Didsbury; ◇ Feng and Uhlig; △ Ohmae and Tsukizoe; ○ Feng and Rightmire. (From Lewis and Didsbury, 1977.)

2. Amplitude of Slip

There is general agreement that at amplitudes greater than 70–100 μm the volume of material removed in a given number of cycles is directly proportional to the amplitude of slip. In other words, the wear volume is proportional to the sliding distance and is identical with normal mild wear. At the low amplitudes typical of fretting the damage rate is much lower. Some investigators claim that there is no measurable damage below 100 μm and that the movement is taken up in the elastic deformation of the surfaces (Sasada, 1959; Aoki *et al.*, 1967). This has led Stowers and Rabinowicz (1973) to suggest that this is a manifestation of the difficulty of measuring the small amplitudes actually at the contacting surfaces. In his original work Tomlinson (1927) found that debris was produced by movements of 2 nm. Johnson (1955) was working with amplitudes down to 50 nm. There is ample visual and x-ray evidence of surface damage at these low amplitudes but little quantitative data. Lewis and Didsbury (1977) have presented the available data on steel as specific wear rate versus amplitude curves (Fig. 6). Except for one instance in which the wear rate appears to be independent of amplitude, the remaining curves all show an increase of over two orders of magnitude in the wear rate in the amplitude range 50–150 μm. Back reflection x-ray diffraction photographs of the damaged surface at an amplitude of slip of 80 μm and above showed a strong (211) ring whereas below this amplitude the (211) ring was extremely faint and the Laue spot pattern of the underlying material was clearly visible. This indicates that above this critical amplitude material is involved to a much greater depth below the surface (Ohmae and Tsukizoe, 1974).

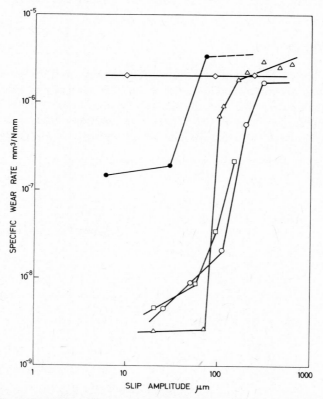

Fig. 6. Comparison of specific wear rates for fretting mild steel surfaces as a function of slip amplitude. ● Lewis and Didsbury; ◇ Feng and Uhlig; △ Ohmae and Tsukizoe; □ Halliday and Hirst; ○ Halliday. (From Lewis and Didsbury, 1977.)

3. NORMAL LOAD

Curves of wear volume against applied load are approximately linear where the movement is forced and occurs over the whole of the contacting area. This implies a constant specific wear rate. If, however, the increased pressure decreases the area of the interface over which slip is occurring and reduces the slip amplitude, then the rate at which material is removed will fall. This fact has been used to reduce fretting damage. Modification of a design to reduce the apparent contact area and thereby increase the pressure at the interface can result in reduced fretting wear. This is not recommended where fatigue is involved since fatigue cracks initiate in the slip/nonslip boundary and this is merely moved by increasing the pressure, and the stress concentration at the boundary is increased.

4. FREQUENCY

Figure 7 shows how frequency may affect the rate of fretting wear if it is assumed that the process is the removal and regrowth of an oxide film that is growing according to a logarithmic law. Figure 8, which is derived from Fig. 7, shows the form of the curves if a constant time or a constant number of cycles is considered. The specific wear rate curve versus fre-

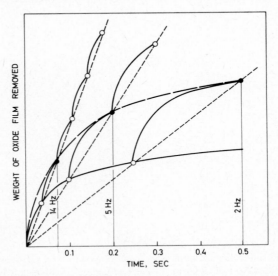

Fig. 7. Schematic representation of the effect of frequency, assuming a mechanism based on scraping off of an oxide film that is growing according to a logarithmic law.

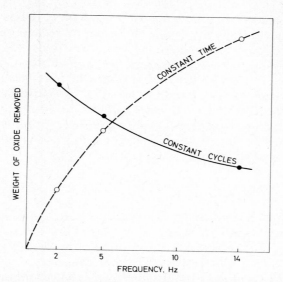

Fig. 8. Schematic representation of accumulation of oxide debris as a function of frequency of vibration, determined for a fixed time and a fixed number of cycles.

quency will have the same form as the curve for a constant number of cycles. The little experimental evidence available conforms to this type of curve and is the basis of the fretting mechanism proposed by Uhlig (1954). Frequency effects become negligible at frequencies above 17 Hz because the only part of the oxidation curve that is involved is the near-linear portion at the origin.

Amplitude and frequency effects are also considered in electrochemical observations of fretting (see Section III).

B. *Environmental Variables*

1. ATMOSPHERE

Fretting of steel in a nonoxidizing atmosphere such as nitrogen or in a vacuum reduces the wear, as assessed by the permanent removal of material from the surfaces (Feng and Uhlig, 1954). Under such conditions it is to be expected that local welding of contacting asperities, which is the basis of Bowden and Tabor's theory of wear (Bowden and Tabor, 1950), would be enhanced. This is probably so, but it results in the transfer of material back and forth from one surface to the other without its removal as loose debris. This strongly suggests that the corrosive nature of the environment plays a role in the actual production of debris and is not a secondary effect. Humidity has been shown to reduce the amount of wear, a fact which is attributed to the hydrated oxides of iron being less hard than their dehydrated counterparts (Feng and Uhlig, 1954). A similar result has been found with pure nickel where the fretting wear was some five times greater in dry air than in air of 25% relative humidity (Bill, 1974). Instances of fretting of steel surfaces in electricity generators operating in hydrogen are known, but the oxygen and water vapor content of the hydrogen is sufficiently great for the debris and volume of damage to be similar to that obtained in air.

2. TEMPERATURE

Since fretting involves both the mechanical properties of the material and its chemical reactivity, it is to be expected that changes in temperature will have considerable effect on fretting wear. Published quantitative data are scarce, but with the possibility of fretting occurring in gas turbine engines and nuclear power installations there is a need for more active research in this field.

To deal with low temperature first, on mild steel the volume of fretting wear increases as the temperature is lowered below 0°C. There is a sharp increase at this temperature followed by a slower increase as the temperature is further lowered (Feng and Uhlig, 1954). It has been suggested that the sudden increase at 0°C is related to the ductile–brittle transformation, but it is more likely to be associated with the lubricant effect of moisture.

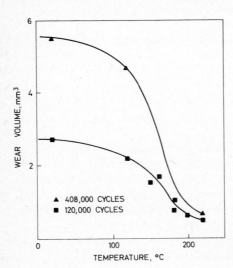

Fig. 9. Volume of wear damage produced in fretting mild steel as a function of temperature. (From Hurricks, 1972.)

TABLE II

Fretting Wear of Nickel-Base Alloys Fretted in Dry Air at
Temperatures up to 816°C[a]

Material	Fretting wear volume ($10^{-5} \times mm^3$)			
	23°C	540°C	650°C	816°C
99.9% Ni	153	—	68	—
Ni–10% Cr–2% Al	180	17	19	24
Ni–10% Cr–5% Al	93	7.2	8	7.7
Ni–20% Cr–2% Al	100	13	29	—
Ni–20% Cr–5% Al	70	4.7	14	—

[a] Amplitude of slip 75 μm, frequency 80 Hz, 288,000 cycles.

Raising the temperature produces a dramatic reduction in the wear rate on mild steel in air in the temperature region 150–200°C (Fig. 9) (Hurricks, 1972). This is due to the greater thickness and adherence of the oxide at temperatures above 200°C, which prevents the initial adhesion and local welding stage of the process, the oxide film acting as a solid lubricant. Table II shows results obtained on nickel and nickel-based alloys (Bill, 1974). In this case a marked drop in wear rate occurs in the temperature interval between room temperature and 540°C. Increasing the chromium content produced a small improvement in wear resistance at 540°C, but at 810°C the deterioration in mechanical properties caused buildup of material on the higher chromium alloys. Increasing the aluminum content has a much greater effect by virtue of its contribution to the formation of a more protective film with more rapid self-repairing properties.

Figure 10 shows a curious rumpling of the surface oxide film observed in fretting 18% Cr–10% Ni austenitic steel surfaces in an oxidizing atmosphere at 650°C. Peeling off of the oxide also occurs in the contact region (Fig. 11). Figures 12 and 13 show chromium and iron distribution maps of this area. The removal of the chromium-rich oxide film leaves a roughened chromium-depleted surface below (Taylor, 1977). The potential dangers of this situation have been discussed by the author (Waterhouse, 1975a). The lower chromium content steels tend to form the faster growing nonprotective iron–chromium spinel in oxidizing atmospheres at elevated temperatures instead of the protective Cr_2O_3 oxide. If the diffusion rate of chromium in the bulk material is too low to maintain the bulk concentration in the surface, then the disruption caused by the fretting action is likely to promote rapid oxidation. Much will depend on the mechanical and lubricant properties of the resultant oxide film. Stott *et al.* (1973) have found that nickel-base high-temperature alloys develop a smooth glazed surface in reciprocating sliding at high temperatures in an

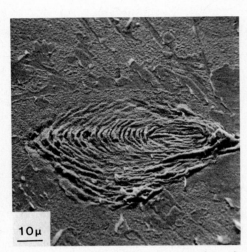

Fig. 10. Fretting damage on austenitic stainless-steel surfaces fretted in an oxidizing atmosphere at 650°C showing rumpling of oxide film (amplitude 25 μm, load 1000 kg, frequency 185 Hz, 1.8 × 10⁶ cycles). (From Taylor, 1977.)

oxidizing atmosphere, which results in a low coefficient of friction and low wear rate. A similar observation has been made on a cobalt-base alloy: the wear rate drops by an order of magnitude as the temperature is raised from 250 to 400°C (Thiéry and Spinat, 1975). Whether this type of behavior is maintained with the small-amplitude movement of fretting is at present being investigated.

Fig. 11. Fretting damage on austenitic stainless-steel surfaces fretted in an oxidizing atmosphere at 650°C, showing peeling of oxide film (amplitude 25 μm, load 1000 kg, frequency 185 Hz, 1.8 × 10⁶ cycles). (From Taylor, 1977.)

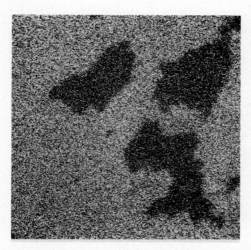

Fig. 12. X-ray distribution map of chromium for the field shown in Fig. 11 (same scale). Chromium-rich areas are light. (From Taylor, 1977.)

Some work has been done on the fretting behavior of high-temperature materials in liquid sodium, stimulated by problems in reactor technology. Campbell and Lewis (1977) have measured specific wear rates for a number of ferritic chromium steels, nickel, and cobalt-base alloys fretting against a 9% Cr–1% Mo steel in liquid sodium at 530°C. Both impact and

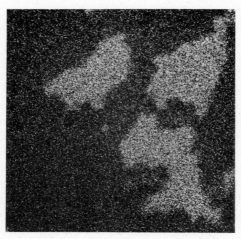

Fig. 13. X-ray distribution map of iron for the field shown in Fig. 11 (same scale). Iron-rich areas are light. (From Taylor, 1977.)

Fig. 14. Fretting damage on Ti–6% Al–4% V fretted in air at 600°C (amplitude 12 μm, clamping pressure 60 MN/m^2, frequency 50 Hz, 24,000 cycles).

rubbing fretting were investigated with a slip amplitude of 100 μm and a frequency of 25 Hz. The specific wear rates were very similar for all the materials, lying in the range 3×10^{-9}–3×10^{-8} mm^3/N mm. The impact fretting rates were generally higher than the rubbing fretting. The similarity in wear rates is due to the transfer of the 9% Cr–1% Mo material to the opposing surface. A treatment that reduced the wear rate by some three orders of magnitude under these conditions was aluminizing, which prevented metal transfer.

The author, in studying the fretting fatigue behavior of titanium- and nickel-base alloys at temperatures up to 600°C in air has found massive metal transfer, often in the form of piled-up platelets (Fig. 14).

III. Fretting in Aqueous Electrolytes

A. *Electrochemical Measurement*

Events at a metal electrolyte interface are usually controlled by the oxide layer on the metal surface and depend very much on its perfection and electrical conductivity. Disruption of the film by fretting can produce large changes in electrode potential, particularly on the baser metals such as aluminum and titanium. With aluminum in sodium chloride solution the solution rapidly becomes black with minute suspended particles of colloidal dimensions. The debris is dispersed immediately as it is produced. The potential–time trace shows a rapid fall in potential as fretting commences, followed by fluctuating behavior characteristic of rapidly changing points of disruption and regrowth of the oxide film. On terminating the fretting the potential immediately rises but takes some time before it achieves its starting value, indicating that the disturbance to the surface causes increased chemical activity which does not immediately disappear (Fig. 15). The noble metals silver and copper show no fall in potential, but the baser metals tantalum, chromium, and aluminum show falls of several hundred millivolts (Taylor and Waterhouse, 1974). Vijh (1976) has correlated the fall in potential with the difference in hardness of the oxide and the underlying metal, i.e., the harder the oxide and the softer the substrate, the greater the potential drop. Other factors such as stirring of the electrolyte and the increased reactivity of deformed metal have little or no effect.

Potentiostatically controlled experiments, in which the potential of the fretting surfaces is kept constant, allow the corrosion current to be measured. A linear relation is found between corrosion current and amplitude of slip at a fixed frequency, and also between corrosion current and frequency at a fixed amplitude. This is to be expected since the number of metal ions going into solution per unit time, which constitutes the corrosion current, depends on the area and the rate at which the underlying metal is being exposed (Sherwin *et al.*, 1971). Calculations show that the bulk of the material removed when metals are fretted in an electrolyte is the result of mechanical action rather than chemical dissolution (Waterhouse, 1972). However this could be of considerable significance in cases of fretting in surgical implants within the body since even a very slow rate of liberation of metal ions can have toxic effects.

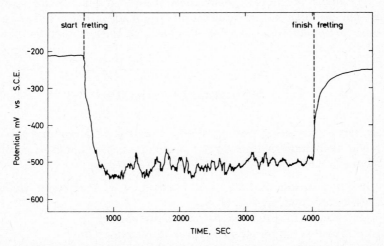

Fig. 15. Potential–time curve for chromium surfaces fretted in 0.5*M* NaCl (amplitude 50 μm, load 300 g, frequency 120 Hz).

TABLE III

<small>Fatigue Lives of 18% Cr–8% Ni Stainless Steel in Air and 0.1 M NaCl
with and without Cathodic Protection by Applying a Potential of
−1200 mV versus SCE[a]</small>

Fatigue in air	Fretting fatigue in air	Corrosion fatigue in 0.1M NaCl		Fretting corrosion fatigue in 0.1 M NaCl	
		Free potential	Cathodic protection	Free potential	Cathodic protection
>10^7	463,000	314,000	2,046,000	286,000	2,279,000

[a] Mean stress 246 MN/m² and alternating stress ±154 MN/m², slip amplitude 16 μm, frequency 50 Hz.

B. Electrochemical Control

Corrosion reactions can be controlled by impressing a cathodic potential (or on metals which display passivation, an anodic potential) on the system. Cathodic and anodic protection are also capable of reducing, and in some cases completely nullifying, the effects of corrosion fatigue, which involves conjoint mechanical and chemical action as does fretting corrosion. It was of interest to investigate their effects in fretting. Table III gives results obtained on an austenitic stainless steel and shows that cathodic protection can eliminate the effects of fretting as assessed by its effect on fatigue life. Visual inspection showed a reduction in surface damage but no quantitative assessment was made. This result suggests that surface fatigue is a contributory factor in the production of wear debris by fretting.

IV. Mechanism of Fretting Wear

When first considering fretting nearly a quarter of a century ago, the author was of the opinion that three possible processes could be involved (Waterhouse, 1955):

(1) the production of loose metal particles by local cold welding, which were subsequently oxidized;
(2) abrasion of the surfaces by hard oxide debris;
(3) continual scraping off and regrowth of oxide films.

If (2) were an important contribution to the process, then an accelerating wear rate would be expected as more debris was produced, which from

Fig. 5 is plainly not the case and so this can be discounted, or at least assigned a minor role. Indeed the debris may have a beneficial lubricant effect (Halliday and Hirst, 1956). Of the two remaining possibilities, (1) would apply in cases of fretting in which oxidation was not a contributory factor, e.g., in the fretting of a noble metal such as gold, or fretting of baser metals in a nonoxidizing atmosphere. In fretting of a material such as steel or aluminum in air it might apply in the early stages of the process but buildup of oxide debris would soon result in loss of metal-to-metal contact, which is known to occur from observation of the rise in contact resistance. At that time (3) seemed the more likely process and Uhlig (1954) produced an expression based on this supposition but included a minor term covering possible plowing action by asperities. His expression agreed well with available data. In the intervening years the advent of the scanning electron microscope has removed much of the speculation concerning surface phenomena at sliding contacts. Hurricks (1970), in a painstaking review of all the evidence, considered three stages to be involved:

(1)   adhesion and metal transfer,
(2)   production of wear debris by mechanicochemical action, and
(3)   steady-state production of debris by fatigue action rather than abrasion.

The author, in a recent paper, based on his observations of the last few years (Waterhouse, 1977b), believes that adhesion and welding occur in the early stages with steel and the more noble metals, causing material to be raised above the level of the original surface (Fig. 16), followed by

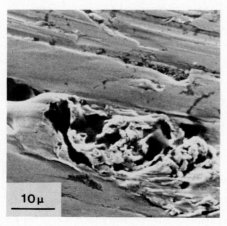

Fig. 16.   Welds formed in the initial stages of fretting between 0.2% C steel surfaces (amplitude 18 μm, clamping pressure 60 MN/m^2, frequency 25 Hz, 500 cycles).

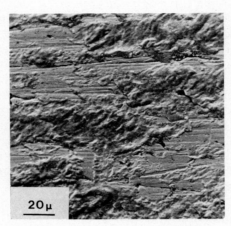

Fig. 17. Smeared material produced in the fretting of 0.2% C steel surfaces (amplitude 18 μm, clamping pressure 60 MN/m², frequency 25 Hz, 5000 cycles).

smearing (Fig. 17), and finally removal of material by a process of delamination (Fig. 18). These stages can be identified in successive Talysurf surveys of the surface (Fig. 19). With extremely chemically active materials such as titanium the adhesion stage is not so apparent and the delamination process operates from the inception of fretting. The delamination theory of wear (Suh, 1973) envisages the coalescence of subsurface

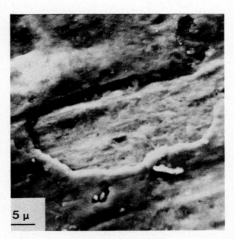

Fig. 18. Removal of material in the form of platelets from the surface of 0.7% C steel by fretting (amplitude 37.5 μm, load 300 g, frequency 150 Hz, 8 × 10⁶ cycles). (From Waterhouse and Taylor, 1974.)

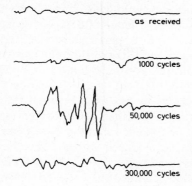

as received

1000 cycles

50,000 cycles

300,000 cycles

Fig. 19. Talysurf surveys of fretting damage on mild steel surface as fretting proceeds (ratio of vertical to horizontal scale is 100). (From Waterhouse and Taylor, 1971a.)

voids produced by dislocation pileups at obstacles such as inclusions. Whether the subsurface cracks originate in this manner or whether they arise as a result of surface fatigue is not clear. However, the evidence that the process is very sensitive to the chemical nature of the environment suggests that corrosion fatigue may be playing a part.

The delamination theory explains the platelike appearance of the debris, and subsurface cracks have been detected in sections through fretted surfaces (Fig. 20). The process is probably more complicated since other effects have been observed, e.g., a rise in temperature of several hundred degrees very locally (Waterhouse, 1961; Alyab'ev *et al.*, 1970a) and the emission of exoelectrons by the Kramer effect (Grunberg and Wright, 1955; Evdokimov, 1968). The rise in temperature locally can result in overaging of an age-hardened alloy (Bethune and Waterhouse, 1968) and in the appearance of nonequilibrium phases such as martensite

5 μ

Fig. 20. TEM photograph of section through fretted surface of Ti–6% Al–4% V alloy, thinned by ion bombardment, showing subsurface crack. The original surface is in the top right-hand corner.

in steels (Waterhouse, 1975b). These changes would have considerable effect on the fatigue properties of the material affected. Local work hardening in the fretted region has also been observed, particularly with materials of low stacking fault energy, e.g., austenitic steels (Bethune and Waterhouse, 1965; Alyab'ev *et al.*, 1970b). Work-hardened material displays a greater notch sensitivity and this would also adversely influence the fatigue behavior. The emission of exoelectrons has been invoked to explain the production of α-Fe_2O_3 on steels through the agency of hydrogen peroxide (Wright, 1952–1953) and there is evidence that their emission is much enhanced in reciprocating sliding as compared with unidirectional sliding (Evdokimov, 1968). Such phenomena appear to be of secondary importance in the fretting process as also are the thermoelectric effects that have been reported (Golego *et al.*, 1971).

To sum up, the initial damage arises from adhesion and welding at points of real contact between the surfaces, resulting in material being plucked up and raised above the original surface. The extent and severity of this part of the process depend on the reactivity of the metal and the corrosivity of the environment. Raised material is smeared out, the surface becomes smoother, and the smeared material is removed by a delamination to produce platelike particles which are essentially metallic but oxide covered. The loose-wear particles may be further comminuted by grinding between the surfaces or rolling up into cylinders or spheres. Delamination proceeds by the spreading of subsurface cracks arising from coalescence of voids (Suh, 1973) or by surface fatigue. The compaction of debris in the contact zone allows the transmission of the alternating shear stresses across the interface, and the delamination continues. Other effects such as work hardening or work softening have adverse effects on the fatigue properties and accelerate the process. In cases in which the movement is due to one of the surfaces being cyclically stressed, propagating fatigue cracks may be initiated in the early stage of the process when there is a boundary between slip and nonslip areas of the surface. Although in the initial stages the wear rate is comparable with rates experienced in adhesive wear, it falls by several orders of magnitude when steady-state conditions are established and is more akin to mild wear. There are those who disagree with this interpretation and hold that the wear rate is characteristic of rates obtained in unidirectional adhesive wear, the very low wear rates in fretting being attributed to errors in measurement of the small amplitudes of slip involved (Stowers and Rabinowicz, 1973).

In the case of fretting between a polymer and a metal the initial stage of cold welding and intense surface disruption is absent. After an induction period damage results to a steel surface by adhesion of the iron oxide to the polymer (Higham *et al.*, 1978b). As described previously, the oxide particles are platelike and could result from delamination.

V. Preventive Measures

It is not intended to give an exhaustive account of the very numerous and diverse methods that have been recommended to prevent the onset of fretting since these have been dealt with in detail in the two monographs mentioned in the introduction, but rather to indicate the basic ideas behind the methods. There are three lines of attack:

(1) Prevent the relative movement from occurring by a change in design. This is not always possible since redesigning is a costly and time-consuming process.

(2) If movement cannot be prevented, treat the contacting surfaces so they are not degraded by cold welding or delamination. This is not always practicable since even if a suitable treatment can be found it may be difficult to apply, particularly to very large components.

(3) Encourage the movement by reducing the coefficient of friction and minimize wear by use of lubricants. This is usually difficult to achieve with conventional liquid lubricants since the relative velocity of the moving surfaces is low and pressures are high so that the regime is entirely within that of boundary lubrication.

Nevertheless many practical cases of fretting have been overcome by fairly simple measures, some of which are outlined below.

A. *Design*

The ultimate in redesigning is to do away with interfaces by construction in one solid piece or, more usually, by using a welded structure. The author has seen the latter remedy used in a gas turbine engine in which the rotor disk was bolted to the main shaft via a flange. Fretting occurred under the bolt heads and resulted in a fatigue failure of the disk. By producing a welded assembly of shaft and disk, interface problems were overcome. In most instances the components must be demountable and, therefore, less drastic modifications of design are possible. Slip occurs at an interface because the shear force is greater than the friction force opposing movement. This gives the possibility of two approaches: (1) reduce the shear stress, or (2) increase the friction force. Reduction of the shear stress can be achieved by ensuring that there is not a stress concentration

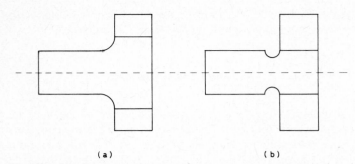

(a)   (b)

Fig. 21. (a) Preferred design of wheel-on-axle assembly to minimize stress concentration at the interface. (b) Stress-relieving groove in wheel-on-axle assembly.

at the interface. A simple case is that of a wheel on an axle. There should be an increase in section at the wheel seat with a generous fillet between the seat and the shaft (Fig. 21a). If an increase in shaft radius is impracticable, a radiused groove may be used (Fig. 21b) (Ørbeck, 1958). Such measures are the normal practice in designing to combat fatigue failure. The friction force may be increased by reducing the apparent area of contact while keeping the applied normal load the same, thus increasing the pressure. This method has been successfully applied to a tapered connection (Rolfe, 1952). It is sometimes claimed that plating the surfaces with softer metals such as cadmium results in an increased coefficient of friction and promotes seizure. The author has produced a similar effect by introducing hard particles between steel surfaces, which key into the surfaces and lock them together (Waterhouse and Allery, 1966). No doubt modern adhesives could also be applied to this effect. Such methods would obviously apply only to permanent joints.

B. Surface Treatments

Surface treatments can either be applied coatings such as electrodeposited metals, sprayed metals, anodized or phosphate coatings, or be the result of diffusion or implantation such as carburizing, nitriding, chromizing or ion implantation. The aim is usually to increase surface hardness and thus reduce cold welding or achieve the same result by rendering the surface nonmetallic. A secondary effect with diffusion or implantation treatments is the development of a compressive stress in the surface which improves the resistance to surface fatigue. Both hardening and compressive stress can also be achieved by surface rolling or shot peening. Most of the quantitative assessment of such treatments have been based on the improvement in fretting-fatigue properties, and evaluation of surface damage has been merely visual. Treatments that the author has found effective when applied to plain carbon steel are sprayed molybdenum (Taylor and Waterhouse, 1972) and a sulfidized coating (Waterhouse and Allery, 1965). Carefully controlled shotpeening also has beneficial effects when applied to carbon steel, but the effect is much more pronounced on an austenitic steel, which has a much greater work-hardening capacity (Waterhouse and Saunders, to be published). Some coatings, e.g., phosphating and hard chromium plating, can have a damaging effect on fatigue properties so caution should be exercised if the fretting is a result of fatigue.

The development of ion implantation provides a new method of wear prevention which has been tested under fretting conditions. Ion-implanted gold films on 0.25% C steel showed much greater resistance to fretting than vacuum evaporated or sputtered gold films. Boron carbide applied by ion implantation gave the best resistance to fretting (Ohmae *et al.,* 1974).

So far there are few available data on the behavior of materials with these surface treatments at high temperatures. Coatings that are nonmetallic and stable at the temperature concerned should still give protection, but work-hardened surfaces are likely to be annealed and surface compressive stresses relieved, the extent depending on the temperature involved and the annealing characteristics of the material.

One method of alleviating fretting damage which is allied to surface treatment is that of introducing a nonmetallic insert between the surfaces which may or may not be bonded to one or both surfaces. This is sometimes used in riveted or bolted joints. A 75-μm-thick layer of terylene completely eliminated fretting damage when placed between contacting surfaces of a high-strength aluminum alloy (Bowers *et al.,* 1967–1968). The nature of the polymer is of importance in this application. Polytetrafluoroethylene produced no fretting wear on mild steel, whereas polythene caused damage at amplitudes of slip greater than 7 μm and polyvinylidenefluoride, polychlorotrifluoroethylene, polysulfone, polyvinylchloride, polymethylmethacrylate, polycarbonate, and nylon 66 all caused wear at amplitudes above 3 μm (Higham *et al.,* 1978b).

Metal plating, phosphating, sulfidising, anodizing, chromizing, and ion implantation are all conducted in baths or enclosures and are therefore unsuitable for applying to large pieces of equipment. The much simpler and more mobile equipment used in metal spraying can be taken to a large installation and used on the spot. The use of nonmetallic inserts is also unrestricted in this sense.

Before closing this section mention should be made of titanium and its alloys, which are particularly susceptible to fretting damage. Anodizing, nitriding, carburizing, chromosiliciding, and plating with electroless nickel have all been applied with some success in this case (Waterhouse and Wharton, 1974).

C. Lubricants

As mentioned above, liquid lubricants are not usually applicable in fretting situations because of the low relative velocities of the surfaces and the high pressures. Improved lubricant feeding has proved satisfactory, however, in a lubricated flexible claw coupling (Rolfe, 1952). Greases are not so easily squeezed out of the contact zone and may provide a remedy if they can be replenished at regular intervals. Shear susceptible greases, i.e., greases whose viscosity falls on being sheared, are the most efficient for machine applications (Herbek and Strohecker, 1953). In locked coil steel ropes the lubricant has to survive for the life of the rope, and the properties required of the grease are quite different. The author has found that a shear-resistant grease with high viscosity and high softening temperature is most suitable (Waterhouse and Taylor, 1971b).

Probably the most suitable lubricant system for the general treatment of fretting is the solid lubricant. Additions of materials such as molybdenum disulfide have not proved particularly successful when added to a grease but have shown considerable promise when bonded to a steel surface (Godfrey and Bisson, 1950). Bonded coatings are analogous to certain of the treatments dealt with in Section V,B.

VI.   Conclusion

Fretting corrosion can arise in any situation in which repeated relative movement occurs between contacting surfaces. The form and extent of damage depends on the chemical nature of the environment and on whether or not the debris can escape. On metals the debris is oxide with possibly some metal content. As initially produced it has a platelike form and seems to arise from a process of delamination, either by the coalescence of subsurface voids or the propagation of fatigue cracks parallel to the surface. Methods of prevention are grouped under three headings: prevention of movement by improved design, protection of the surfaces by coatings, and lubrication. Some of these methods are contradictory and it is necessary to analyze the particular practical situation before opting for a particular solution.

References

Alyab'ev, A. Ya., Kazimirchik, Yu. A., and Onoprienko, V. P. (1970a). *Sov. Mater. Sci.* **6,** 284.

Alyab'ev, A. Ya., Shevelya, V. V., and Rozhkov, M. N. (1970b). *Sov. Mater. Sci.* **6,** 672.

Andrew, J. F., Donovan, P. D., and Stringer, J. (1968). *Br. Corros. J.* **3,** 85.

Aoki, S., Fujiwara, T., and Furukawa, I. (1967). *Bull. Jpn. Soc. Proc. Eng.* **2,** 238.

Bethune, B., and Waterhouse, R. B. (1965). *Wear* **8,** 22.

Bethune, B., and Waterhouse, R. B. (1968). *Wear* **12,** 289.

Bill, R. C. (1974). NASA Tech. Note TN D-7570.

Bowden, F. P., and Tabor, D. (1950). "Friction and Lubrication of Solids." Oxford Univ. Press, London and New York.

Bowers, J. E., Finch, N. J., and Goreham, A. R. (1967–1968). *Proc. Inst. Mech. Eng.* **182,** Pt. L, 703.

Campbell, C. S., and Lewis, M. W. J. (1977). *Proc. BNES Conf. Ferritic Steels for Fast Reactor Steam Generators, London, 30 May –2 June* Paper 55.

Eden, E. M., Rose, W. N., and Cunningham, F. L. (1911). *Proc. Inst. Mech. Eng.* **4,** 839.

Evdokimov, V. D. (1968). *Sov. Mater. Sci.* **4,** 546.

Feng, I. M., and Rightmire, B. G. (1956). *Proc. Inst. Mech. Eng.* **170,** 1055.

Feng, I. M., and Uhlig, H. H. (1954). *J. Appl. Mech.* **21,** 395.

Field, J. E., and Waters, D. M. (1967). N. E. L. Rep. No. 275.

Godfrey, D., and Bisson, E. E. (1950). NACA Tech. Note No. 2180.

Golego, N. L., Alyab'ev, A. Ya., Shevelya, V. V., and Kulagin, N. S. (1971). *Sov. Mater. Sci.* **7,** 54.

Golego, N. L., Alyab'ev, A. Ya., and Shevelya, V. V. (1974). "Fretting Corrosion of Metals." Tekhina, Kiev.

Grunberg, L., and Wright, K. H. R. (1955). *Proc. R. Soc. London Ser. A* **232,** 403.

Halliday, J. S. (1957). *Proc. Conf. Lub. and Wear, London 1958* p. 640. Institution of Mechanical Engineers, London.

Halliday, J. S., and Hirst, W. (1956). *Proc. R. Soc. London Ser. A* **236,** 411.

Higham, P. A., Stott, F. H., and Bethune, B. (1978a). *Corros. Sci.* **18,** 3.

Higham, P. A., Stott, F. H., and Bethune, B. (1978b). *Wear* **47,** 71.

Hurricks, P. L. (1970). *Wear* **15,** 389.

Hurricks, P. L. (1972). *Wear* **19,** 207.

Hurricks, P. L. (1974). *Wear* **27,** 319.

Johnson, K. L. (1955). *Proc. R. Soc. London Ser. A* **230,** 535.

Lewis, M. W. J., and Didsbury, P. B. (1977). Nat. Centre of Tribology, unpublished results.

Markworth, M. (1976). *Aluminium* **52,** 481.

Mindlin, R. D. (1949). *J. Appl. Mech.* **16,** 259.

Ohmae, N., and Tsukizoe, T. (1974). *Wear* **27,** 281.

Ohmae, N., Nakai, T., and Tsukizoe, T. (1974). *Wear* **30,** 299.

Ørbeck, F., (1958). *Proc. Inst. Mech. Eng.* **172,** 953.

Rabinowicz, E. (1961). *J. Appl. Phys.* **32,** 1440.

Rolfe, R. T. (1952). *Allen Eng. Rev.* No. 28, 2.

Rollins, C. T., and Sandorff, P. E. (1972). *J. Aircr.* **9,** 581.

Sasada, T. (1959). *J. Jpn. Soc. Lub. Eng.* **4,** 127.

Sherwin, M. P., Taylor, D. E., and Waterhouse, R. B. (1971). *Corros. Sci.* **11,** 419.

Stott, F. H., Lin, D. S., and Wood, G. C. (1973). *Corros. Sci.* **13,** 449.

Stowers, I. F., and Rabinowicz, E. (1972). *J. Appl. Phys.* **43,** 2485.

Stowers, I. F., and Rabinowicz, E. (1973). *J. Lub. Tech.* **95,** 65.

Suh, N. P. (1973). *Wear* **25,** 111.

Taylor, D. E. (1977). Private communication.

Taylor, D. E., and Waterhouse, R. B. (1972). *Wear* **20,** 401.

Taylor, D. E., and Waterhouse, R. B. (1974). *Corros. Sci.* **14,** 111.

Thiéry, J., and Spinat, R. (1975). AGARD Conf. Proc. No. 161, Fretting in Aircraft Systems, p. 6.1.

Tomlinson, G. A. (1927). *Proc. R. Soc. London Ser. A* **115,** 472.

Toth, L. (1972). *Wear* **20,** 277.

Uhlig, H. H. (1954). *J. Appl. Mech.* **21,** 401.

Vijh, A. K. (1976). *Corros. Sci.* **16,** 541.

Waterhouse, R. B. (1955). *Proc. Inst. Mech. Eng.* **169,** 1157.

Waterhouse, R. B. (1961). *J. Iron Steel Inst. London* **197,** 301.

Waterhouse, R. B. (1972). "Fretting Corrosion." Pergamon, Oxford.

Waterhouse, R. B. (1975a). *Wear* **34,** 301.

Waterhouse, R. B. (1975b). AGARD Conf. Proc. No. 161, Fretting in Aircraft Systems, p. 8.1.

Waterhouse, R. B. (1977a). *Proc. Eur. Congr. Corros., 6th London, September.*

Waterhouse, R. B. (1977b). "Wear of Materials," p. 55. ASME, New York.

Waterhouse, R. B., and Allery, M. (1965). *Wear* **8,** 112.

Waterhouse, R. B., and Allery, M. (1966). *Trans. Am. Soc. Lub. Eng.* **9,** 179.

Waterhouse, R. B., and Taylor, D. E. (1971a). *Wear* **17,** 139.

Waterhouse, R. B., and Taylor, D. E. (1971b). *Lub. Eng.* **27,** 123.

Waterhouse, R. B., and Wharton, M. E. (1974). *Ind. Lub. Tribol.* **25,** 20, 56.

Wright, K. H. R. (1952–1953). *Proc. Inst. Mech. Eng.* **1B,** 556.

Erosion Caused by Impact of Solid Particles

G. P. TILLY†

Transport and Road Research Laboratory
Crowthorne, Berkshire
England

I. Introduction

This chapter is concerned with the type of wear that occurs when solid particles impact against a target material at speed. Solid particle erosion is distinct from rain erosion caused by impact of water drops and exhibits

† Formerly at the National Gas Turbine Establishment, Pyestock, Farnborough, Hampshire, England.

different types of behavior. Erosion is commonly measured as a specific weight loss and expressed as ϵ, the weight of material removed by unit weight of impacting particles. Alternatively, when considering the performance of target materials having different densities, it is more appropriate to use volumetric erosion ϵ_v, which is given by ϵ/ρ. Erosion commonly involves impact velocities of up to 550 m/sec and particle sizes of up to 1000 μm. Larger velocities and particles can cause a different type of damage, commonly termed foreign object damage.

Erosion is a form of wear that occurs in several industries and has been reported for situations as diverse as rocket motor tail nozzles (Neilson and Gilchrist, 1968), helicopter rotors and gas turbine blading (Hibbert, 1965), gas turbine engines operated by pulverized brown coal (Lheude and Atkin, 1963) and by exhaust from supercharged fluid bed boilers, transportation of airborne solids through pipes (Bitter, 1963), boiler tubes exposed to fly ash (Raask, 1968), and fluid cat crackers (Clarke, 1953). An interesting but rare example of erosion occurs for space vehicles landing on Mars, where 44–105μm dust can be blown by winds of up to 60 m/sec. In the 1976 landing the weather was calm, but evidence of erosion on rocks could clearly be seen. Erosion can be very expensive; for example, ingestion of dust clouds can reduce lives of helicopter engines by as much as 90% (Tilly, 1972). For a 1000 shaft horse power (SHP) engine, serious damage can be caused by ingestion of about 6 kg of sand. Local stall can be caused by removal of as little as 0.05 mm of material from leading edges of compressor blades. In pneumatic transportation of material through pipes, the erosive wear at bends can be up to 50 times more than that in straight sections. An extreme example has been reported where steel bends had to be replaced every three or four months in a pneumatic plant carrying wood chips (Lehrke and Nonnen, 1975). In analysis of failures of boiler tubes, it was found that about one-third of all occurrences were due to erosion and total service lives were as low as 16,000 hours (Raask, 1968).

Erosion testing is used for assessment of the performance of paint. In the ASTM Standard D658-44 the time is measured for a controlled jet of 74–88-μm silicon carbide particles to erode through the surface and expose a 2-mm diameter spot of the base material (ASTM, 1965).

II. Types of Erosion Test

There has been a wide variety of tests developed for studying mechanisms and assessing the extent of erosion. These tests can broadly be divided into those that are designed to simulate a specific type of erosion and those that are intended to be used for fundamental studies.

A. Simulative Tests

In the gas turbine industry there have been several investigations in which full-scale tests were made on production engines to determine the general susceptibility to erosion (see Table I). In such tests, sand having a controlled size distribution was fed into the intake and engine performance was measured at predetermined intervals. Some of the tests have been for acceptance and were terminated after ingestion of a predetermined quantity of sand, but in some cases engines failed prematurely or were deliberately taken to failure. In pneumatic transportation, there have been programs of work to determine the erosion of bends in pipe circuits. One of the most significant points to emerge has been that erosion measured as a specific-weight loss only gives a partial measure of the performance of bends. The criterion for life to failure is determined by when a hole is eroded through the pipe wall. However, small particles that produce lower levels of specific erosion ϵ cut a sharper wear profile so that failure occurs when substantially less material has been removed from the bend (Mason and Mills, 1977). Thus conventional use of specific erosion can give an incorrect assessment of pipe bend erosion.

B. Laboratory Tests

There has been a variety of tests developed to study the basic mechanisms of erosion. The main requirements are to have a cheap method of studying a wide range of parameters and to be able to produce erosion under precisely controlled conditions.

The type of test most commonly used involves blasting a stream of airborne particles against the target. A weighed quantity of abrasive particles is fed into an airstream, accelerated through a nozzle, and directed onto the target. The most important parameter to control is velocity, and there have been many investigations in which the acceleration length was inadequate. This is particularly important in cases for which a range of particle sizes is involved because the smaller particles can be accelerated more readily. A major weakness of airblast tests is monitoring of the impact conditions, and it is essential to measure particle speed using an established technique such as high-speed cine photography or streak photography. An ingenious mechanical technique has been developed in

TABLE I

COMPARISONS BETWEEN ESTIMATED AND MEASURED EROSION LOSSES OF GAS TURBINE ENGINES
TESTED UNDER CONTROLLED CONDITIONS

Reference	Engine	Test condition	Erosion			
			Weight loss (mg/g)		Power loss (%)	
			Measured	Estimated	Measured	Estimated
Montgomery and Clark (1962)	4 SHP Solar Mars	0–5-μm quartz	0.65	0		
		0–10-μm quartz	1.3	1.0		
		0–15-μm quartz	1.6	1.4		
		0–43-μm quartz	7.0	4.0		
		0–74-μm quartz	7.9	6.8		
Duke (1968)	Rover 1 S/60 rotor	14-μm quartz at 180 m/sec	0.48	0.45		
		14-μm quartz at 275 m/sec	0.82	0.63		
		14-μm quartz at 370 m/sec	1.12	1.01		
Unpublished Allison report (1969)	T56	0–180-μm quartz			0.15	0.11
		105–210-μm quartz			0.84	0.75
Skinner (1962)	T58	0–40-μm quartz			0.53	0.46
		0–200-μm quartz			1.82	1.41
Bianchini and Koschman (1966)	T63	0–20-μm quartz			0.78	0.55
		0–80-μm quartz			5.15	2.2

which speed is measured using two rotating disks fixed to a common shaft (Ruff and Ives, 1975). A radial slit cut in one disk permits particles to pass through and erode the second disk. Erosion exposures are made with the disks stationary and rotating at a known velocity. A pair of marks eroded on the second disk permits the particle speed to be calculated.

As an alternative to air blasting, low velocities can be produced precisely by dropping particles under gravity down an evacuated tower. This technique has been used for a 5-m height of drop to produce erosion that simulates the damage in pipe bends (Bitter, 1963).

A variety of whirling arm rigs has been developed to enable tests to be made at precisely controlled velocities over a wide range of impact conditions. The target specimens are attached to the tips of a rotor arm and whirled through a curtain or a narrow band of erosive particles. One of the first whirling arm rigs was designed for studies of rain erosion and was operated in air (Fyall and Strain, 1962). However, such rigs are very noisy and consume a lot of power. Smaller rigs have been used for studies of solid particle erosion. These have been operated in evacuated chambers so that very little power is required and noise is minimal. The National Gas Turbine Establishment (NGTE) whirling arm rig incorporates a range of rotor arms 200–920 mm in diameter. The specimens can be whirled at velocities of up to 550 m/sec. Velocities are measured very precisely by using a capacitive pickup which gives a signal to the Y plates of an oscilloscope. The output of a signal generator is applied to the X plates and adjusted to give a Lissajous figure when the two frequencies are matched (Tilly et al., 1969).

C. Single Impacts

In order to study the mechanisms of material removal in detail, it is necessary to examine the behavior of individual particles. There has been a number of investigations of single-impact behavior, all tending to use relatively large particle sizes, for experimental convenience and to enable well-defined high-speed photographs to be taken. In most studies, a gas gun is used to fire a particle against the target. Unfortunately gas guns require spherical or cylindrical projectiles and are difficult to adapt for the irregular particles, such as quartz, that are more relevant to erosion. One solution to this problem is to fire a suitably shaped specimen at a particle suspended on wires (Winter and Hutchins, 1974). In this apparatus, velocity is measured by photoelectric equipment. Single-impact studies have also been conducted using a whirling arm rig in vacuum (Tilly and Sage, 1970). Here, a small number of irregular quartz particles, sieved into the range 500–850 μm, was impacted against a 10-mm square specimen attached to a rotor arm and moving at 180 m/sec. Photographs were taken using two processes: a gas-driven rotating mirror camera and an image converter camera. In the same investigation, 3000-μm glass spheres were fired by a gas gun against a stationary target. Here, particle speed was measured by recording the time interval between the breaking of two wires positioned in the trajectory a fixed distance apart.

High-speed photography is used to study the behavior of impacting particles immediately after the impact process. The damage pits can be examined by optical and stereo scan microscopy to identify the removal mechanisms of material under different impact conditions.

Single impact studies have been successful in establishing the modes of material removal and the role of the erosive particles during and after impact.

D. Correlation of Different Tests

The pros and cons of testing in air and in vacuum have been examined and it has been shown that the presence or absence of air has no effect on erosion (Tilly and Sage, 1970). Comparisons of ten types of laboratory test (Table II) show that there is remarkable consistency in the data. The standard for comparison, derived from the NGTE whirling arm rig, was for 125–150-μm quartz impacted against an 11% chromium steel at 90° and 250 m/sec.

Erosion testing of gas turbines is an essential requirement for civilian aeroengines that may be operated in dusty environments and for all military engines. It has been found that small axial flow engines are more susceptible because blading is physically smaller and removal of very little material can impair the aerodynamic performance. Results for erosion tests on five types of engine are given in Table I. For the 45 SHP Solar Mars and the Rover IS 60 engines, it is possible to compare measured losses from the rotors with estimates derived from laboratory data. It can be seen that the data correlate very well and give confidence to predictions based on laboratory scale work. Very good estimations of the erosion of a 1½-stage compressor were made by Grant and Tabakoff (1974). The predictions were within 20% of observed values for the inlet guide vanes, 10% for the rotor, and 7.5% for the discharge stator.

III. Aerodynamic Effects

It is important to be able to assess the aerodynamic behavior of particles approaching the target because this influences the angle and veloc-

TABLE II

Reference	Test type	Erosion (mg/g)	
		Measured	Predicted
Wood and Espenschade (1964)	Steel eroded by 0–74-μm quartz at 40° and 255 m/sec	3.1	1.7
Kleis (1969)	Steel eroded by 200–300-μm quartz at 90° and 107 m/sec	0.95	1.06
Millington and Spooner Unpublished	Nimonic 90 eroded by 60–120-μm quartz at 90° and 130 m/sec	2.1	1.2
Work reported by Tilly (1972)	Titanium alloy eroded by 60–120-μm quartz at 90° and 130 m/sec	1.0	1.2
Sheldon and Finnie (1966)	Aluminum eroded by 60–130-μm SiC at 20° and 150 m/sec	1.5	2.3
Neilson and Gilchrist (1968)	Aluminum eroded by 210-μm SiC at 20° and 280 m/sec	1.7	2.2
	Glass eroded by 210-μm alumina at 90° and 110 m/sec	40	25
Sheldon (1969)	Glass eroded by 170-μm SiC at 90° and 150 m/sec	55	67
	Aluminum eroded by 330-μm SiC at 20° and 90 m/sec	0.43	1.1
Tadolder (1966)	Iron eroded by quartz at 82 m/sec	0.0052	0.006
Williams and Lou (1974)	Graphite/epoxy eroded by 77–125-μm sand at 70 m/sec	5	5.6
Mills (1977)	Mild steel pipe bend eroded by 70-μm quartz at 26 m/sec	0.011	0.026

ity of impact. In air blast tests the air flow passes around the target, and particles are deflected from the free stream trajectory. Thus, if the target is a flat plate normal to the air blast, particles are deflected so that impacts are at angles of less than 90°. Small particles are deflected most, and some may miss altogether. The number of particles that strikes the target, related to the number initially aimed, is the strike efficiency η_s. The particle trajectories can be obtained using equations suggested by Taylor (1940):

$$m \frac{dV_x}{dt} = \frac{1}{2} \rho (V_x - U_x)^2 C_d A \tag{1}$$

$$m \frac{dV_y}{dt} = \frac{1}{2} \rho (V_y - U_y)^2 C_d A \tag{2}$$

The drag coefficient C_d can be derived from the Reynolds number Re,

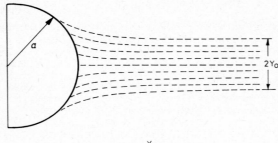

$$\eta_s = \frac{Y_0}{a} 100$$

Fig. 1. Trajectories of 5-μm particles quartz approaching a cylinder at 100 m/sec. (After Tilly, 1969.)

using an expression proposed by Langmuir and Blodgett (1946) and an equivalent value of particle diameter related to sieve diameter d_s:

$$d = 0.75 d_s \tag{3}$$

suggested by Bagnold (1960). Calculations using a digital computer have been made for particles in sizes from 5 to 60 μm (mean sieve diameter) approaching a cylinder, and for flat plates inclined at different angles to the air flow. Typical trajectories are given in Figs. 1 and 2. Values of strike angle, velocity, and efficiency are given in Figs. 3–6. For a free stream velocity of 100 m/sec, quartz particles bigger than 20 μm are not significantly affected by curved airflow. Smaller particles are deflected so that strike conditions are modified. Data for erosion due to 60–125-μm quartz against flat plates can be used to derive the erosion of a cylinder, using the expression

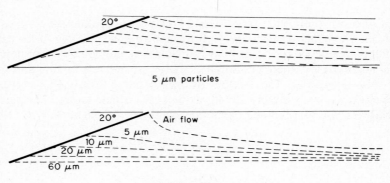

Fig. 2. Particle flow approaching an inclined plate at 100 m/sec. (After Tilly, 1969.)

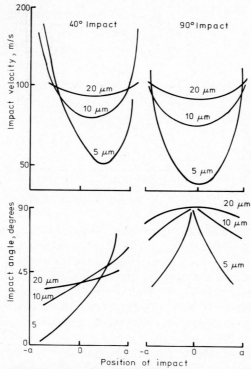

Fig. 3. Influence of airflow on impact velocities and angles against inclined plates at 100 m/sec nominal velocity. (After Tilly, 1969.)

$$\epsilon = \int_0^\pi \frac{f(\alpha)\ \sin\alpha\ a\ d\alpha}{2a} \tag{4}$$

The interaction between particles and airflow is also important in the free stream because this is usually where the particles are accelerated. It is necessary to have an adequate acceleration length, and this is a function of particle size. For a length of 0.6 m it was calculated that 60-μm particles reach 104 m/sec, 90-μm particles reach 102 m/sec, and 120-μm particles reach 90 m/sec with an air speed of 104 m/sec (Tilly, 1969).

IV.   The Impacting Particles

The properties of the impacting particles have received comparatively little attention and there have been few controlled measurements of the

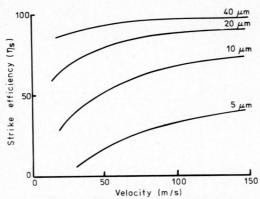

Fig. 4. Influency of velocity on strike efficiency against a cylinder. (After Tilly, 1969.)

influence of the main characteristics. Many investigations have used quartz because it is the main erosive constituent in natural sands and fly ash. However, others have used irregular silicon carbide or alumina particles, partly because they can be obtained in closely defined size ranges and partly because they have a high erosiveness, which gives convenient testing times. Glass and steel spheres have also been used in work where features such as single-impact behavior or comparative erosion are studied.

It is generally considered that erosiveness is a function of hardness of the impacting particles. Hardness is commonly expressed as a Mohs number obtained by scratch testing. It can also be measured by the

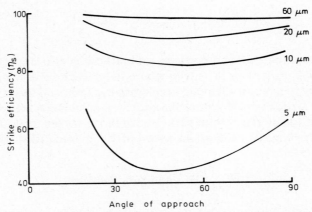

Fig. 5. Variation of strike efficiency with angle of approach for flat plates at 100 m/sec nominal velocity. (After Tilly, 1969.)

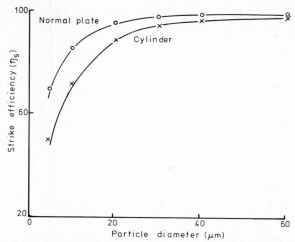

Fig. 6. Variation of strike efficiency with particle size for 100 m/sec nominal velocity. (After Tilly, 1969.)

Vickers microhardness test in which a diamond pyramid is pressed into sectioned samples of the particles and the size of indentation measured. The erosiveness has been measured for a number of natural and industrial materials having hardnesses varying over an order of magnitude. The test was defined by using a close size range 125–150 μm and a velocity of 130 m/sec. The resulting erosion (Fig. 7) is related to hardness by the expression

$$\epsilon = \text{const} \cdot H^{2.3} \qquad (5)$$

For naturally occurring materials the erosiveness increases with quartz content and particle size (Tilly *et al.*, 1969). It is not necessary for the particles to be harder than the target material to cause erosion. Head *et al.* (1973) showed that fluorite (Mohs hardness 4.0) caused more erosion than alumina (Mohs hardness 9.0). In addition to hardness, it is necessary to consider shape. Clearly, particles having sharp corners are potentially more erosive than those that are rounded or spherical. In tests on mild steel, using relatively low velocities, 100-μm quartz was found to be ten times more erosive than glass spheres of the same size (Raask, 1968). However, such tests are complicated by the fact that glass is usually much softer than natural quartz and the relative behavior differs at glancing and normal angles of impact (Tilly and Sage, 1970). Shape is influenced by hardness so that soft particles tend to be more rounded. It is not possible

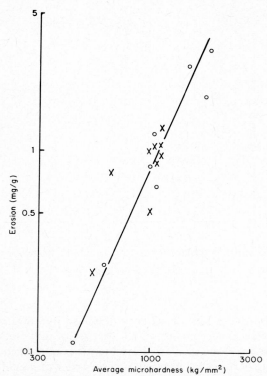

Fig. 7. Effect of particle hardness on erosion. 125–150-μm particles against an 11% chromium steel at 130 m/sec and 90° impact angle. ○ industrial material; × natural sand. (After Tilly *et al.,* 1969).

to give precise relationships, particularly for quartz, which may be relatively rounded if it is wind blown or very sharp if freshly crushed. Other potentially important features that have not been studied in detail include particle density, friability, ductility, and elasticity.

For naturally occurring materials, particle size is very important because it determines the liability to become airborne as well as strongly influencing intrinsic erosiveness. Particles smaller than 70 μm are defined as dust and can be blown to heights of up to 20 m. Increase in particle size can increase the erosiveness but in a complicated manner. There is an aerodynamic effect, mentioned in Section III, because particles smaller than 20 μm may be deflected around the target without impacting. Thus interaction with the airflow can falsify studies of the effects of particle size. There is an intrinsic size effect which varies with the target material. Erosion of engineering alloys and resilient plastics initially increases with particle size until the onset of a saturation level $\hat{\epsilon}$ where it is independent of size (see Fig. 8). The relationship is of the form

$$\epsilon = \hat{\epsilon}[1 - (d_0/d)^{3/2}]^2 \tag{6a}$$

The onset of saturation is itself dependent on velocity. Erosion of a brittle material such as glass exhibits a power law dependence on particle size:

$$\epsilon = \text{const } d^m \tag{6b}$$

For tests on annealed glass, values of $m = 2$ have been obtained (Sage and Tilly, 1969). Values of 3.14–5.12 have been reported by Sheldon and Finnie (1966) for a range of brittle materials eroded by different sizes of silicon carbide and steel spheres. Erosion of resins and composite materials such as fiber glass exhibits a logarithmic relationship:

$$\epsilon = \text{const} \cdot \log d \tag{6c}$$

Examples of the three types of erosion are given in Fig. 9.

V. The Impact Parameters

The primary impact parameters: velocity, angle of attack, and duration of exposure, strongly influence the erosion process and comparatively small changes can produce tenfold variations in damage. The secondary

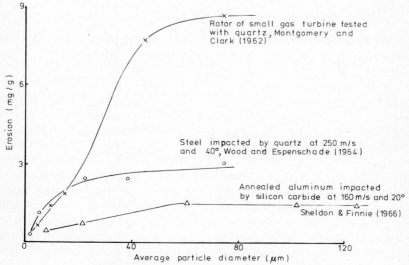

Fig. 8. Effect of particle size on erosion of metals.

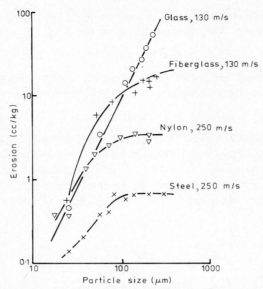

Fig. 9. Effect of particle size on erosion of ductile and brittle materials. (After Tilly and Sage, 1970.)

parameters: particle concentration, temperature, and tensile stress, have comparatively little effect.

A. Velocity

There have been numerous investigations of the effects of velocity, all of which have shown that there is a relationship of the form

$$\epsilon = \text{const} \cdot V^n \tag{7}$$

In early work on an SAE 1020 steel, Finnie (1960) reported a value of 2 for the exponent n. However, he and his co-workers subsequently reported work on different materials (Finnie et al., 1967) which gave values of n between 2.05 and 2.44. Other workers have reported values of 2.0–2.5, and the consensus of opinion is that erosion of ductile materials involves an exponent of 2.3–2.4. Higher values have been reported for brittle materials; Finnie (1960) obtained 6.5 for 575-μm steel spheres against glass. This is in contrast to a value of 2.3 reported by Tilly et al. (1969) for 125–150-μm quartz against glass and 2.4 by Behrendt (1970) for 25–200-μm quartz also against glass. Typical curves for effects of velocity, with two types of rig and eight materials, are given in Fig. 10.

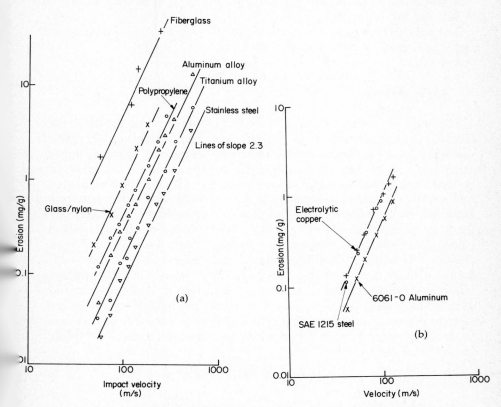

Fig. 10. Effects of velocity on erosion. (a) 125–150 μm quartz at 90°. (After Tilly *et al.*, 1969). (b) 0–250 μm silicon carbide at 20°. (After Sheldon, 1969.)

Most studies have been at relatively high velocities because they have been associated with applications such as aircraft, which involve values in excess of 250 m/sec. In contrast, transport of airborne materials through pipes involves velocities of around 25 m/sec and there is comparatively little information about this range. In one of the few relevant investigations, Bitter (1963) reported tests using 300-μm iron spheres against glass at 9.9 m/sec and postulated that there is a threshold value below which deformation is elastic and no damage occurs. A threshold velocity of 0.67 m/sec was calculated for iron spheres against a low-carbon steel. In a later investigation, Tilly (1973) suggested that the threshold velocity is also dependent on particle size and derived a value of 2.7 m/sec for 225-μm quartz against an 11% chromium steel. These low values are in contrast to rain erosion for which high thresholds have been estimated (Fyall and Strain, 1962).

The velocity exponent is also influenced by particle size, and for quartz tends to a value of 2.0 for small sizes (Tilly *et al.*, 1969).

B. Angle of Impact

The angle of impact is defined as the inclination between the target surface and the particle trajectory. The effect of angle depends on the type of material; ductile materials exhibit maximum damage for glancing impact at an angle of 20–30°, whereas brittle materials exhibit a maximum for normal impact; typical curves are given in Fig. 11. The effect of angle is interrelated with other parameters and it has been shown that erosion of glass, although brittle for large particles, exhibits ductile characteristics for 9-μm particles at 152 m/sec (Fig. 12). Typically brittle materials include glass, silicon nitride, silicon carbide, hard coatings, and some plastics such as fiber glass. Most metals and resilient plastics behave in a ductile manner. Very soft materials can exhibit deposition for 90° impacts but this can vary with duration of test (see Section V,C). Tough materials such as alloy steels are predominantly ductile but exhibit significant erosion for normal impacts, which can be one-third to one-half the maximum value for glancing impact. Erosion can be expressed in terms of impact angle by the equation

$$\epsilon = A'\cos^2 \alpha \sin n\alpha + B'\sin^2 \alpha \qquad (8)$$

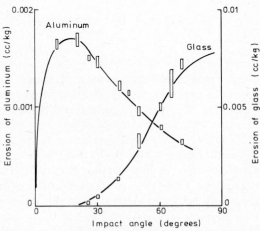

Fig. 11.   Effect of impact angle for 300-μm iron spheres at 10 m/sec. (After Bitter, 1963.)

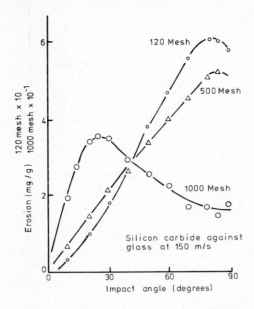

Fig. 12.   Effects of particle size and impact angle. (After Sheldon and Finnie, 1966.)

If the material is fully brittle, $A' = 0$. If fully ductile, $B' = 0$ and $n = \pi/2\alpha$ for $\alpha > \alpha_0$. Many materials exhibit a combination of brittle and ductile erosion so that the ductile term predominates at glancing angles and the brittle term at high angles.

C.   Duration of Exposure

In rain erosion, materials usually exhibit an incubation phase during which there is comparatively little damage, followed by accelerating erosion until the process stabilizes. Erosion caused by solid particles only involves an incubation phase for soft and ductile materials such as annealed light alloys and resilient plastics. The common engineering materials, including steels, do not have a discernible incubation phase and the process stabilizes immediately. The incubation can be caused by surface roughening and material work hardening, which changes the erosion rate, or by deposition, which increases until the surface becomes saturated and degraded. Materials commonly exhibit increasing deposition at higher angles of impact, as shown for aluminum (Fig. 13). The initial state of the material can be restored by light application of abrasive paper to remove the surface layer.

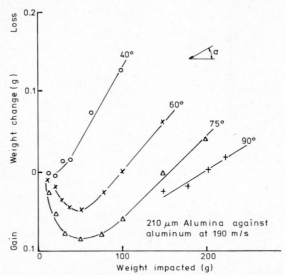

Fig. 13. Variation of weight change with duration of erosion for different impact angles. (After Neilson and Gilchrist, 1968.)

D. Particle Concentration

There have been several investigations of the influence of particle concentration, including tests on pipe bends (Mills, 1977) and gas turbines (Montgomery and Clarke, 1962). The consensus of opinion is that erosion reduces with increased concentration, a typical reduction being 50% for a fortyfold increase in concentration (Fig. 14). In applications such as aero engine intakes, it is convenient to express concentration as mass of dust per unit volume of air. In pneumatic transportation it is more relevant to use phase density, the ratio of mass flow rate of material conveyed to mass flow rate of air.

While pipe tests confirm that there is a small reduction in specific erosion with increased phase density, the shape of the eroded channel changes so that less material is removed when failure occurs. Thus at higher concentrations failure occurs sooner than would be predicted from specific erosion, assuming constant weight loss at failure (Mills, 1977).

E. Temperature

The effects of temperature have been examined by several investigators. With one exception (Neilson and Gilchrist, 1968) they have

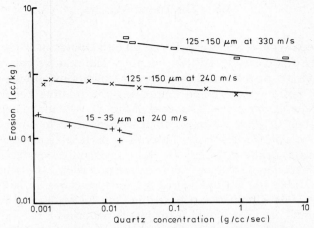

Fig. 14. Effect of concentration on erosion. (After Tilly and Sage, 1970.)

shown that erosion increases with temperature. Various theories have been advanced to account for the effect of temperature. Wellinger and Uetz (1957) suggested that the increased resistance is caused by enhancement in the development of a protective oxide layer which forms again and again. Tilly (1969) suggested that erosion is proportional to the fracture ductility for four engineering alloys. The effect of temperature, however, is complicated by the fact that the erosion process itself involves a substantial rise in temperature; Neilson and Gilchrist (1968) measured an average temperature rise of 170°C at a point 0.4 mm from the eroded surface. Jennings *et al.* (1976) have reported metallographic evidence of local melting. Tilly (1969) found that for tests on nylon and polypropylene in which deposition occurred, the surface appearance blackened and there was charred material due to high local temperatures.

F. Tensile Stress

In applications such as erosion of helicopter rotor blades and aero engines, the target material is subjected to centrifugal stresses of up to 200 N/mm². Simplified tests have been conducted in which a plate was loaded in tension and exposed to erosion by 60–125-μm quartz at 100 m/sec. The resulting data showed that there is a small decrease in erosion with increased stress, but the difference barely exceeds normal experimental scatter and is not considered to be significant (Tilly, 1969).

VI. The Target Material

It has been seen in Section V,B that material behavior is characterized by whether it is basically brittle or ductile. The relative behavior of different materials also depends on impact conditions, principally strike angle and particle size. When comparing the relative performance of brittle and ductile materials, as shown in Fig. 11, a brittle material may be more resistant to glancing impacts but less resistant to normal impacts. There can be a similar crossover in terms of particle size; for erosion at a glancing angle, brittle materials are more resistant to small particles but less resistant to large particles. The erosion of a wide variety of materials is given in Table III, but the figures are for impact conditions that favor ductile materials. Comparisons based on data for a glancing angle of impact and small particles would favor brittle materials.

In studies of the types of material commonly used for compressor blading in aero gas turbine engines, it has been found that as a "rule of thumb" it can be assumed that the erosion expressed as specific weight loss ϵ is the same for these materials. This simplification includes steels and the common types of aluminum, titanium, and nickel alloy. The figure that can be used for reference purposes is 4.7 mg/g erosion for impact by

TABLE III

EROSION VALUES OF COMMON MATERIALS TESTED WITH
125–150-μm QUARTZ AT 250 m/SEC AND 90°

Material	Diamond pyramid hardness (kg/mm²)	Erosion (cm³/kg)
Hot pressed silicon nitride	1800	0.2
11% chromium steel	355	0.6
Forged nickel alloy	207	0.6
0.05-mm plated tungsten carbide	1050	0.85
Titanium alloy, sheet	340	1.0
Aluminum alloy, sheet	155	2.0
Cast cobalt alloy	600	2.2
Type 66 nylon	—	2.3
Polypropylene	—	3.1
30% glass-reinforced nylon	—	7.0
25% carbon-reinforced nylon	—	7.0
Extruded magnesium alloy	37	13.0
70% glass-reinforced epoxy	—	53.0
Annealed soda glass	450	100

125–150-μm quartz at 90° and 250 m/sec. For these materials, erosion can be calculated for different particles and velocities using the equation

$$\epsilon = 1.4 \times 10^{-12} H^{2.3} V^{2.3} \quad \text{mg/g} \tag{9}$$

from Eqs. (5) and (7), where H is kg/mm^2 and V is m/sec. (It is fortuitous that hardness and velocity have the same exponent.) Effects of impact angle and particle size can be handled by equations such as (6) and (8). When expressed as specific volume loss to give a more relevant form of erosion, the relative performance of the lighter alloys is less satisfactory. This can sometimes be used to advantage if heavier materials can be tolerated; for example, titanium has successfully been substituted for aluminum in compressor blading for a civilian version of the Rolls Royce Spey engine. This gave a 100% improvement in erosion resistance for a total increase in weight of 8.3 kg for the first three stages of blading.

Considerable attention has been given to searches for materials having improved erosion resistance but without finding anything that is wholly acceptable in terms of other criteria such as resistance to large objects, fatigue, or long-term stability. Cobalt alloys have reportedly given good performance in rain erosion but proved to be less satisfactory than chromium steel or nickel alloys in solid particle erosion. One of the best available flame-sprayed tungsten carbine coatings was found to improve the resistance of chromium steel to glancing impacts but was very poor for 90° impacts (Fig. 15a). In most engineering applications an increase in the thickness of the base material is as effective, possibly more so, than an application of a coating. It is also substantially cheaper. However, protection of more susceptible materials is a viable proposition and some good systems have been developed. In Fig. 15b the behavior of a nickel coating on fiber glass is shown in comparison with the parent material and it can be seen that a very substantial improvement has been achieved. This erosion resistance is similar to that of aluminum alloys but inferior to chromium steel.

There have been several studies of the influence of hardness on erosion and there is some evidence to show that there is a relationship (Finnie *et al.*, 1967). However, heat treatment of two steels to increase hardness by factors of up to 4, produced no significant change in erosion (Fig. 16).

VII. Mechanisms of Erosion

Several theories have been developed in attempts to explain erosion phenomena in relation to mechanisms of material removal. Although

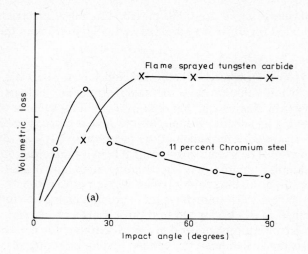

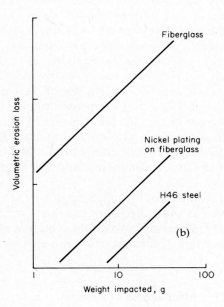

Fig. 15. Comparative behavior of parent materials and coatings. (a) Hand coating on steel 60–125 μm quartz at 100 m/s, (b) ductile coating on fiberglass 125–150 μm quartz at 130 m/s.

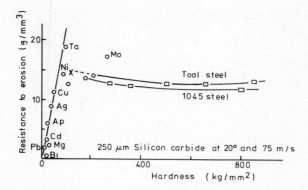

Fig. 16.   Effect of hardness for different metals. (After Finnie *et al.*, 1967.)

these differ in detail, there is general agreement that brittle and ductile behaviors are due to fundamentally different mechanisms. In the following sections some of the main theories are briefly summarized.

A.   *Brittle Erosion*

Materials such as glass fracture with little or no deformation. Tillet (1959) showed that impacts by steel spheres cause ring cracks having radii greater than the area of contact. Finnie (1960) analyzed the behavior in terms of the Hertzian stresses developed in the impact process, which have maximum tensile values at the surface and in radial directions, i.e.,

$$\sigma_R = 0.187(1 - 2\gamma_2)\rho^{1/5}V^{1/5}\left(\frac{1 - \gamma_1^2}{E_1} + \frac{1 - \gamma_2^2}{E_2}\right)^{-4/5} \tag{10}$$

Finnie suggested that in the early stages of brittle erosion each impact produces a ring crack which flares out beneath the surface. A typical brittle crack in glass is shown in Fig. 17. With further impacts, the cracks interact and fragments of material are removed. Variations of the strike angle produce Hertzian crack diameters that are proportional to the vertical component of strike velocity.

Bitter (1963) argued that damage can occur only if the stresses exceed the elastic limit. Repeated loads cause work hardening and raise the elastic limit until the stage is reached in which the stresses exceed the strength of the material and cracking occurs. The erosion was termed deformation wear, given by

$$W_D = M(V \sin \alpha - K)^2/2e \tag{11}$$

_10 μm_

Fig. 17.   Brittle crack caused by 150-μm quartz particle against glass at 100 m/sec and 90°
to the surface. (After Tilly, 1969.)

Adler (1976) considered the material removal mechanisms for glass beads impacted against glass plates at 61–130 m/sec. It was confirmed that the erosion is due to the production and interaction of ring fractures.

B. Ductile Erosion

1. Cutting Wear

In contrast to brittle behavior, erosion of truly ductile materials is low for 90° impacts, and is a maximum at glancing strike angles of about 20°. Finnie (1960) proposed that ductile erosion is caused by sharp particles removing chips by an action similar to a tool cutting process. A stereo scan photograph showing this type of chip partially removed is given in Fig. 18. It was assumed that the particle remains intact and that the ratio of forces parallel and perpendicular to the target face remains constant. Two types of cutting wear were postulated:

(i) For glancing angles of impact, the particle leaves the surface when its perpendicular velocity is zero:

$$W_1 = \frac{mV^2}{p\psi K} \left(\sin 2\alpha - \frac{6}{K} \sin^2 \alpha \right) \qquad \text{for} \quad \tan \alpha \le \frac{K}{6} \qquad (12)$$

50 μm

Fig. 18.   Metal chip partly removed by cutting erosion.

(ii) For angles of impact exceeding α_0, the cutting action stops when the kinetic energy of the particle is zero:

$$W_2 = \frac{mV^2 \cos^2 \alpha}{6p\psi} \tag{13}$$

It can be shown that Finnie's equations correlate with ductile erosion for glancing angles but predict zero loss for 90° impacts whereas there is invariably substantial erosion.

Bitter (1963) also postulated that ductile erosion is caused by a cutting action but considered the energy balance for the process and produced the equations

$$W_{c1} = \frac{2MC(V \sin \alpha - K)^2}{(V \sin \alpha)^{1/2}} \left(V \sin \alpha - \frac{C(V \sin \alpha - K)^2 q}{(V \sin \alpha)^{1/2}} \right)$$

$$\text{for} \quad \alpha \leq \alpha_0 \tag{14}$$

$$W_{c2} = \frac{M}{2q} [V^2 \cos^2 \alpha - K_1(V \sin \alpha - K)^{3/2}] \quad \text{for} \quad \alpha \geq \alpha_0 \tag{15}$$

where α_0 is the angle at which $W_{c1} = W_{c2}$. Bitter extended the analysis to include brittle and ductile terms for the same material:

$$W_T = W_c + W_D \tag{16}$$

Thus, the erosion for 90° impacts not accounted in Finnie's theory is said to be due to a brittle component. The equations for cutting wear are very similar to those of Finnie; if K_1 is neglected and $3p\psi$ made equal to q, W_2 and W_{c2} are identical.

Neilson and Gilchrist (1968) simplified Bitter's approach but retained the same basic treatment to produce the equations

$$W_1 = \frac{M(V^2 \cos^2 \alpha - v_p^2)}{2\phi} + \frac{M(V \sin \alpha - K)^2}{2\beta} \quad \text{for} \quad \alpha \leq \alpha_0 \tag{17}$$

$$W_2 = \frac{MV^2 \cos^2 \alpha}{2\phi} + \frac{M(V \sin \alpha - K)^2}{2\beta} \quad \text{for} \quad \alpha \geq \alpha_0 \tag{18}$$

where the term involving ϕ is ductile erosion and that involving β is brittle erosion, and α_0 is the angle at which v_p is zero. Equations (17) and (18) have the advantage that they can very simply be adjusted to describe differing behavior and form the basis of the simplified equation (8).

These theories are based on models of brittle cracking and ductile cutting. The resulting equations correlate with erosion in terms of impact angle but require empirical calibration. They incorrectly predict that erosion depends on the square of the velocity. Metallographic evidence confirms that ductile cutting and brittle cracking can occur but shows that there are other modes of material removal. This has led to development of models based on extrusion of lips of material followed by damage caused by fragments of the particles scouring across the surface.

2. EXTRUSION AND FRAGMENTATION

In studies of single impacts, it has been noticed that in most cases material is extruded to the edges of the damage scar to form "lips" that are vulnerable to subsequent impacts (Fig. 19). Also, hard particles such as quartz and silicon carbide invariably break up and the fragments scour across the surface in a manner similar to the "splat" of a water drop (Fig. 20). Evidence of fragmentation was obtained by high-speed photography (Fig. 21) and by measurement of particle sizes before and after impact (Fig. 22). These observations led to the suggestion of the two-stage mechanism of ductile erosion by Tilly (1973). In this theory it is accepted that other processes take place, such as local melting and removal of chips by a machining action. Nevertheless, a major role is played by extrusion of target material and fragmentation of the particles. The process involves threshold values of velocity and particle size below which damage cannot occur. In the primary stage, the impacting particles produce indentations with extruded lips that may be removed if the impact conditions are favorable. The primary erosion is given by

50 μm

Fig. 19.   Ductile scar caused by a glancing impact.

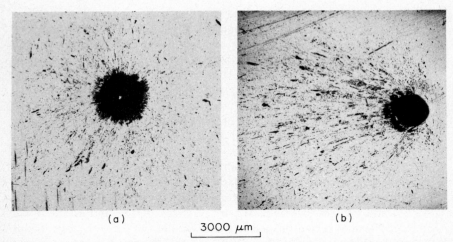

(a)   (b)

|_____ 3000 μm _____|

Fig. 20. Erosion scars caused by single 3000-μm glass spheres at 270 m/sec. (a) Normal impact at 90°; (b) Glancing impact at 30°.

$$\epsilon_1 = \hat{\epsilon}_1 \left(\frac{V}{V_r}\right)^2 \left[1 - \left(\frac{d_0}{d}\right)^{3/2} \frac{V_0}{V}\right]^2 \tag{19}$$

where $\hat{\epsilon}_1$ is the maximum or plateau value of first-stage erosion. In the secondary stage, fragments of the particles scour across the surface causing damage given by

$$\epsilon_2 = \hat{\epsilon}_2 (V/V_r)^2 F_{d,v} \tag{20}$$

where V_r is a reference velocity. Total erosion, given by $\epsilon_1 + \epsilon_2$, has been correlated with experimental data and shown to give the correct expo-

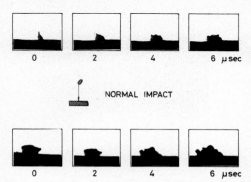

Fig. 21. High-speed photographs of particle fragmentation during impact of 500–850-μm quartz against steel at 180 m/sec.

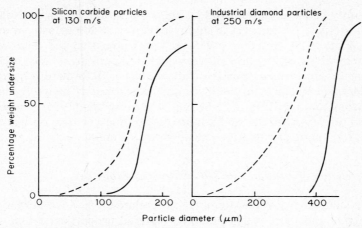

Fig. 22.   Particle size distributions before (————) and after (— — — —) impact at 90° impact angle. (After Tilly, 1973.)

nents for velocity and to fit the shapes of curves for different sizes (Fig. 23).

In a series of carefully planned impact experiments, Hutchins *et al.* (1975) showed that the extruded lips may be removed if impact exceeds a critical velocity. For impacts at 30° to the surface, the material removed by steel spheres, expressed nondimensionally in relation to mass impacted, is given by $5.82 \times 10^{-10} V^{2.9}$. This correlates remarkably well with data produced by Sheldon and Kanhere (1972), giving $6.2 \times 10^{-10} V^{2.9}$ and by Kleis (1969), giving $6.6 \times 10^{-10} V^{2.9}$. A distinction is drawn between

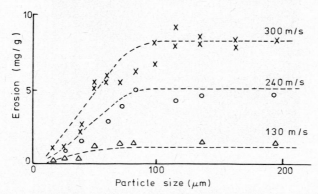

Fig. 23.   Comparisons of estimated and experimental data for different particle sizes and velocities. Quartz at 90° against 11% chromium steel. — — — — curves derived from Eqs. (19) and (20). (After Tilly, 1973.)

the quantity of material removed as opposed to that which is displaced; the fraction is $\frac{1}{12}$ for impacts at 50 m/sec and $\frac{1}{4}$ at 400 m/sec. In a later paper, Winter and Hutchins (1975) demonstrated that the deformation process involves production of very high temperatures, and adiabatic shear bands are formed. Bands of intense shear deformation have also been identified by Harrison (1970) for impact of irregular quartz against an 11% chromium steel (Fig. 24). Winter and Hutchins (1974) studied the behavior of glass particles and confirmed that they shatter on impact so that the fragments can remove extruded lips of metal.

C. Correlation of Erosion with Other Material Properties

There have been several attempts to correlate erosion with other readily available and more easily obtained material properties. Finnie *et al.* (1967) claimed some success in relating erosion to hardness, but there are a number exceptions such as the behavior of the relatively hard cobalt

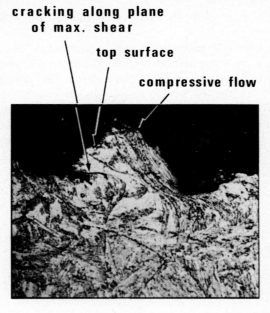

**cracking along plane
of max. shear**

top surface

compressive flow

50 μm

Fig. 24. Section through erosion scar showing intense shear deformation of an 11% chromium steel.

alloys, which have lower erosion resistance than several much softer materials such as the nickel-based alloys (see Table III). Furthermore, heat treatment can change the hardness but has no effect on erosion (Fig. 16). Tilly (1969) suggested that for a given material it is necessary to consider the ductile and brittle components separately. In this way, erosion of the common types of engineering alloys was correlated with ductility at fracture. Tuitt (1974) had some success correlating erosion with the elastic modulus of certain alloys. Ascarelli (1971) used a term which he called "thermal pressure." This was correlated with erosion data published by Finnie *et al.* (1967). Hutchins (1975) examined Ascarelli's correlation and showed that further improvement can be obtained by correlating volumetric erosion of a wide range of metals with the product $C\rho$ ΔT. It was argued that the factors that influence thermal softening are a low heat capacity per unit volume $C\rho$. A high rate of thermal softening is favored by a low value of ΔT, the difference between the temperature of the metal and its melting point. This approach gives a very good correlation with erosion and is based on reasonable assumptions of the mechanisms involved. However, it must be noted that although it refers to erosion in general, it is only relevant to ductile materials.

List of Symbols

ϵ	erosion	a	target dimension
$\hat{\epsilon}$	maximum or plateau erosion	H	particle hardness
ρ	density	σ	stress
V	particle velocity	E	Young's modulus
v_p	residual parallel component of velocity	C, K, K_1	A', B' constants
U	air velocity	e	deformation wear factor
m	mass of a particle	$q\phi$	cutting wear factor
M	total mass of impacting particles	W	erosion caused by particles of total mass M
C_d	drag coefficient	p	plastic flow stress
Re	Reynolds number	ψ	constant, ratio of depth of contact to depth of cut
A	particle cross sectional area perpendicular to air flow	F	degree of particle fragmentation
η_s	strike efficiency	γ	Poisson's ratio
d	particle diameter	C	specific heat
d_0, V_0	threshold values of particle diameter and velocity	ΔT	difference between metal temperature and melting point
α	impact angle		

References

Adler, W. F. (1976). *Wear* **37** (2), 353–364.
Ascarelli, P. (1971). U. S. Army Mater. and Mech. Res. Centre Tech. Rep. 71–47.

ASTM (1965). Standard Method of Test Designation: D658-44.

Bagnold, R. A. (1960). "The Physics of Wind Blown Sands and Desert Dunes." Methuen, London.

Behrendt, A. (1970). *Proc. Int. Conf. Rain Eros. Assoc. Phenomena, 3rd, RAE, England* **II,** 797–820.

Bianchini, G. V., and Koschman, R. B. (1966). *Proc Conf Environ. Aircraft Propulsion Syst.* Paper 66-ENV-20.

Bitter, J. G. A. (1963). *Wear* **6,** 5–21, 169–190.

Clarke, J. S. (1953). *Oil and Gas J.* 262–276.

Duke, G. A. (1968). Note No. ARL/ME 297, Aero Res Labs Melbourne, Australia.

Finnie, I. (1960). *Wear* **3,** 87–103.

Finnie, I., Wolak, J., and Kabil, Y. H. (1967). *ASTM J. Mater.* **2,** 682–700.

Fyall, A. A., and Strain, R. N. C. (1962). *J. Roy. Aero. Soc.* **66,** 447–453.

Grant, G., and Tabakoff, W. (1974). *AIAA Aero Sci. Meeting, 12th, Washington,* paper 74–16.

Harrison, G. F. (1970). Unpublished work reported by Tilly (1972).

Head, W. J., Lineback, L. D., and Manning, C. R. (1973). *Wear* **23,** 291–298.

Hibbert, W. A. (1965). *J. Roy. Aero. Soc.* **69,** 769–776.

Hutchins, I. M. (1975). *Wear* **35,** 371–374.

Hutchins, I. M., Winter, R. E., and Field, J. E. (1975). *Proc. Roy. Soc. London* **A348** 379–392.

Jennings, W. H., Head, W. J., and Manning, C. R. (1976). *Wear* **40,** 93–112.

Kleis, I. (1969). *Wear* **13,** 199–215.

Langmuir, I., and Blodgett, K. (1946) U. S. AAF Tech. Rep. 5418.

Lehrke, W. D., and Nonnen, F. A. (1975). *Int. Conf. Protect. of Pipes, 1st, BHRA, Durham, England,* Paper G2.

Lheude, E. P., and Atkin, M. L. (1963). Rep. ARL/ME III, Dept. Supply, Aust. Defense Sci. Serv.

Mason, J. S., and Mills, D. (1977). *Int. Powder and Bulk Solids Handling and Proc. Conf., 2nd, Chicago, Illinois.*

Mills, D. (1977). The Erosion of Pipe Bends by Pneumatically Conveyed Suspension of Sand, PhD Thesis, Thames Polytechnic, London.

Montgomery, J. E., and Clarke, J. M. (1962). Paper 538A, SAE Summer Meeting, New York.

Neilson, J. H., and Gilchrist, A. (1968). *Wear* **11,** 111–122, 123–143.

Raask, E. (1968). *Wear* **13,** 301–313.

Ruff, A. W., and Ives, L. K. (1975). *Wear* **35,** 195–199.

Sage, W., and Tilly, G. P. (1969). *Aero. J.* **73,** 427–428.

Sheldon, G. L. (1969). *Trans. ASME* **92D,** 619–628.

Sheldon, G. L., and Finnie, I. (1966). *Trans. ASME* **88B,** 387–392.

Sheldon, G. L., and Kanhere, A. (1972). *Wear* **21,** 195–209.

Skinner, R. M. (1962). General Electric Rep., Schenectady, New York.

Tadolder, Y. A. (1966). *Tr. Tallin. Polytech. Inst.* **A237,** 15.

Taylor, G. I. (1940). R and M 2024, ARC Tech. Rep.., HMSO, London.

Tillet, J. P. A. (1959). *Proc. Phys. Soc.* **69B,** 47–54.

Tilly, G. P. (1969). *Wear* **14,** 63–79.

Tilly, G. P. (1972). *Congr. ICAS 8th,* Amsterdam.

Tilly, G. P. (1973). *Wear* **23,** 87–96.

Tilly, G. P., and Sage, W. (1970). *Wear* **16,** 447–465.

Tilly, G. P., Goodwin, J. E., and Sage, W. (1969). *Proc. Inst. Mech. Eng.* **184** (15), 279–292.

Tuitt, D. (1974). *Proc. Int. Conf. Rain Eros. 4th, Meersburg.*

Wellinger, K., and Uetz, H. (1957). *Wear* **1**, 225–231.

Williams, J. H., and Lou, E. K. (1974). *Wear* **29**, 219–230.

Winter, R. E., and Hutchins, I. M. (1974). *Wear* **29**, 181–194.

Winter, R. E., and Hutchins, I. M. (1975). *Wear* **34**, 141–148.

Wood, C. D., and Espenschade, P. W. (1964). *SAE Summer Meeting, New York.* Paper 880A.

Rolling Contact Fatigue

D. SCOTT†

National Engineering Laboratory
East Kilbride, Glasgow
Scotland

I. Introduction

The useful life of lubricated rolling elements such as rolling bearings, gears, and cams is limited by surface disintegration, pits being formed by a fatigue process. The phenomenon is characterized by the sudden removal of surface material and, in some instances, catastrophic fracture due to repeated alternating stresses. The process has three phases: pre-

† Present address: Paisley College of Technology, Paisley, Scotland.

conditioning of the material prior to cracking, crack initiation, and crack propagation.

Theoretically, anelastic rolling elements may be considered to have point or line contact, but in practice a certain degree of elasticity is always present, and even the hardest materials will deform until the contact area is sufficient to support the load. The deformation caused by pressure, combined with rolling movement, results in material pileup ahead of the rolling contact. This departure from the theoretical idea by continuous distortion and relaxation of surface material requires energy and manifests itself as a source of rolling friction.

It is now generally considered that fatigue crack initiation requires plastic deformation, which may occur only on a submicroscopic scale. In rolling elements plastic deformation occurs because the surfaces are not perfect. The high spots or small protuberances on even the smoothest surfaces are the first points on the contacting surface to be subjected to pressure, and they are deformed by very small loads. This deformation has little effect on the functioning of the rolling elements and can usually be observed only by the difference in optical reflectivity of a bearing track or contact area, but the plastic work is of importance in the onset of surface fatigue.

To satisfy modern engineering requirements a high degree of reliability is demanded from rolling mechanisms, and in some specialized applications such as aircraft engines, atomic reactors, and space vehicles the occasional premature failure and scatter in life causes considerable concern. The great diversity in the rolling contact fatigue life of apparently identical rolling elements even when run under closely controlled conditions may be compared with that in human life and electric light bulb life (Heindlhofer and Sjovall, 1923). Load is the dominant factor determining the life of rolling elements. The incidence of failure, however, has been found to be influenced by many factors besides load, the most important being the properties of the material, the nature of the lubricant, and the environment (Scott, 1964).

II. Theoretical Considerations

When solid elastic bodies roll together, contact takes place over a finite area defined by the classical theory of Hertz (1881) (see Timoshenko and Goodier, 1970), who investigated mathematically the distribution of stresses over the contact area and the shape and size of the pressure area. Although his theoretical expressions are subject to assumptions not strictly applicable to rolling bearings, much subsequent work (Stribeck,

1901; Goodman, 1912; Allan, 1946) has shown that they may be accepted as a basis for research and as accurate enough for practical purposes. The theories of Reynolds (1876) and Heathcote (1921) expressed the origin of rolling friction as being the forces required to overcome interfacial slip. It is generally recognized that the friction at the interface results in regions of microslip and regions of adhesion within which the surfaces roll without relative motion (Johnson, 1960). While Palmgren (1959) confirmed that loss of energy due to interfacial slip contributes a large portion of rolling resistance, work by Tabor (1956) indicated that the greater portion of basic rolling resistance results from elastic hysteresis losses in the metal rather than friction attributable to interfacial slip. Developments that have increased the understanding of rolling friction have been reviewed by Johnson and Tabor (1967/1968).

Theories of rolling contact stresses have been summarized by Anderson (1964a) and Johnson (1966). Three principal lines of approach have been adopted; a study of elastic stresses due to slip and friction at the contact interface, pressures developed in elastohydrodynamic lubricating films, and the influence of inelastic (viscostatic or plastic) behavior of the material of the rolling solids. The influence of tangential forces on rolling contact stresses has been extensively studied (Heathcote, 1921; Poritsky, 1950, 1970; Johnson, 1958; Vermeulen and Johnson, 1964; Haines and Ollerton, 1963; Kalker, 1964; Halling, 1964). The effect of spin has been considered (Lutz, 1955; Wernitz, 1958). There has been considerable activity in the field of elastohydrodynamic lubrication (Dowson, 1970; Dowson *et al.*, 1962, 1964; Dowson and Whitaker, 1965; Grubin, 1949; Petrusevich, 1961; Sternlight *et al.*, 1961). The subject has been reviewed (Theyse, 1966; Dowson, 1967/1968) and elastohydrodynamics in practice have been outlined (Bolton, 1977), the most complete work relevant to the stress distribution being by Dowson and Higginson (1966). Inelastic deformation plays a significant part in the mechanism of rolling motion (May *et al.*, 1959; Flom and Bueche, 1959), the importance of the complex strain cycle has been demonstrated (Greenwood *et al.*, 1961), and theories for viscoelastic solids have been developed (Morland, 1962). The theory of plastic deformation and shakedown has led to a theoretical capacity for cylinders (Johnson, 1964). The mode of plastic deformation has been studied experimentally (Crook, 1957; Hamilton, 1963) and theoretically (Merwin and Johnson, 1963). The correlation of theory and experiment in research on fatigue in rolling contact has been discussed (Johnson, 1964). A simple theory of asperity contact in elastohydrodynamic lubrication has been formulated by Johnson *et al.* (1972), and a more detailed theory of partial elastohydrodynamic contacts has been presented by Tallian (1972).

III.   Material Requirements for Rolling Elements

In mechanisms the principal rolling elements are gears, cams, and rolling bearings. Material selection for these is important to ensure a rolling contact fatigue life adequate for the design requirements. For economic reasons the cheapest possible material is usually selected (Scott, 1967/1968, 1968a,b,c).

Most types of gears transmit load by line contact and in service may be subjected to rolling, sliding, abrasive, chemical, vibratory, and shock-loading action. Besides pitting failure due to rolling contact fatigue their useful lives may also be terminated by scuffing, fretting, abrasive wear, corrosion, and fracture. Gear materials must be chosen to resist these phenomena (Brown, 1958; Chesters, 1958; Frederick and Newman, 1958; Merritt, 1958; Woodley, 1977). The most extensively used gear steels are the carbon steels; manganese contributes markedly to strength and hardness, its effect depending upon carbon content. It also enhances hardenability, and fine-grained manganese steels attain good toughness and strength. For more stringent gear applications alloy steels, heat treated to provide the optimum properties, are required. Nickel imparts solid solution strengthening and increases toughness and resistance to impact, particularly at low temperatures, reduces distortion in grinding, and improves corrosion resistance. With nickel, given strength levels are obtained at lower carbon content, thus enhancing toughness and fatigue resistance. Chromium increases hardenability and has a strong tendency to form stable carbides which retard grain growth to provide fine-grained tough steels. Vanadium forms stable carbides that do not readily go into solution and that are not prone to agglomeration by tempering. It inhibits grain growth, thus imparting strength and toughness to heat treated steels. Molybdenum and vanadium are generally used in combination with other alloying elements. Lead may be added to gear steels to attain faster machining rates, increased production, and longer tool life.

Surface hardening to reduce wear is extensively applied to gear steels. Carbon and alloy steels can be carburized; nitriding is usually applied to aluminum-containing steels. Flame and induction hardening and surface treatments such as Sulphinuz, phosphating, and soft nitriding may be used (Scott. 1967/1968; Woodley, 1977; Neale, 1973).

Besides pitting due to rolling contact fatigue, accelerated normal wear, scuffing, and burnishing can cause distress of cams and tappets of high-speed high-output small automotive units. For the cams for such engines chromium- and molybdenum-containing irons are generally used with individual cams surface hardened (Metals Handbook, 1961). Tappet mate-

rials are usually through hardened, high-carbon, chromium, or molybdenum types or carburized low-alloy steels. The most common tappet materials in automotive applications are gray hardenable cast iron, containing
chromium, molybdenum, and nickel and chilled cast iron (Metals Handbook, 1961). Hardenable gray cast iron is the most widely used camshaft
material. Water-quenched high-carbon or oil-quenched alloy steels of
high carbon content are used in the through hardened condition. Carburized steels may also be used with selective flame or induction hardening of surface areas. Salt bath nitriding treatment can be beneficial to
cam followers.

Cam follower systems, although extensively used in engineering, do not
have an extensive literature of their own, but work on pitting failure of
cams and friction drives has been reviewed by Naylor (1967/1968) and design and manufacture by Jensen (1965).

Although ball and roller bearings are basically rolling elements, in
operating mechanisms they may also be subjected to some wear by sliding
and to chemical attack by lubricant and environment. Besides good
fatigue resistance to contend with high alternating stresses, the principal
qualities of ball-bearing steels are dimensional stability, high hardness to
resist abrasion and wear, and high elastic limit to minimize plastic deformation under load. A high-carbon steel satisfies the requirements if a
carbide-forming element is incorporated to allow oil quenching to minimize distortion during heat treatment. A 1.0% C, 1.5% Cr steel is generally used for conventional ball bearings (Neale, 1973; Scott, 1964, 1967,
1968c), but because of through hardening and heat treatment difficulties,
for large races and rollers, case-hardening materials containing nickel and
molybdenum may be used (Neale, 1973; Zaretsky and Anderson, 1970).

For elevated temperature applications conventional ball-bearing steel is
not satisfactory due to loss of hardness and fatigue resistance at temperatures above the tempering temperature of the steel. Materials for elevated
temperature use must possess adequate hot hardness, good oxidation resistance, and dimensional stability. High-speed tool steels that contain
principally tungsten and molybdenum and have high tempering temperatures have been found to be superior to other materials used at elevated
temperatures (Scott and Blackwell, 1967a; Scott, 1967/1968, 1968c,
1969a,b, 1970). Under elevated temperature conditions in a corrosive
environment chromium-containing types of stainless steels are used.

For specific applications under arduous conditions of rolling contact
other types of hard materials such as sintered carbides (Scott and Blackwell, 1964a, 1966a, 1967a,b; Scott, 1968d, 1969a, 1970, 1975a, 1977a;
Baughmann and Bamberger, 1963; Parker *et al.*, 1964), alumina (Parker *et
al.*, 1964b), and ceramics (Taylor *et al.*, 1961, 1963; Carter and Zaretsky,

1960; Zaretsky and Anderson, 1961; Parker *et al.*, 1965) are potentially suitable. Hot pressed silicon nitride (Scott and Blackwell, 1973a; Scott *et al.*, 1971; Dalal *et al.*, 1975) and some chemical vapor deposition (CVD) coatings (Scott, 1977a,b, 1978) appear attractive for use with in situ lubrication effected by reinforced polymeric cage materials (Scott and Blackwell, 1975; Scott and McCullagh, 1975a; Scott, 1974a; Scott and Mills, 1973, 1974a).

IV. Effect of Load and Geometry on Rolling Contact Fatigue

The life of rolling elements is that period of service which is limited by rolling contact fatigue, and the dominant factor controlling life is load. It is generally agreed that for ball bearings there is an inverse cube relationship relating life to load (Palmgren, 1959), but that in roller bearings, because of roller geometry, this may vary between values of 3 and 4, and the figure 10/3 has been adopted (Palmgren, 1959). Most experimental evidence to date (Anderson, 1964b; Scott, 1957, 1958) indicates that the inverse cube relationship between life and load for point contact in conventional ball-bearing steels with mineral oil lubricants is approximately true for other materials and lubricants; it has been established that fatigue life varies inversely as the third to fourth power of the load for a variety of materials, including glass, ceramics, and high-speed tool steels (Butler and Carter, 1957; Carter and Zaretsky, 1960) and lubricants including nonflammable and synthetic fluids (Barwell and Scott, 1956; Cordiano *et al.*, 1956, 1958). From the evidence available it is suggested (Palmgren, 1959) that the relationship will exist for new bearing materials and lubricants.

Contact angle can affect life and load-carrying capacity (Palmgren, 1959); as the contact angle is increased, ball spin velocity and relative sliding between ball and race is increased with increase of heat generation and an increase in contact temperature. Increased friction and heat generation can increase the shearing stresses in the mating elements. Experimental results (Zaretsky *et al.*, 1962a) agree with theory up to 30°, but at contact angles above 30° the experimental load-carrying capacity is less than that predicted and the deviation increases as the contact angle increases. It has been shown (Sibley and Orcut, 1961) that the geometry of the contact area in rolling contact changes with lubricant shear rate and viscosity. Bearing design considerations have been reviewed (Wellons and Harris, 1970).

Full-scale tests to study cam and tappet failure as influenced by design and other factors have been carried out by Ayers *et al.* (1958) and with test rigs by Andrew *et al.* (1962, 1965) and Taylor and Andrew (1963). The effect of tappet rotation has been reported by Bona and Ghilardi (1965).

Niemann and Rettig (1958) have studied the effects of design parameters on pitting failure of gears. With small tooth errors increase in speed can extend pitting life, but if tooth errors are great, an increase in speed reduces life because of an increase in dynamic tooth loads. Coleman (1967/1968, 1970) has presented formulas and charts to show the effect of various geometry and design parameters on pitting of gears.

V. Effect of Material Properties on Rolling Contact Fatigue

A. *Effect of Surface Finish*

Fatigue is a surface phenomenon and is greatly influenced by the physical structure and chemical nature of the surface and the surface finish (Scott *et al.*, 1962). Since the surface/volume ratio of the amount of metal stressed is much greater in rolling contact, where only the surface layer is involved, surface effects may be expected to be of greater importance in rolling contact fatigue than in ordinary fatigue failure. Helbig (1949) has shown that the life of finely ground rollers was 11% higher and buffed and polished rollers 22% higher than that of turned rollers. Way (1935) reported that rough surfaces tend to pit sooner. Baughmann (1960) obtained the best rolling contact fatigue life with the best surface finish. Accelerated service simulation testing of bearing balls (Scott, 1964) has shown that rolling contact fatigue life is greatly dependent upon surface finish, there being a trend toward increased life with improved surface finish (Fig. 1). However, the method of obtaining the surface finish and the stresses induced are also important; a heavily mechanically worked surface layer, for instance, produced by lapping to aid dimensional accuracy can considerably reduce fatigue resistance (Fig. 1). Light etching or electrolytic polishing can remove a worked surface layer deleterious to fatigue resistance, but if too great a depth of material is removed beneficially compressive stressed material is also removed and there is no increase in life. With flat washer-type specimens, surface flatness is as important as surface finish.

Surface finish and the method of surface finishing can influence gear performance (Fletcher and Collier, 1966). The surface roughness and oil film thickness control the extent of metallic contact and significantly influence the incidence of failure (Dawson, 1962, 1964). Increased surface

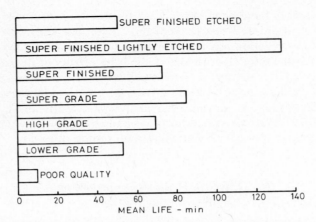

Fig. 1. Comparative rolling four-ball tests on bearing balls of different surface finish.

roughness reduces fatigue life, while increased speed and increased lubricant viscosity, which tend to increase oil film thickness, increase fatigue life. The life of disks and gears appears to correlate with the D value, the ratio of the initial combined peak-to-valley roughness of both surfaces and the theoretical film thickness; a large value of D results in a short life and vice versa. Friction appears to increase with surface roughness (Shotter, 1958; Niemann and Rettig, 1958). Johnson (1964) found little variation of the measured coefficient of traction [(friction force)/(normal load)] throughout a range of experimental conditions with both smooth and rough disks. Comparison of the friction measurement and Dawson's fatigue results shows no correlation between the bulk shear stress introduced by the traction force and the fatigue life.

Residual stresses affect the fatigue performance of gear materials (Mattson and Roberts, 1959). Shot peening is a well-established method of introducing mechanically compressive surface residual stresses (Kirk *et al.*, 1966), and it has been established that the surface residual stress level is the important factor in determining fatigue life rather than the structural changes due to shot peening action (Mattson and Roberts, 1959). Residual compressive stresses introduced by some surface treatments may be more beneficial than the more obvious structural and hardness changes (Chesters, 1964). Sulfurized coatings afford fatigue protection owing to compressive stresses introduced which can raise the fatigue limit; phosphating has an adverse effect on fatigue (Waterhouse and Allery, 1965). Soft nitriding can considerably increase the load-carrying capacity of carbon-steel gears.

Nitriding has been reported to produce a cam surface that is highly resistant to pitting (Naylor, 1967/1968). Surface coatings have been reviewed (Wilson, 1973; Scott, 1977a, 1978).

B. Effect of Material Hardness

Hardness is one of the simplest parameters to determine, and the effect of material hardness on rolling contact fatigue life has been studied (Scott and Blackwell, 1966a; Zaretsky *et al.*, 1965a,b). There appears to be an optimum hardness range for rolling elements (Fig. 2). To ensure maximum life the hardness of both race and rolling elements should be within the optimum hardness range or be of similar hardness, or the rolling elements should be up to 10% harder than the races. The highest measured residual compressive stress, beneficial in rolling contact, is obtained when there is no difference in hardness between the contacting surfaces (Zaretsky *et al.*, 1965b). The fatigue strength of ball-bearing steel has been found to be unchanged over the range 58–64 HRC (Styri, 1951) and it has been found that increasing the hardness above 750 HV did not improve fatigue strength (Frith, 1954). Anderson (1964a) has summarized the mass of rolling contact fatigue data accumulated with materials of various hardness. Maximum fatigue life appears to be obtained at the maximum hardness obtainable with the optimum heat treatment procedure. Since a certain amount of plastic flow and work hardening takes place in rolling contact, the largest possible contact area consistent with plastic flow and

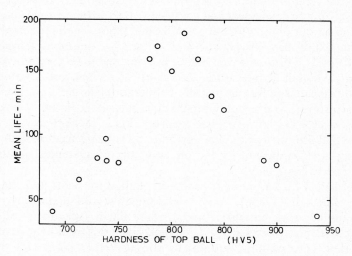

Fig. 2. Life–hardness curves of top balls from rolling four-ball tests.

work hardening that is not sufficiently great to accelerate fatigue, should be provided to reduce unit stress (Scott and Blackwell, 1966a).

The effect of hardness on the selection of gear materials has been summarized by Woodley (1977).

C. Effect of Heat Treatment and Structure

Since material hardness greatly affects rolling contact fatigue resistance, heat treatment and material structure must also be important factors. Lightly tempered EN 31 ball-bearing steel contains a considerable proportion of retained austenite. Transformation of retained austenite to martensite during service would involve dimensional instability (Mikus *et al.*, 1960), and for this reason a low austenite content is an important criterion for precision instrument bearings. Also, the force necessary to roll a ball decreases as the retained austenite decreases from 18 to 3.9% (Drutowski and Mikus, 1960). Frankel *et al.* (1960) report that fatigue strength is reduced by increasing austenite content and also that the transformation to martensite during alternating stressing is deleterious since large localized stresses resulting from the austenite/martensite reaction may initiate submicroscopic cracks which lower the fatigue strength. It has been suggested (Castleman *et al.*, 1952; Averback *et al.*, 1952) that the austenite phase may be sensitive to strain and may decompose to martensite during mechanical testing. Alternatively, it has been reported that contrary to general opinion (Prenosial, 1966) fatigue strength and resistance to pitting of case-hardened gears increased with increased austenite content to an optimum effect at 15% (Razin, 1966). The percentage of retained austenite may be reduced and the service operating temperature range of EN 31 ball steel extended by use of a stabilizing treatment, tempering at a higher temperature. Alternatively, a subzero treatment to below the marten site formation (Mf) temperature should allow the breakdown of austenite to proceed with falling temperature in the normal manner. Jackson (1959), however, has reported that a freeze treatment has no effect on ball life; the removal of austenite by subzero treatment in EN 31 steel may be complicated by the phenomenon of austenite stabilization (Harris and Cohen, 1949).

The effect of heat treatment has been studied by Scott and Blackwell (1966a), and Scott (1968b). To eliminate the effects of chemical composition, manufacturing variables, and surface finish, specimens manufactured from a single bar of EN 31 steel and given different heat treatments were tested. The optimum hardness for rolling contact fatigue life resulted from a tempering treatment which would reduce the percentage of retained austenite. A double tempering treatment was more effective than a single treatment. Deep freezing did not seem to be beneficial when using the tempering treatment to give optimum hardness, but was beneficial when lower tempering temperatures were used, possibly due to reduction of retained austenite. Slightly higher hardness was obtained by subzero treatment.

D. Effect of Manufacturing Variables

Makers of bearings and aircraft and missile manufacturers and users, concerned with tightening design limits, are constantly demanding better quality materials to meet the more stringent demands and stricter inspection specifications; steelmakers are endeavoring to provide cleaner ball-bearing steels with improved mechanical properties. Besides modifications in conventional steelmaking processes, newer techniques of melting and refining are being developed (Scott, 1968b, 1969c,d, 1975b, 1977b; Scott and Blackwell, 1964b, 1966a, 1971a). Published data on EN 31 steels show a range of properties, but a summary of results (Child and Harris, 1958) indicates that, generally, open hearth material is superior to basic electric arc material and that vacuum melting is beneficial. Frith (1954) has reported similar findings. Johnson and Sewell (1960) and Johnson and Blank (1964), on the other hand, report that good basic electric arc steel can match open hearth quality and that modifications in the steelmaking process can be effective in reducing the scatter of results.

Much work remains to be done in the comparison, under service conditions, of bearings made from different steels and the correlation of bearing tests with fatigue results. Kjerrman (1952) has stated that rolling bearings made from acid open hearth material were superior to those made from basic electric arc material. From other bearing tests (Scott, 1964, 1969c,d; Scott and Blackwell, 1964a) it has been shown that vacuum melting is beneficial, although some premature failures still occurred. Morrison *et al.* (1959) have shown that vacuum remelting, up to five times, progressively improves ball-bearing steel and has a great influence on postponing substantially early failures. Hustead (1962) found that steel processed by the vacuum consumable arc process gave improved bearing performance as compared with air-cast steel, but that vacuum degassed material in full-scale bearing tests proved inferior to similar air-cast material.

Laboratory accelerated service simulation tests have shown that the steelmaking procedure can greatly influence the performance of steel under conditions of rolling contact (Scott 1964, 1969c,d; Scott and Blackwell, 1964b, 1968a, 1969, 1971a). While there is a difference in performance among different casts of steels made by the same steelmaker and similar types of steel produced by different steelmakers, there is generally an overall trend; acid material appears to be superior to basic material, and modifications within each process, which reduce factors found to be detrimental, can improve material properties. Vacuum remelting is beneficial, possibly because of the better distribution of smaller inclusions and reduction in the number of deleterious-type inclusions; improving the vacuum used in melting improved the bearing life (Scott and Blackwell, 1968b; Scott, 1969c,d). Vacuum degassing appears to be an economical way of improving steel quality. There seems to be a trend toward improved rolling contact fatigue life with improved fatigue strength as measured by rotating bending (Scott, 1964) (Fig. 3). Although recent results still show a variation from cast to cast, the overall results indicate that modern EN 31 steels, following considerable development work, are superior to those of earlier manufacture (Scott, 1969c). Further tests have shown that all newer methods of steel refining improve the performance of steels used for ball bearings (Scott, 1969d; Scott and Blackwell, 1968b), (Fig. 4).

Because of the method of manufacture, upsetting slugs of bar material between hemispherical dies, bearing balls exhibit diametrically opposed polar areas having end grain perpendicular to the surface and areas between the poles having fiber orientation parallel to the surface (Fig. 5). It has been shown that fiber orientation is an important factor in rolling contact fatigue of balls (Scott 1964, 1968b). Carter (1958a) and Anderson

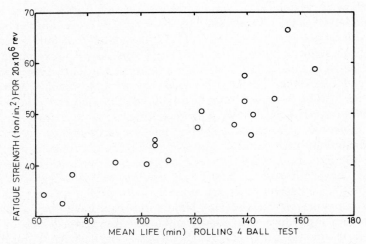

Fig. 3. Correlation of fatigue strength with rolling contact fatigue resistance.

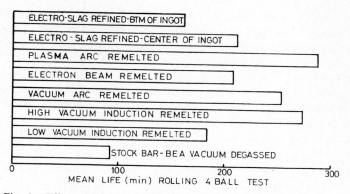

Fig. 4. Effect of steel rerefining process on rolling contact fatigue life.

and Carter (1959) report that life is reduced as the angle of fiber orientation approaches the perpendicular, but other test results suggest that the worst condition occurs when the angle of fiber orientation in relation to the bearing surface corresponds with the maximum shear plane (Scott, 1964, 1968b).

The position of material in the bar or billet of conventionally manufactured EN 31 steel is important (Scott, 1964). Specimens manufactured

Fig. 5. Section through a bearing ball deeply etched to show fiber orientation (×6).

from the center of a rolled bar have a lower rolling contact fatigue life than those manufactured from material away from the center line, possibly attributable to a greater inclusion concentration at the center of the bar and the fact that the outer material of the bar is subjected to a greater degree of working (Scott, 1964). An almost opposite effect is found with electroslag refined material; material from the top of the ingot is better than that from the middle and bottom of the ingot; material from the center of the ingot is superior to that from the outside of the ingot (Scott, 1968b). Rotating bending fatigue limits of the same material showed a similar trend.

For high-speed tool steels, ingot size has an important effect on the rolling contact fatigue resistance of rolling bearings (Scott and Blackwell, 1971b), better results being obtained with the smaller sized ingots. It appears that deleterious carbide segregation and large carbides are unavoidable in ingots larger than 10 in.

One method that is potentially attractive for the elimination of carbide segregation and the production of a homogeneous structure is to manufacture the high-speed tool steel via the powder method (Scott and Blackwell, 1974; Scott, 1976b, 1977b). Sound homogeneous material with uniform carbide distribution and a gas content comparable with conventionally cast material was found to be potentially suitable for rolling bearings. However, interstitial nitrogen content (Scott and Blackwell, 1974; Scott and McCullagh, 1973a) appears to have a dominant effect on rolling contact fatigue resistance and must be kept to a minimum.

In the quest for steels of higher strength with acceptable levels of ductility, considerable interest has been shown in ausforming (Lips and Van Zuilen, 1954), which produces a more uniform distribution of smaller carbides than conventionally produced material. Such a thermomechanical treatment of high-speed tool steels appears to be beneficial for resistance to rolling contact fatigue (Scott and Blackwell, 1977a), but metallurgical investigations have revealed that the process is difficult to control to obtain the optimum benefits even with small sized specimens. Larger specimens would present more severe difficulties. Breakup of large carbides and the movement of fractured carbides through the matrix material produce structural features and surface defects that result in premature fatigue failure (Fig. 6).

With the development of cleaner steels of improved physical properties and lower inclusion content, carbide content (an inherent feature of ball-bearing steel), carbide size, and carbide segregation have become important because large nondeforming carbides can fracture to become a source of rolling contact fatigue initiation and premature failure. A lower carbon and chromium-type steel with increased silicon and manganese contents

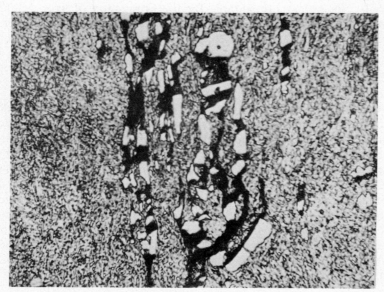

Fig. 6. Section through an ausformed specimen showing cavities formed by the breakup of large carbides by the deformation process. Etched nital (× 1500).

to provide matrix strengthening and to maintain hardness offers a method of reducing the carbide content and suppressing deleterious structure features such as carbide segregation, carbide stringers, and carbide network (Scott and Blackwell, 1977b, 1978).

Such modified composition ball-bearing steel has compared favorably in rolling contact fatigue resistance with steels of conventional analysis. The increased silicon content to counteract reduced carbon effects by matrix strengthening also controls deleterious interstitial nitrogen content by forming silicon nitride precipitates (Scott and Blackwell, 1977b, 1978).

Trends in design indicate that future aircraft engines may require bearings to operate in the range of high *DN* [bearing bore (in millimeters) times shaft speed (in revolutions per minute)] values. At such high speeds analyses predict a significant reduction in rolling contact fatigue life because of the high centrifugal forces developed at the outer race contact (Jones, 1952). The use of low-mass balls, either spherically hollow (Coe *et al.*, 1971, 1975) or cylindrically drilled (Coe *et al.*, 1971; Holmes, 1972a,b), is one approach to reduction of stresses (Harris, 1968). It has been shown that heat treatable electrodeposited materials with hard carbides codeposited, free from deleterious nonmetallic inclusions and having rolling contact fatigue resistance comparable with currently used materials, are potentially suitable for the production of light-weight or hollow rolling elements and races for high-speed applications (Scott *et al.*, 1975; Scott and McCullagh, 1972, 1976; Scott, 1976a).

E. *Effect of Nonmetallic Inclusions*

Nonmetallic inclusions that break the metallic continuity in the stressed area of rolling elements can act as stress raisers to initiate fatigue cracks. The nature and type of the inclusions, characteristic of the different steel-making processes, appear to be more important than the number and size of the inclusions (Frith, 1954; Scott and Blackwell, 1964b; Johnson and Sewell, 1960; Murray and Johnson, 1963; Uhrees, 1963). Crack initiation is facilitated by the presence of nondeforming brittle-type nonmetallic inclusions which crack during deformation in the stressed area. Such inclusions are more characteristic of basic electric arc steel than the older open hearth steel, which contains mainly pliable sulfide-type inclusions that deform without cracking. Trace elements can be important (Morrison *et al.*, 1964).

F. *Effect of Residual Gas Content*

Because of rolling and sliding action, rolling bearing materials undergo metallographic changes which have been well documented (Scott, 1968b; Scott and Scott, 1960; Tallian, 1967a). Metallurgical investigation (Scott and McCullagh, 1973b) has shown that the metallographic changes did not differ for materials of similar composition and heat treatment, but from extensive hardness exploration it was concluded that in the AOH steels investigated, the spread of strain hardening was less inhibited than in the BEA steels, and, therefore, crack initiation and propagation are delayed in AOH steels. Since rolling contact fatigue resistance correlates with cavitation erosion resistance (Tickler and Scott, 1970), it was further suggested that some metallurgical factor controlling strain hardening, crack initiation, and crack propagation, such as gas content, is important in both phenomena and could be the dominant factor in material performance. Oxygen content is mainly associated with nonmetallic inclusions; the morphology of the inclusion appears to be the dominant feature. While hydrogen content can be important, modern steels are generally low in hydrogen and it is hydrogen pickup in service that can be deleterious. Interstitial nitrogen and nitrides can pin dislocations to accelerate crack initiation. Thus nitrogen content is important and can account for the difference in performance of otherwise identical ball-bearing steels (Scott and McCullagh, 1973a, 1975b). There is a trend toward reduced rolling contact fatigue life with increased nitrogen content (Fig. 7). It appears that a principal factor in the production of high-quality ball-bearing steel is the reduction of residual nitrogen. Vacuum remelting does this but electroslag refining does not; therefore, the latter process should use low-nitrogen steel.

G. Effect of Material Combination

Resulting from uneven stressing, lubricant, temperature, or environmental effects, one element of a rolling mechanism may be prone to premature failure and in some instances the apparently simple solution of replacing the failing EN 31 steel element with one manufactured of a superior material has not been successful. It has been shown (Scott and Blackwell, 1966b) that the combination of materials in rolling contact is important and that the elastic and plastic properties of materials in contact have a significant effect on rolling contact fatigue resistance. Figure 8 summarizes test results which show that the heavily loaded rolling contact fatigue lives of various high-speed tool steels can be reduced by rolling against EN 31 steels. The life of tungsten carbide under similar conditions can be increased by rolling against EN 31 steel (Scott and Blackwell, 1966a, 1967b).

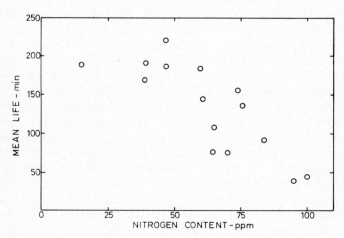

Fig. 7. Correlation between mean rolling contact fatigue life and nitrogen content.

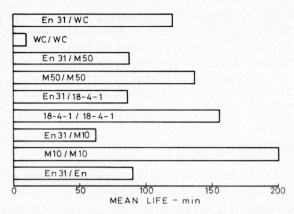

Fig. 8. Effect of material combination on mineral oil lubricated rolling contact fatigue life.

VI. Effect of the Lubricant on Rolling Contact Fatigue

The lubricant used and the environment it creates can have a significant influence on the life of rolling mechanisms (Scott, 1977a). The effect of lubricant on rolling contact fatigue of ball bearing materials has been reviewed (Scott, 1964, 1968b, 1977a; Anderson, 1964a) and the correlation among accelerated testing machines and among accelerated tests, full-scale tests, and practice have been reported (Scott and Blackwell, 1964a; Murphy *et al.*, 1977; Krivoshein, 1960; Bezborodko and Krivoshein, 1962). There are now I.P. methods to assess the effect of lubricant on rolling contact fatigue (I.P., 1975a,b; Yardley, 1977; Kenny, 1977).

Barwell and Scott (1956) and Scott (1957) showed that the nature of the lubricant had a dominant effect on the incidence of pitting failure of rolling bearings, some of the poorer nonflammable fluids yielding one-fifth of the life expected from a conventional mineral oil lubricant. Cordiano *et al.* (1956, 1958) reported similar results and proposed lubricant factors of 1.2, 1.5, and 2.6 for phosphate esters, phosphate ester based fluids, and water–glycol based fluids, respectively, for correcting design loads determined from bearing handbooks. Other workers (Macks, 1953; Baugham, 1958; Anderson and Carter, 1958; Carter, 1958b, 1960; Zaretsky *et al.*, 1962b; Rounds, 1962a,b; Hobbs and Mullet, 1969; Gustafson, 1964; Kenny and Yardley, 1972; Yardley *et al.*, 1974; March, 1977; Tokuda *et al.*, 1977) have reported significant variation in bearing life with different base stock types of lubricant. Parker (1977) found that all synthetic fluids tested gave rolling contact fatigue lives less than that of a paraffinic mineral oil. It has been established (MacPherson, 1977) that surface temperature exerts a powerful influence on fatigue life so that testing at different temperatures is essential to evaluate oil/steel combinations. Lubricant material compatibility is as important as material compatibility (Scott, 1969a).

A. Effect of Lubricant Viscosity

Viscosity is probably the most important property of a lubricant. Barwell and Scott (1956), however, found that lubricant viscosity as measured at room temperature and pressure was not a dominant factor in rolling contact fatigue since some lubricants of low viscosity gave as long a life as others of high viscosity. With certain types of lubricants such as mineral oils from the same crude source, there is a continuous trend toward longer life with increasing viscosity. Proportionality with (viscosity)$^{0.3}$ (Scott, 1960) and with (viscosity)$^{0.2}$ (Carter, 1960; Anderson, 1964b) has been suggested. Scott (1960) found that with mineral oils from the same crude stock the viscosity index (VI) decreased with increasing viscosity and that the lubricant with the lowest VI gave the longest life, confirming earlier results (Barwell and Scott, 1965). Since low VI is equivalent to a high pressure viscosity coefficient and also a high temperature viscosity coefficient (Hartung, 1957), life appears to correlate with the pressure–viscosity coefficient. Thus viscosity under pressure may be more important in rolling contact fatigue than viscosity measured at room temperature. Polyalkylene glycol fluids, diesters, and polyphenyl ethers also show a trend toward longer life with increased viscosity, but with silicone fluids rolling contact fatigue life appears to be independent of viscosity. It is possible that the more viscous oils of low VI are almost solid under contact pressures and do not readily enter the fatigue crack to accelerate crack propagation, whereas all silicone fluids are probably of similar viscosity under the contact conditions. Sternlicht et al. (1960) reported that fatigue life with mineral oils appeared to increase proportionally to VI. Baughman (1958) considered that fatigue appeared to be related to the lubricant viscosity in the high pressure region, while Carter (1960) and Anderson (1964a) found that fatigue life increased with increasing lubricant pressure–viscosity coefficient. The rise in viscosity with pressure for phosphate ester base and phosphate ester fluids is similar to that for petroleum oils, but the rate of rise is greater (Cordiano et al., 1956, 1958). Water glycols have relatively flat pressure viscosity curves, indicating a relatively low viscosity of such fluids at high pressure, which may be a contributory factor to the poor rolling contact fatigue life obtained with such fluids. Rounds (1962a) reported consistently better fatigue life with naphthenic oils of lower VI than with paraffinic oils of high VI, thus confirming the findings of Barwell and Scott (1956), Scott 1957), and later the findings of Scott and Blackwell (1964a, 1966a, 1967a, 1977a) and Scott (1974a).

B. Effect of Lubricant Additives

Most modern lubricants contain additives which reinforce or impart some desired property. The addition of conventional amounts of mild antiwear additives such as tricresyl phosphate do not cause any significant change in rolling contact fatigue resistance (Scott, 1957); elemental sulfur and lead naphthenate appear to increase resistance. Increased concentration of the more active extreme pressure (EP) additives used to prevent metal-to-metal contact between heavily loaded contacts results in reduced rolling contact fatigue life; increased concentration of the milder type EP additives has little effect (Scott, 1957). Rounds (1962a) reported that a reactive chlorine additive reduced bearing fatigue life, phenyl ether and methyl phenyl silicone additives increased life, and antiwear additives increased specimen surface distress in many material/additive tests (Rounds, 1971). Parker and Zaretsky (1973) found the presence of several surface-active additives detrimental to rolling element fatigue life. The effect of zinc dithiophosphate appears still to be unclear (Carter, 1958a; Rounds, 1975; Lowenthal and Parker, 1976). Various results have been obtained with the use of zinc dialkyl dithiophosphate, depending upon concentration and other additives present (Rounds, 1967).

The buildup of acid species in an oil through additive reaction can lead to reduced rolling element fatigue life (Mould and Silver, 1975, 1976). Strong acids such as dibutyl phosphate and dichloroacetic acid even in small concentration cause considerable reductions in bearing life, but large concentrations of a weak acid such as myristic acid are required before any noticeable effect on fatigue life is obtained.

Solid lubricants can be beneficial as additives if correctly used in oils and greases (Scott and Jamieson, 1962; Jost, 1958; Scott, 1972) for rolling elements. To be effective they must be compatible with other additives present and must adhere strongly to the surface. Failure to do so may lead to sedimentation, which is deleterious in most lubricating systems. Acid surfactants present in the lubricant tend to prevent bonding of solid lubricant particles to the rolling surfaces but alkaline surfactants appear to aid bonding (Scott and Jamieson, 1962). Particle size appears to be important; submicron sized particles are the most effective. While additives have generally been found not to be beneficial for rolling contact fatigue resistance with diester-type lubricants, tungsten disulfide and molybdenum disulfide have been found to be beneficial in polyalkylene glycol and in a silicone fluid for aircraft engines and to a lesser extent PTFE was also beneficial (Scott and Blackwell, 1973b). A fluoride additive was effective in greases.

VII. Effect of Environment

Environment is known to play an important role in metal fatigue (Hoyt, 1952; Likhtman *et al.*, 1954). A corrosive environment can also adversely affect fatigue behavior (Gilbert, 1956). Fatigue is a surface phenomenon and since the surface/volume ratio of the amount of material stressed is greater in rolling contact fatigue, where only the surface layer is involved, this factor may be expected to be of greater importance in pitting of rolling elements than in ordinary fatigue failure. The presence of water in mineral oil lubricants can accelerate rolling contact fatigue failure of conventional ball-bearing steel (Grunberg and Scott, 1960). Even small quantities of dissolved water are detrimental. The deleterious effect of water in the lubricant of rolling bearings has been extensively studied (Shatzberg and Felsen, 1968; Murphy *et al.*, 1977), and additives have been found to counteract the detrimental effects of water contamination (Grunberg and Scott, 1962; Scott, 1962). A synergistic effect was found with a combination of additives. Surface active materials form an adsorbed monolayer on the surface of rolling elements to give fatigue lives that are independent of the extent of water contamination and may be regarded as acting in the sense of adsorption fatigue (Likhtmann *et al.*, 1954). Corrosion fatigue plays some part in pitting failure but is not inhibited by recognized water-soluble anticorrosive additives. The lubricating film thickness is reduced when water is present and is increased by the presence of additives effective in counteracting water accelerated fatigue failure (Grunberg and Scott, 1960). The effect of environment on bearing performance has been reviewed by Scott (1968a).

Water in the lubricant can accelerate the propagation of fatigue cracks, resulting in premature failure by complete fracture of ball bearings (Figs. 9 and 10). It has been hypothesized that vacancy-induced diffusion of hydrogen into highly stressed surface material causes hydrogen embrittlement and accelerated failure (Grunberg and Scott, 1960; Grunberg *et al.*, 1960). Use of a radioactive tracer investigational technique (Grunberg *et al.*, 1963) together with electron microfractography of failed specimens (Scott and Scott, 1960; Scott *et al.*, 1967, 1969/1970) has verified the hy-

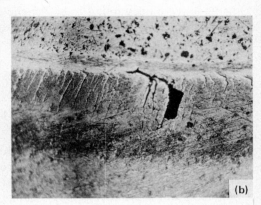

Fig. 9. (a) Single small failure pit formed in a mineral oil lubricant (×3.9). (b) Crack propagation toward a single small failure pit at the inner edge of a bearing track of a mineral oil lubricated test ball (×65).

pothesis (Figs. 11 and 12). A similar explanation for premature fatigue failure has been offered by Morgan (1958) and for the premature failure of water cooled rolling mills by Bardgett (1961) and Das *et al* (1962). Petch (1956) has suggested that hydrogen gas adsorption reduces the surface energy and enables crack propagation at a lower stress. Acidic oxidation during rolling contact can liberate hydrogen through electrochemical

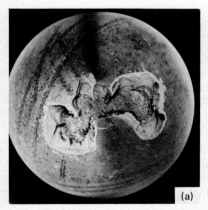
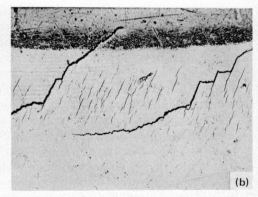

Fig. 10. (a) Fractured ball by test in a water-contaminated mineral oil (×3.9). (b) Rapid crack propagation toward fracture of a test ball lubricated with a water-contaminated lubricant (×65).

Fig. 11. Electron microfractograph of the fracture surface of a failed ball tested in water-contaminated mineral oil showing intergranular fracture indicative of hydrogen embrittlement. Normal rolling contact failure is by a transcrystalline fatigue mechanism.

reaction with steel with traces of water in mineral oils as the conducting medium (Cirana and Szieleit, 1973).

High-temperature oil degradation in highly stressed bearing regions leads to a buildup of acidic species deleterious to rolling contact fatigue life (Mould and Silver, 1975). The addition of naphthanic acid to a base oil

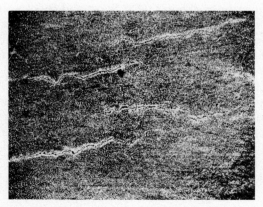

Fig. 12. Hairline cracks in the surface of a steel ball, indicative of hydrogen embrittlement due to testing in a water-contaminated mineral oil (×70).

produces a reduction in bearing life similar to that found with acids formed by oil oxidation (Rounds, 1971). Fatty acids give much lower bearing lives than mineral oils (Rounds, 1962a).

Galvin and Naylor (1965) found that the environment formed by the lubricant affected fatigue life by either influencing crack initiation or crack propagation. Oleic acid in white oil considerably reduced fatigue resistance as compared to white oil alone. A corrosion mechanism has been proposed to explain the effects of fatty acid blends on lubrication (Goldblatt, 1972). The rolling contact fatigue control of ball-bearing steel through lubrication engineering has been reviewed by Tallian (1967/1968).

In some specific applications, problems of operating rolling bearings with liquid metals as lubricants arise (Anderson, 1964c; Markert and Ferguson, 1957), and in high-vacuum applications the role of oxide films and surface contaminants on rolling elements becomes important (Johnson and Buckley, 1967/1968). Space environment has a dominant effect on lubricants and poses specific problems for rolling elements (Anderson and Glen, 1967/1968; Scarlett, 1967/1968).

VIII. Effect of Temperature

Elevated temperatures and low temperatures constitute arduous conditions under which some rolling mechanisms are expected to operate. Increasing temperature reduces lubricant viscosity, which in many cases caused a reduction in both lubricating film thickness and rolling contact fatigue life (Anderson, 1964a). Investigations of materials and lubricants at elevated temperatures (Scott, 1958; Scott and Blackwell, 1964a, 1965, 1966b, 1967a, 1973a) have shown that for all rolling bearing materials tested life varied with different lubricants and for all lubricants tested rolling contact fatigue life varied with different materials; thus material/lubricant combination in rolling contact is important (Scott, 1969a; MacPherson, 1977).

While excessive degradation of a lubricant is deleterious to rolling contact fatigue life, some lubricant degradation products can be beneficial (Scott and Blackwell, 1965), and thus it is an advantage for elevated temperature use if lubricant degradation products are not deleterious to the materials used and if they possess some lubricating ability. The success or failure of an elevated temperature lubricant may depend largely on this latter property.

Special atmospheres may be used to delay lubricant degradation (Bailey *et al.*, 1958; Nemeth and Anderson, 1955; Sorem and Caltaneo,

1956); oxygen plays an important role (Nemeth and Anderson, 1956). Otterbein (1958) found that rolling bearing fatigue life decreased steadily with increase of temperature from 120 to 300°F with a diester lubricant. Scott (1958) found that there was no reduction in fatigue life of EN 31 steel up to 160°C but that fatigue resistance rapidly deteriorated above this to 200°C; high-speed tool steels suffered no reduction in life up to 200°C when lubricated with diester-type lubricants (Scott and Blackwell, 1973a; Scott, 1969a). Some improvement in the rolling contact fatigue resistance of high-speed tool steels may be found at elevated temperatures because of metallurgical changes controlling deleterious nitrogen content (Scott and McCullagh, 1975a). Further data have been reported on the effects of lubricant temperature on bearing life (Morrison *et al.,* 1959; Zaretsky and Anderson, 1961).

IX. Mechanism of Failure

Controversy exists as to whether cracks initiating rolling contact fatigue originate at the surface and propagate into the material or originate below the surface and propagate toward the surface. Metallurgical investigations have revealed that both mechanisms are operative (Barwell and Scott, 1956; Scott, 1964; Scott *et al.,* 1967; Scott, 1967; Davies and Day, 1964, Moyar and Morrow, 1964; Tallian, 1967b; Littmann, 1970; Littmann and Widner, 1965; Way, 1935), the more dominant depending upon prevailing localized circumstances. Propagation of surface cracks that appear to initiate at the edges of the contact pressure area seems to be controlled by the nature of the lubricant and the environment. With mineral oils surface cracks appear to propagate into the material at an acute angle to the rolling direction and then turn parallel to the surface at a depth associated with the subsurface area of calculated maximum shearing stress to detach a piece of surface material (Fig. 9). With some poorer lubricants such as water-based nonflammable fluids, crack propagation is rapid and goes deeply into the material so that fracture ensues (Fig. 10). The mutual effects of corrosive attack and stressing can accelerate failure and produce a continuous type of surface pitting (Fig. 13). Way (1935) proposed that lubricant under pressure enters a surface crack and that contact under load then seals the crack. The trapped liquid is compressed and the pressure buildup extends the crack toward removal of surface material. Increasing lubricant viscosity probably increases life since it is more difficult for the more viscous lubricant to enter the fatigue crack. The chemical nature of the lubricant, such as molecule size and surface tension, may

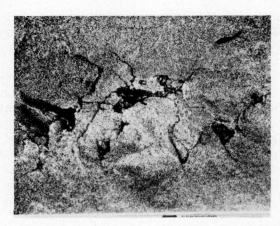

Fig. 13. Corrosive attack by lubricant and continuous-type pitting of the surface of a test ball (× 70).

also influence lubricant impingement into the crack by capillarity. A liquid must be present for surface pitting to occur (Scott and Blackwell, 1964a, 1966a, 1967a; Way, 1935). Pitting can be prevented if the viscosity of the lubricant used is above a certain load-dependent value (Way, 1935). Dawson (1961, 1962, 1964, 1965/1966) has reported that surface pitting only occurs if the lubricant film thickness is less than the combined asperity heights of the mating surfaces.

All materials undergo change due to rolling contact. The changes manifest themselves by hardness changes (Scott and McCullagh, 1973b), plastic deformation, and metallographic change of surface material. The surface finishing marks of rolling elements are smoothed over by rolling and sliding action. Electron microscopic examination of sections has revealed metallographic change of surface material and evidence of plastic deformation (Scott and Scott, 1960; Scott *et al.,* 1967). Hardened and tempered EN 31 ball-bearing steel consists of dispersed fine carbides in a martensitic matrix (Fig. 14). Metallographically changed surface material can be devoid of carbides. Such white etching material results from high contact stresses, plastic deformation, and high-temperature flashes, causing solution of the carbides followed by rapid quenching under pressure by bulk material and lubricant. Absorption of gases from lubricant breakdown in the contact zone may contribute to hardening of the white etching material (Studenok and Makhon'Ko, 1957; Seleznev, 1957; Grozin, 1957).

In the subsurface area of contact below the carbide-free zone, stringer-type carbides are found (Fig. 15), possibly by annihilation of coarse carbides by plastic deformation, causing internal extrusion and in-

Fig. 14. Electron micrograph of a section of a bearing ball, etched nital to show structure.

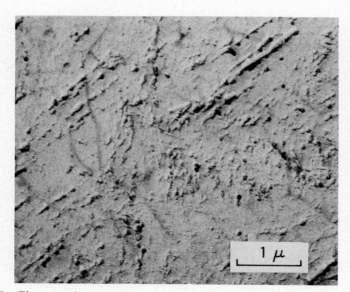

Fig. 15. Electron micrograph of a section of the subsurface bearing track material showing stringer-type carbides.

trusion (Bush *et al.*, 1960) similar to an external extrusion and intrusion mechanism on the surfaces of cyclically stressed specimens (Forsyth and Stubbington, 1954/1955). Metallographic changes also occur in the surface material of high-speed tool steels subjected to rolling contact. Electron microscopic investigations have revealed that the worked surface material still retains carbides, although these are much smaller than the carbides in the original structure (Scott and Blackwell, 1965). It may be that the local high-temperature flashes do not reach the austenitizing temperature of the high-speed tool steel to effect carbide solution.

Rolling contact causes microstructural changes in the subsurface area of contact in the region of calculated maximum Hertzian stress. Sections through bearing tracks, transverse to the rolling direction, reveal the presence of localized areas of tempered martensite (Scott, 1964a; Barwell and Scott, 1956; Scott and Scott, 1960) or mechanical troostite (Jones, 1946) that are softer than the original material. This may be a tempering phenomenon, heat being generated by elastic hysteresis attributable to cyclic stressing or a strain-induced transformation. Subsurface cracks are found in such areas (Fig. 16). The darker etching soft material may be incapable

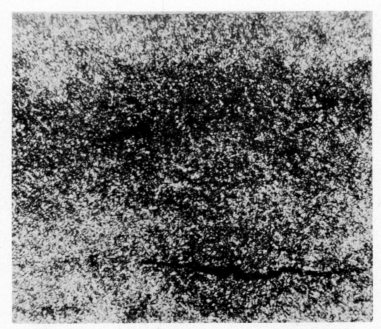

Fig. 16. Transverse section through the bearing track of a tested ball showing subsurface metallographic change and cracks—etched nital. The white material at the top of the micrograph is nickel plating to protect the surface during specimen preparation (×510).

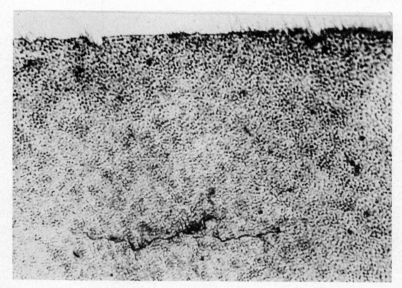

Fig. 17. Transverse section through the bearing track of a tested ball showing a subsurface crack—etched nital (×500).

of supporting, without plastic deformation, the stresses imposed, and cracks initiate and propagate parallel to the bearing surface. Similar subsurface cracks occur without the associated metallographic change in lightly loaded rolling elements (Fig. 17), and thus there appears to be a threshold stress level above which subsurface metallographic change occurs, indicating that such change is a plastic flow or yielding phenomenon rather than a tempering effect. Brittle-type nonmetallic inclusions which break the metallic continuity can act as stress raisers to initiate cracks in the subsurface region of calculated maximum Hertzian stress. The stress concentration at nonmetallic inclusions has been suggested as the cause of dark etching microstructural alteration (Gentile *et al.,* 1965), but the dark etching constituents form in steels of low nonmetallic inclusion content (O'Brien and King, 1965).

Similar metallographic changes and subsurface cracks occur in sections through the bearing track that are longitudinal to the rolling direction, but white etching areas may be associated with them (Fig. 18). Many white etching areas have a smooth appearance, some contain fine structural features, and some contain markings suggestive of deformation (Fig. 19). Stringer-type carbides are also associated with the edges of the white etching areas and occur within the white etching areas. Davies and Day (1964) suggested that local areas of untempered martensite may be produced around cracks initiated at stress raisers, such as inclusions, by in-

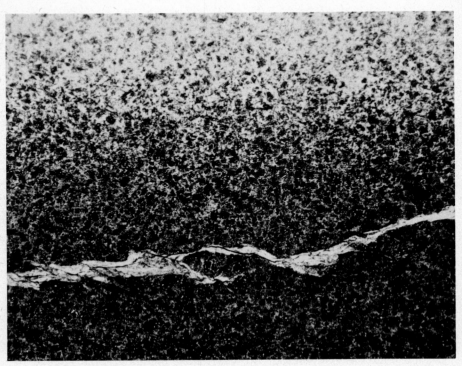

Fig. 18. Longitudinal section through the bearing track of a tested ball showing subsurface metallographic change with associated white etching material and subsurface cracks—etched nital (×700).

tensive heating brought about by dislocation activity in the vicinity of the cracks. Heating of relatively short duration may be sufficiently intensive to exceed the austenitizing temperature and cause carbide solution, and then rapid quenching by bulk material would produce martensite of up to 1100 HV5. The presence of white etching areas without associated cracks tends to nullify this theory. Microprobe analysis of the hard white etching material revealed an increased alloy content and a reduced iron content compared with the matrix of unchanged material indicative of carbide solution. The large carbides in the original microstructure are usually mixed carbides, and the stringer-type carbides are generally iron carbides; thus some stress-induced diffusion of elements may occur. It has been suggested (O'Brien and King, 1965) that diffusion of carbon can account for a volume of stringer-type iron carbide greater than that of the original mixed carbides. The use of special thermal etching techniques has indicated that some white etching material may be a form of austenite and the high hardness may be caused by the almost colloidal dispersion of fine carbides. In situ transformation of any retained austenite could give rise to stress concentration attributable to a volume increase and thus initiate cracks. Thin foils prepared by electropolishing subsurface white etching material showed a small cell structure with no resemblance to the normal platelike and carbide structure of martensite. Such a cell structure has been reported between 0.05 and 0.1 μm (O'Brien and King, 1965; Martin et al., 1965). Selected area diffraction of thin foils of white etching material indicates a lattice parameter of ferrite. Rolling contact produces similar elongated white material associated with subsurface cracks in high-speed tool steels after prolonged running, but little evidence has been found of carbide elongation in or adjacent to the white etching material.

It appears that several distinctly different modes of rolling contact fatigue can cause cracks to nucleate and propagate independently at various rates; thus the phenomenon appears to be influenced by highly localized conditions as well as by the general properties of the bulk material. When geometry, materials, and lubrication are satisfactory, failure is

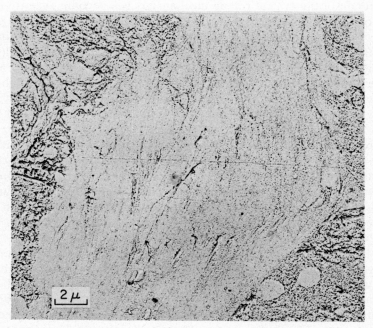

Fig. 19.   Electron micrograph of white etching material showing fine structural features and deformation.

probably by means of a subsurface stress-induced mechanism. However, scatter in life and the absence of a fatigue limit in service and laboratory tests (Styri, 1951) indicate that stress can influence the nucleation of various modes of failure. Surface failure mechanisms that are influenced by sliding action, lubricant and environmental effects, and subsurface failure mechanisms associated with microstructural changes and stress raisers appear to be dominated by prevailing circumstances.

To ensure an adequate life for rolling elements, all causes of failure other than fatigue must be avoided and all factors which influence fatigue life should be controlled carefully by design, manufacture, and material and lubricant selection.

X. Failure Detection

The complex machines created by modern technology often depend upon rolling mechanisms for their successful operation. The practice of withdrawing such equipment from service periodically to dismantle for inspection of the critical rolling mechanisms has become expensive, and economic pressures are causing the switch from such maintenance procedure to failure prevention maintenance. This involves the use of machine-condition monitoring to avoid failure in service and to allow remedial action to be taken before the breakdown point is reached. A convenient method of doing this is by contaminant analysis of the lubricant by ferrography (Scott *et al.*, 1974, 1975; Seifert and Westcott, 1972; Bowen *et al.*, 1976; Scott, 1975c; Scott and Smith, 1973; Scott and Westcott, 1977a,b, 1978).

Wear particles are unique, having individual characteristics that bear evidence of the conditions under which they were formed. Thus careful examination of their morphology and determination of their composition can yield specific information about the condition of the moving surfaces of the machine elements from which they were produced, the mechanism of their formation, and the operative wear mode in the system from which they were extracted (Scott, 1975c,d, 1976b; Bowen and Westcott, 1976).

Rolling elements such as ball and roller bearings and gears generate characteristic wear particles which are particularly useful in the elucidation of element condition (Bowen and Westcott, 1976; Scott and Mills, 1974b,c). Three distinct particle types are found with rolling bearing fatigue. Thin laminar particles that frequently contain holes are generated throughout the life of a bearing, and only an increase in their size and concentration indicates some initial change in bearing operation. Spherical particles (Scott and Mills, 1974b) (Fig. 20) are a characteristic feature associated with a fatigue crack so that their detection indicates a propagating fatigue crack in a rolling element; the concentration of spherical particles indicates the extent of crack propagation (Middleton *et al.*, 1974). Chunky particles up to about 100 nm in size indicate the initiation of surface disintegration (Bowen and Westcott, 1976). Fatigue particles produced from gear pitch lines have much in common with rolling bearing fatigue particles (Bowen and Westcott, 1976). A high proportion of large chunky particles to small particles found when surface deterioration is occurring (Fig. 21).

Particle tribology (Scott, 1975d, 1976b) by the ferrographic oil analysis procedure (Scott and Westcott, 1978) is a potentially attractive method of incipient failure detection and catastrophic failure prevention in rolling mechanisms and is the basis of a powerful machinery monitoring technique.

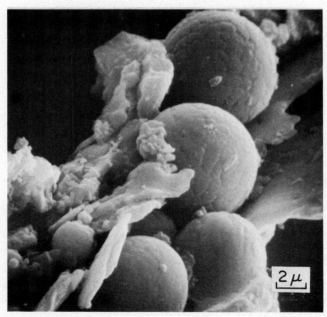

Fig. 20. Electron micrograph of spherical particles on a ferrogram prepared from the lubricant of an aircraft engine. The spherical particles of bearing material were produced by a propagating rolling contact fatigue crack.

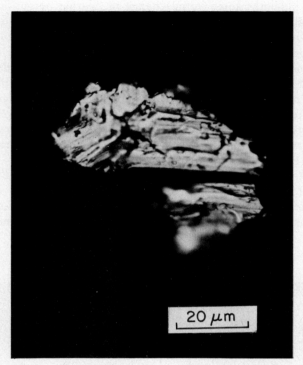

Fig. 21. Electron micrograph of a chunky fatigue particle.

References

Allan, R. K. (1946). "Rolling Bearings." Pitman, London.
Anderson, W. J. (1964a). *In* "Advanced Bearing Technology" (E. E. Bissen and W. J. Anderson, eds.), NASA SP-38, pp. 139–173.
Anderson, W. J. (1964b). *In* "Advanced Bearing Technology" (E. E. Bissen and W. J. Anderson, eds.), NASA SP-38, pp. 371–450.
Anderson, W. J. (1964c). *In* "Advanced Bearing Technology" (E. E. Bissen and W. J. Anderson, eds.), NASA SP-38, pp. 469–496.
Anderson, W. J., and Carter, T. L. (1958). *ASLE Trans.* **1,** 266–272.
Anderson, W. J., and Carter, T. L. (1959). *ASLE Trans.* **2,** 108–120.
Anderson, W. J., and Glen, D. C. (1967/1968). *Proc. Inst. Mech. Eng. London* **182** (3A), 506–519.
Andrew, S., Taylor, R. G., and Lane, R. J. (1962). (MIRA) Rep. 7.
Andrew, S., Wykes, F. C., and Taylor, R. G. (1965). MIRA Rep. 4.
Averback, B. L., Lorns, S. G., and Cohen, M. (1952). *Trans. Am. Soc. Met.* **44,** 746–757.
Ayers, V., Bidwell, J. B., Pilger, A. C., and Witham, R. K. (1958). *SAE Trans.* **66,** 242.
Bailey, G. H., Sorem, S. S., and Caltaneo, A. E. (1958). SAE Paper 24A.

Bardgett, W. E. (1961). *NML Tech. J.* **3**, 4–14.

Barwell, F. T., and Scott, D. (1956). *Engineering London* **182** (4713), 9–12.

Baughman, R. A. (1958). ASME Paper 58-A235.

Baughman, R. A. (1960). *J. Basic Eng.* **82**, 287–294.

Baughman, R. A., and Bamberger, E. N. (1963). *J. Basic Eng.* **85**, 265.

Bezborolko, M. D., and Krivoshein, G. S. (1962). *In* "Friction and Wear in Machinery," Collection 16.5. Akad. Nauk. SSSR, Moscow.

Bolton, W. K. (1977). *In* "Rolling Contact Fatigue—Performance Testing of Lubricants" (R. Tourret and E. P. Wright, eds.), pp. 17–26. Institute of Petroleum, London.

Bona, F. C., and Ghilardi, F. G. (1965). *Proc Inst. Mech. Eng. London* **180**, 269.

Bowen, E. R., and Westcott, V. C. (1976). "Wear Particle Atlas." Foxboro Transonics, Burlington, Massachusetts.

Bowen, R., Scott, D., Seifert, W. W., and Westcott, V. C. (1976). *Tribol. Int.* **9**, 109–115.

Brown, R. J. (1958). *Proc. Int. Conf. Gearing* pp. 225–231. Institution of Mechanical Engineers, London.

Bush, J. J., Grube, W. L., and Robinson, G. H. (1960). *Proc. Symp. Rolling Contact Phenomena* (J. Bedwell, ed.), pp. 363–399. Elsevier, Amsterdam.

Butler, R. H., and Carter, T. L. (1957). NASA TN 3030.

Carter, T. L. (1958a) NASA TN 4216.

Carter, T. L. (1958b). NASA TN 4161.

Carter, T. L. (1960). NASA TR-60.

Carter, T. L., and Zaretsky, E. V. (1960). NASA TN D259.

Castleman, L. S., Averback, B. L., and Cohen, M. (1952). *Trans. Am. Soc. Met.* **44**, 240–263.

Chesters, W. T. (1958). *Proc. Int. Conf. Gearing* pp. 91–98. Institution of Mechanical Engineers, London.

Chesters, W. T. (1964). *Proc. Symp. Fatigue Rolling Contact* pp. 86–94. Institution of Mechanical Engineers, London.

Child, H. C., and Harris, G. T. (1958). *J. Iron Steel Inst. London* **190**(4), 414–431.

Circena, I. R., and Szieleit, H. S. (1973). *Wear* **24**, 107.

Coe, H. H., Scribbe, H. W., and Anderson, W. J. (1971). NASA TN D7007.

Coe, H. H., Parker, R. J., and Scribbe, H. W. (1975). NASA TN D7869.

Coleman, W. (1967/1968). *Proc. Inst. Mech. Eng. London* **182** (3A), 191–204.

Coleman, W. (1970). NASA SP-237 (P. Ku, ed.), pp. 551–589.

Cordiano, H. V., Cochrane, E. P., and Wolfe, R. J. (1956). *Lub. Eng.* **12**, 261–266.

Cordiano, H. V., Cochrane, E. P., and Wolfe, R. J. (1958). *Lub. Eng.* **14**, 33–39.

Crook, A. W. (1957). *Proc. Inst. Mech. Eng. London* **171**, 187.

Dalal, H. M., Chui, Y. F., and Rabinowicz, E. (1975). *ASME/ASLE Lub. Conf., Montreal.*

Das, B. N., Basak, R., and Nijhawan, B. N. (1962). *NML Tech. J.* **4**, 4.

Davies, W. J., and Day, K. L. (1964). *Proc. Symp. Fatigue Rolling Contact* pp. 23–39. Institution of Mechanical Engineers, London.

Dawson, P. H. (1961). *Power Transm. London* **30**, 268.

Dawson, P. H. (1962). *J. Mech. Eng. Sci.* **4**, 16.

Dawson, P. H. (1964). *Proc. Symp. Fatigue Rolling Contact* pp. 41–45. Institution of Mechanical Engineers, London.

Dawson, P. H. (1965/1966). *Proc. Inst. Mech. Eng.* **180** (3B), 95–100.

Dowson, D. (1967/1968). *Proc. Inst. Mech. Eng.* **182** (3A), 151–167.

Dowson, D. (1970). NASA SP-237 (P. Ku, ed.), pp. 27–76.

Dowson, D., and Higginson, G. R. (1966). "Elastohydrodynamic Lubrication." Pergamon, Oxford.

Dowson, D., and Whitaker, A. V. (1965). *ASLE Trans.* **8**, 224.

Dowson, D., Higginson, G. R., and Whitaker, A. V. (1962). *J. Mech. Eng. Sci.* **4**, 121.

Dowson, D., Higginson, G. R., and Whitaker, A. V. (1964). *Proc. Symp. Fatigue Rolling Contact* pp. 66–75. Institution of Mechanical Engineers, London.

Drutowski, R. C., and Mikus, E. B. (1960). *J. Basic Eng.* **82**, 302.

Fletcher, H. A. R., and Collier, P. (1966). NEL Rep. 223, National Engineering Laboratory, East Kilbride.

Flom, D. G., and Bueche, A. N. (1959). *J. Appl. Phys.* **30**, 1725.

Forsyth, P. J. E., and Stubbington, C. A. (1954/1955). *J. Inst. Met.* **83**, 395–399.

Frankel, H. E., Bennet, J. A., and Pennington, W. A. (1960). *Trans. Am. Soc. Met.* **52**, 257–276.

Frederick, S. H., and Newman, A. D. (1958). *Proc. Int. Conf. Gearing* pp. 91–98. Institution of Mechanical Engineers, London.

Frith, P. H. (1954). Iron and Steel Inst. Special Rep. 50. Iron and Steel Institute, London.

Galvin, G. D., and Naylor, H. (1965). *Proc. Inst. Mech. Eng. London* **179**, 857–875.

Gentile, A. J., Jordon, E. G., and Martin, A. D. (1965). *Trans. Metall. Soc. AIME* **233**, 1085–1093.

Gilbert, P. T. (1956). *Metall. Rev.* **1**, 379–417.

Goldblatt, I. L. (1972). *ASLE Trans.* **16**, 150.

Goodman, J. (1912). *Proc. Inst. Civ. Eng.* **189**, 82–166.

Greenwood, J. A., Minshall, H., and Tabor, D. (1961). *Proc. R. Soc. London Ser. A* **259**, 48a.

Grozin, D. B. (1957). *Moscow Akad. Nauk. SSSR* 158.

Grubin, A. N. (1949). Department of Scientific and Industrial Research Trans. 337. HM Stationery Office London.

Grunberg, L., and Scott, D. (1960). *J. Inst. Pet. London* **44** (419), 406–410.

Grunberg, L., and Scott, D. (1962). *J. Inst. Pet. London* **48** (440), 259–266.

Grunberg, L., Jamieson, D. T., Scott, D., and Lloyd, R. A. (1960). *Nature London* **188** (4757), 1182–1183.

Grunberg, L., Jamieson, D. T., and Scott, D. (1963). *Philos. Mag.* **8**, 1553.

Gustafson, J. H. (1964). *Lub. Eng.* **20**, 65–68.

Haines, D. J., and Ollerton, F. (1963). *Proc. Inst. Mech. Eng. London* **177**, 195.

Halling, J. (1964). *J. Mech. Eng. Sci.* **6**, 64.

Hamilton, G. M. (1963). *Proc. Inst. Mech. Eng. London* **177**, 667–675.

Harris, T. A. (1968). *ASLE Trans.* **11**, 290–294.

Harris, W. J., and Cohen, M. (1949). *Trans. Am. Inst. Min. Met. Eng.* **180**, 447–470.

Hartung, H. A. (1957). ASME Paper 57-A-277.

Heathcote, H. L. (1921). *Proc. Inst. Mech. Eng. London* **15** (AD), 569–702.

Heindlhofer, K., and Sjovall, H. (1923). *Trans. ASME.* **45**, 141–150.

Helbeg, F. (1949). *Werkstattatech. Maschinenbau* **39**, 111–115.

Hertz, H. (1881). *Gesammelt Werke* **1**, 156.

Hobbs, R. A., and Mullet, G. W. (1969). *Proc. Inst. Mech. Eng. London* **183** (3P), 21–27.

Holmes, P. W. (1972a). NASA Contract Rep. CR 2004.

Holmes, P. W. (1972b). NASA Contract Rep. CR 2005.

Hoyt, S. L. (1952). "Metals and Alloys Data Book." Van Nostrand-Reinhold, Princeton, New Jersey.

Husted, T. E. (1962). SAE Paper 509a.

I. P. 300/75T (1975a). "Pitting Failure Tests for Oils in a Modified Four Ball Machine." Institute of Petroleum, London.

I. P. 305/75T (1975b). "The Assessment of Lubricants by Measurement of their Effect on the Rolling Fatigue Resistance of Bearing Steel using the Unisteel Machine." Institute of Petroleum, London.

Jackson, E. G. (1959). *ASLE Trans.* **2**, 121–128.

Jensen, P. W. (1965). "Cam Design and Manufacture." Industrial Press, New York.

Johnson, K. L. (1958). *J. Appl. Mech.* **80**, 339.

Johnson, K. L. (1960). *In* "Rolling Contact Fatigue Phenomena" (J. Bidwell, ed.), pp. 6–28. Elsevier, Amsterdam.

Johnson, K. L. (1964). *Proc. Symp. Fatigue Rolling Contact.* pp. 155–159. Institution of Mechanical Engineers, London.

Johnson, K. L. (1966). *Wear* **9**, 4–19.

Johnson, K. L., and Tabor, D. (1967/1968). *Proc. Inst. Mech. Eng. London* **182** (3A), 168–172.

Johnson, K. L. Greenwood, J. A., and Poon, S. Y. (1972). *Wear* **19**, 91–108.

Johnson, R. F., and Blank, J. R. (1964). *Proc. Symp. Fatigue Rolling Contact* pp. 95–102. Institution of Mechanical Engineers, London.

Johnson, R. L., and Buckley, D. H. (1967/1968). *Proc. Inst. Mech. Eng. London* **182** (3A), 479–490.

Johnson, R. F., and Sewell, J. F. (1960). *J. Iron Steel Inst.* **196**, 414–444.

Jones, A. B. (1946). ASTM Preprint No. 45.

Jones, A. B. (1952). *Trans. ASME* **74**, 695–703.

Jost, H. P. (1958). *Sci. Lub. Suppl. Issue* November 48–54.

Kalker, J. J. (1964). *Proc. K. Ned. Akad. Wet. Ser. B* **67**, 135.

Kenny, P. (1977) *In* "Rolling Contact Fatigue—Performance Testing of Lubricants" (R. Tourret and E. P. Wright, eds.), pp. 59–74. Institute of Petroleum, London.

Kenny, P., and Yardley, E. D. (1972). *Wear* **20**, 105–121.

Kirk, D., Nelms, P. R., and Arnold, H. (1966). *Metallurgia* **74**, 225.

Kjerrman, B. L. (1952). *Proc. World Metall. Congr., 1st* pp. 502–505. American Society for Metals, Cleveland, Ohio.

Krivoshein, G. S. (1960). *Ind. Lab.* **26**, 405.

Likhtmann, V. I., Rebinder, P. A., and Kapenko, G. V. (1954). "Effect of a Surface Active Medium on the Process of Deformation of Metals." Akad. Nauk. SSSR, Moscow. [Department of Scientific and Industrial Research Trans. (1958). HM Stationery Office, London.]

Lips, F. M. H., and van Zuilen, H. (1954). *Met. Progr.* **66**, 103.

Littmann, W. E. (1970). NASA SP-237 (P. Ku, ed.), pp. 309–378.

Littmann, W. E., and Widner, R. L. (1965). ASME Paper 65 WA/CF2.

Lowenthal, S. H., and Parker, R. J. (1976). NASA TN D-8124.

Lutz, O. (1955). *Konstruktion* **7**, 330.

Macks, E. F. (1953). *Lub. Eng.* **9**, 254–258.

MacPherson, P. B. (1977). *In* "Rolling Contact Fatigue—Performance Testing of Lubricants" (R. Tourret and E. P. Wright, eds.). Institute of Petroleum, London.

March, C. N. (1977). *In* "Rolling Contact Fatigue—Performance Testing of Lubricants" (R. Tourret and E. P. Wright, eds.). Institute of Petroleum, London.

Markert, W., and Ferguson, K. M. (1957). *Lubr. Eng.* **13**, 285–290.

Martin, J. A., Borgese, S. F., and Eberhardt, A. D. (1965). ASME Paper 65-WA/CF-4.

Mattson, R. L., and Roberts, J. G. (1959). "Internal Stresses and Fatigue of Metals." Elsevier, Amsterdam.

May, W. D., Morris, E. L., and Atack, D. (1959). *J. Appl. Phys.* **30**, 1713.

Merritt, H. F. (1958). *Proc. Int. Conf. Gear.* pp. 72–76. Institution of Mechanical Engineers, London.

Merwin, J. E., and Johnson, K. L. (1963). *Proc. Inst. Mech. Eng. London* **177**, 676.

Metals Handbook (1961). "Properties and Selection," 8th ed. American Society for Metals, Cleveland, Ohio.

Middleton, J. L., Westcott, V. C., and Wright, R. W. (1974). *Wear* **30**, 275.

Mikus, E. B., Hughel, T. J., Gerty, S. M., and Knudsen, A. L. (1960). *Trans. Am. Soc. Met.* **52**, 307–320.

Morgan, P. G., (1958). *Iron and Steel London* **31**, 609.

Morland, L. W. (1962). *J. Appl. Mech.* **29**, 345.

Morrison, T. W., Walp, H. O., and Removenko, R. P. (1959). *ASLE Trans.* **2**, 129–146.

Morrison, T. W., Tallian, T. E., Walp, H. O., and Baile, G. H. (1964). *Symp. Fatigue Rolling Contact* pp. 78–85. Institution of Mechanical Engineers, London.

Mould, R. W., and Silver, H. B. (1975). *Wear* **31**, 295.

Mould, R. W., and Silver, H. B. (1976). *Wear* **37**, 377–388.

Moyar, G. J., and Morrow, J. (1964). *Univ. of Illinois Bull.* **486**, 62, 1–80.

Murphy, W. R., Armstrong, E. L., and Wooding, P. S. (1977). *In* "Rolling Contact Fatigue—Performance Testing of Lubricants" (R. Tourret and E. P. Wright, eds.). Institute of Petroleum, London.

Murray, J. D., and Johnson, R. F. (1963). Clean Steels, Iron Steel Inst. Special Rep. 77, p. 104. Iron and Steel Institute, London.

Naylor, H. (1967/1968). *Proc. Inst. Mech. Eng. London* **183** (3A), 237–247.

Neale, M. J. (ed.) (1973). "Tribology Handbook." Butterworth, London.

Nemeth, Z. N., and Anderson, W. J. (1955). *Lub. Eng.* **11**, 267–273.

Nemeth, Z. N., and Anderson, W. J. (1956). *Lub. Eng.* **12**, 327–330.

Niemann, G., and Rettig, H. (1958). *Proc. Int. Conf. Gear.* pp. 31–42. Institution of Mechanical Engineers, London.

O'Brien, J. L., and King, A. H. (1965). ASME Paper 65.

Otterbein, M. E. (1958). *ASLE Trans.* **1**, 33–39.

Palmgren, A. (1959). Ball and Roller Bearing Engineering. SKF Ind., Philadelphia, Pennsylvania.

Parker, R. J. (1977). *In* "Rolling Contact Fatigue—Performance Testing of Lubricants" (R. Tourret and E. P. Wright, eds.). Institute of Petroleum, London.

Parker, R. J., and Zaretsky, E. V. (1973). NASA TN D7383.

Parker, R. J., Grisaffe, S. J., and Zaretsky, E. V. (1964a). NASA TN D2459.

Parker, R. J., Grisaffe, S. J., and Zaretsky, E. V. (1964b). NASA TN D2274.

Parker, R. J., Grisaffe, S. J., and Zaretsky, E. V. (1965). *ASLE Trans.* **8**, 205–216.

Petch, N. J. (1956). *Philos. Mag.* **1**, 331–337.

Petrusevich, A. I. (1961). *Izv. Akad. Nauk, SSSR Otd. Tekh. Nauk* **2**, 209.

Poritsky, H. (1950). *J. Appl. Mech.* **17**, 191.

Poritsky, H. (1970). NASA SP-237 (P. Ku, ed.), pp. 77–152.

Prenosial, B. (1966). *Proc. Int. Conf. Heat Treatment, Bremen.* Société Française de Metallurgie and Inst. für Hutten Technik.

Razin, C. (1966). *Proc. Int. Conf. Heat Treatment, Bremen.* Société Française de Metallurgie and Inst. für Hutten Technik.

Reynolds, O. (1876). *Philos. Trans. R. Soc.* **166**, 155–174.

Rounds, F. G. (1962a). *ASLE Trans.* **5**, 172.

Rounds, F. G. (1962b). *In* "Rolling Contact Phenomena" (J. B. Bidwell, ed.), pp. 346–362. Elsevier, Amsterdam.

Rounds, F. G. (1967). *ASLE Trans.* **10**, 243–255.

Rounds, F. G. (1971). *ASLE Trans.* **13**, 236–245.
Rounds, F. G. (1975). *ASLE Trans.* **18**, 79–89.
Scarlett, N. A. (1967/1968). *Proc. Inst. Mech. Eng. London* **182** (3A), 585–593.
Scott, D. (1957). *Proc. Inst. Mech. Eng. Conf. Lub. Wear* pp. 463–468.
Scott, D. (1958). *Engineering London* **185**, 660–662.
Scott, D. (1960). *NEL Rep. LDR 44/60.* National Engineering Laboratory, East Kilbride.
Scott, D. (1962). *J. Inst. Pet.* **48**, 24–25.
Scott, D. (1964). *Proc. Symp. Fatigue Rolling Contact* pp. 103–115. Institution of Mechanical Engineers, London.
Scott, D. (1967). *Proc. Inst. Mech. Eng. London* **181** (3b), 39–51.
Scott, D. (1967/1968). *Proc. Inst. Mech. Eng. London* **182** (3A), 325–340.
Scott, D. (1968a). *Tribology* **1**, 14–18.
Scott, D. (1968b). *In* "Low Alloy Steels," pp. 203–209. Iron and Steel Institute, London.
Scott, D. (1968c). *In* "Tribology," Proc. J. Residential Course, Paper I. Institution of Metallurgists, London.
Scott, D. (1968d). *NML Tech. J.* **111**, 279–285.
Scott, D. (1969a). *Proc. Symp. Lub. Hostile Environ.* **183** (3L), 9–17. Institution of Mechanical Engineers, London.
Scott, D. (1969b). *In* "Substitute Alloys as Bearing Materials" (P. K. Gupta and J. E. Manna, eds.), pp. 221–229. National Metallurgical Laboratory, Jamshedpur, India.
Scott, D. (1969c). *Proc. Iron Steel Inst. Conf. Tribol. Steelworks* pp. 122–126. Iron Steel Inst. SP125.
Scott, D. (1969d). *Vacuum* **19**, 167–169.
Scott, D. (1970). *In* "Journees d'Etude sur L'Usure." GAMI, Paris.
Scott, D. (1972). *Wear* **21**, 155–165.
Scott, D. (1974a). Colloq. Int. CNRS No. 233, 433–439.
Scott, D. (1975a). *Wear* **32**, 309–314.
Scott, D. (1975b). *Proc. Eur. Symp. Powder Metall.* pp. 1–6. Grenoble Soc. Metall. **3**, Paper 4/9.
Scott, D. (1975c). *Wear* **34**, 15–22.
Scott, D. (1975d). *Proc. Inst. Mech. Eng. London* **189**, 623–633.
Scott, D. (1976a). *Tribol. Int.* **9**, 261–264.
Scott, D. (1976b). *NEL Rep. 627.* National Engineering Laboratory, East Kilbride.
Scott, D. (1977a). *Proc. Plansee Semin. 9th,* Vol. 2, Preprint 41. Metallwerk Plansee, Reutte.
Scott, D. (1977b). *Wear* **43**, 71–87.
Scott, D. (1978). *Wear* **48**, 283–290.
Scott, D., and Blackwell, J. (1964a). *Proc. Inst. Mech. Eng. London* **178** (3N), 75–81.
Scott, D., and Blackwell, J. (1964b). *Proc. Inst. Mech. Eng. London* **178** (3N) 81–89.
Scott, D., and Blackwell, J. (1965). *Proc. Ind. Lub. Symp.* pp. 53–71. Scientific Publ., Broseley, Shropshire.
Scott, D., and Blackwell, J. (1966a). *Rev. Metall. Paris* **3**, 325–341.
Scott, D., and Blackwell, J. (1966b). *Proc. Inst. Mech. Eng. London* **180**, (3K), 32–37.
Scott, D., and Blackwell, J. (1967a). *Proc. Inst. Mech. Eng. London* **181** (3O) 70–76.
Scott, D., and Blackwell, J. (1967b). *Proc. Inst. Mech. Eng. London* **181**, (3O), 77–84.
Scott, D., and Blackwell, J. (1968a). *Proc. Inst. Mech. Eng. London* **182** (3N), 63–67.
Scott, D., and Blackwell, J. (1968b). *Proc. Inst. Mech. Eng. London* **182** (3N), 239–242.
Scott, D., and Blackwell, J. (1969). *Proc. Inst. Mech. Eng. London* **183** (3P), 189–193.
Scott, D., and Blackwell, J. (1971a). *Wear* **17**, 72–86.
Scott, D., and Blackwell, J. (1971b). *Wear* **18**, 19–28.
Scott, D., and Blackwell, J. (1973a). *Wear* **24**, 61–67.

Scott, D., and Blackwell, J. (1973b). *Lub. Eng.* **29**, 99–107.
Scott, D., and Blackwell, J. (1974). *Wear* **34**, 149–158.
Scott, D., and Blackwell, J. (1975). *Proc. Eur. Space Symp. Tribol. 1st, Frascoti, Italy.* pp. 175–183. European Space Agency (Noordwyk), Sp III.
Scott, D., and Blackwell, J. (1977a). *Proc. Eurotrib. 77,* Vol. 1, Paper 69, pp. 1–5.
Scott, D., and Blackwell, J. (1977b). *Proc. Int. Conf. Wear Mater., St. Louis* pp. 306–309. ASME, New York.
Scott, D., and Blackwell, J. (1978). *Wear* **46**, 273–279.
Scott, D., and Jamieson, D. T. (1962). *J. Inst. Pet.* **48**, 91–104.
Scott, D., and Mills, G. H. (1973). *Polymer* **14**, 130–132.
Scott, D., and Mills, G. H. (1974a). "Advances in Polymer Friction and Wear," Vol. 5B, pp. 441–550. Plenum, New York.
Scott, D., and Mills, G. H. (1974b). *Wear* **30**, 275.
Scott, D., and Mills, G. H. (1974c). *In* "Scanning Electron Microscopy," pp. 883–888. IIT Res. Inst., Chicago, Illinois.
Scott, D., and McCullagh, P. J. (1972). *Electrodeposition Surf. Treat.* **1**, 21–31.
Scott, D., and McCullagh, P. J. (1973a). *Wear* **25**, 339–341.
Scott, D., and McCullagh, P. J. (1973b). *Wear* **24**, 119.
Scott, D., and McCullagh, P. J. (1975a). *Proc. Solid Lub. Symp., Tokyo* pp. 49–50. JSLE, Tokyo.
Scott, D., and McCullagh, P. J. (1975b). *Wear* **34**, 227–237.
Scott, D., and McCullagh, P. J. (1976). *Proc. ASLE/JSLE Int. Lub. Conf., Tokyo* pp. 484–491. Elsevier, Amsterdam.
Scott, D., and Scott, H. M. (1960). *Proc. Reg. Eur. Conf. Electron Microsc.* **1**, 539–542. Springer-Verlag, Berlin and New York.
Scott, D., and Smith, A. I. (1973). *Proc. Inst. Mech. Eng. Conf. Select. Mater. Machine Des. Publ.* 22.3–14.
Scott, D., and Westcott, V. C. (1977a). *Wear* **44**, 173–182.
Scott, D., and Westcott, V. C. (1977b). *Proc. Eurotrib. 77,* Vol. 1, Paper 70.
Scott, D., and Westcott, V. C. (1978). *Proc. Conf. Tribol., Swansea.* Institution of Mechanical Engineers, London. (in press).
Scott, R. L., Kepple, R. K., and Miller, M. H. (1962). *In* "Rolling Contact Fatigue Phenomena" (J. Bidwell ed.), pp. 301–306. Elsevier, Amsterdam.
Scott, D., Loy, B., and Mills, G. H. (1967). *Proc. Inst. Mech. Eng. London* **181** (3I), 94–103.
Scott, D., Loy, B., and McCallum, R. (1969/1970). *Proc. Inst. Mech. Eng. London* **184** (3B), 25–35.
Scott, D., Blackwell, J., and McCullagh P. J. (1971). *Wear* **17**, 72–86.
Scott, D., Seifert, W. W., and Westcott, V. C. (1974). *Sci. Am.* **230**, 88–97.
Scott, D., Seifert, W. W., and Westcott, V. C. (1975). *Wear* **34**, 251–256.
Scott, D., Seifert, W. W., and Westcott, V. C. (1975). *Wear 34,* 251–256.
Scott, D., McCullagh, P. J., and Hands, B. J. (1975). *Trans. Inst. Met. Finish.* **32**, 309–314.
Seifert, W. W., and Westcott, V. C. (1972). *Wear* **21**, 22–48.
Seleznev, N. N. (1957). *Tr. Donetsk Ind. Inst.* **19**, 97.
Shatzberg, P., and Felsan, I. M. (1968). *Wear* **12**, 331.
Shotter, B. A. (1958). *Proc. Inst. Conf. Gearing* pp. 120–125. Institution of Mechanical Engineers, London.
Sibley, L. B., and Orcutt, F. K. (1961). *ASLE Trans.* **4**, 234–239.
Sorem, S. S., and Caltaneo, A. G. (1956). *Lub. Eng.* **12**, 258–260.
Sternlicht, B., Lewis, P., and Flynn, P. (1960). ASME Paper 60-WA-286.
Sternlicht, B., Lewis, P., and Flynn, P. (1961). *J. Basic Eng.* **83**, 213–226.

Stribeck, R. (1901). *Engineering* Dec. 4th, 463–468.

Studenok, I. A., and Makhon'ko, E. P. (1957). Nature of the white layers on the rolling surface of wagon wheels. *Dokl. Nauch. Konferentsii. Posviashch. 40-letiu Velikoi Oktiabr'sk. Sots. Revoliutsii, 7th, Tomsk, USSR,* No. 2, pp. 52–53. Tomskii Univ., Tomsk, USSR.

Styri, H. (1951). *Am. Soc. Test. Mater. Proc.* **51,** 682.

Tabor, D. (1956). *Lub. Eng.* **12,** 379–386.

Tallian, T. (1967a). ASME Preprint AMIL-3.

Tallian, T. (1967b). *ASLE Trans.* **10,** 418–439.

Tallian, T. (1967/1968). *Proc. Inst. Mech. Eng. London* **182** (3A), 205–236.

Tallian, T. (1972). *Wear* **21,** 49–101.

Taylor, K. M., Sibley, L. B., and Lawrence, J. C. (1961). ASME Paper 61-LUBS-12.

Taylor, K. M., Sibley, L. B., and Lawrence, J. C. (1963). *Wear* **6,** 226.

Taylor, R. G., and Andrew, S. (1963). MIRA Rep. 5.

Theyse, F. H. (1966). *Wear* **9,** 41–59.

Tickler, M., and Scott, D. (1970). *Wear* **16,** 229.

Timoshenko, S. P., and Goodier, J. N. (1970). "Theory of Elasticity." McGraw-Hill, New York.

Tokuda, M., Ito, S., and Muro, H. (1977). *In* "Rolling Contact Fatigue—Performance Testing of Lubricants" (R. Tourret and E. P. Wright, eds.), pp. 107–116. Institute of Petroleum, London.

Uhrees, L. D. (1963). Clean Steels *Iron Steel Inst. Spec. Rep.* 77 p. 104. Iron and Steel Institute, London.

Vermeulen, P. J., and Johnson, K. L. (1964). *J. Appl. Mech.* **86,** 337.

Waterhouse, R. B., and Allery, M. (1965). *Wear* **8,** 112–120.

Way, S. (1935). *J. Appl. Mech.* **2,** A49–A58, A110–A115.

Wellons, F. W., and Harris, T. A. (1970). NASA SP-237 (P. Ku, ed.), pp. 529–549.

Wernitz, W. (1958). *Schriftenr. Autriebstechnik* **19,** Fr. Vieweg, Braunschweig.

Wilson, R. W. (1973). *Proc. Eur. Tribol. Congr., 1st* pp. 165–175. Institution of Mechanical Engineers, London.

Woodley, B. J. (1977). *Tribol. Int.* **10,** 323–331.

Yardley, E. D. (1977). *In* "Rolling Contact Fatigue—Performance Testing of Lubricants" (R. Tourret and E. P. Wright, eds.), pp. 47–58, Institute of Petroleum, London.

Yardley, E. D., Kenny, P., and Sutcliffe, D. A. (1974). *Wear* **28,** 29–47.

Zaretsky, E. V., and Anderson' W. J. (1961). *J. Basic Eng.* **83,** 603–612.

Zaretsky, E. V., and Anderson, W. J. (1970). NASA SP-237 (P. Ku, ed.), pp. 370–408.

Zaretsky, EV., and Anderson, W. J. (1970). NASA SP-237 (P. Ku, ed.), pp. 379–408.

Zaretsky, E. V., Anderson, W. J., and Parker, R. J. (1962a). *ASLE Trans.* **5,** 210–219.

Zaretsky, E. V., Anderson, W. J., and Parker, R. J. (1962b). NASA TN D1404.

Zaretsky, E. V., Parker, R. J., and Anderson, W. J. (1965a). NASA TN D2640.

Zaretsky, E. V., Parker, R. J., and Anderson, W. J. (1965b). NASA TN D2664.

Wear Resistance of Metals

T. S. EYRE

Department of Metallurgy
Brunel University
Uxbridge, Middlesex
England

I. Introduction

Wear resistance is not an intrinsic property of a metal in the same way as is Young's modulus or thermal conductivity although these and other properties may have an appreciable effect on wear resistance. Wear can occur only as a result of friction arising between one surface and another, which surfaces may be solid, liquid, or gaseous, under both load and motion. The amount of wear will depend upon the particular friction couple and the environmental conditions existing at the interacting surfaces.

There is no simple direct relationship between friction and wear, but for a given metallic couple low friction will generally result in low wear, and the converse is also true. Small changes in friction may, however, indicate quite large changes in wear rate. Whereas the coefficient of friction for metal-to-metal contact under adhesive conditions in the unlubricated condition only varies from about 0.4 to 1.5, wear rates may vary by many orders of magnitude; some examples are given in Table I (Hirst, 1965).

Wear is usually expressed as a weight or volume loss per unit of sliding distance (cubic centimeters per centimeter of sliding). Wear resistance is not a well-defined quantitative term and should only be used in a comparative sense where specific figures are quoted under well-defined conditions. In general, wear resistance is inversely related to wear volume, i.e.,

$$\text{wear resistance} = 1/\text{wear volume}$$

This relationship can only be applied to steady-state conditions, i.e., abrasive wear or adhesive conditions after running in has been completed, and then only if transitions from one mechanism to another do not occur.

Under normal engineering conditions a large number of variables may contribute to wear and a systems engineering approach must be adopted if we are to understand the effect of these (Fig. 1) (Czichos, 1974). Even small changes in one of these variables may cause an unacceptable increase in friction and wear or in vibration and noise. It may be that wear as such is not the controlling factor other than for the effect it has on increased noise or vibration level, causing unacceptable working conditions. The development of debris in situ may, by reducing the clearance between the moving surfaces, cause increased friction and ultimate seizure.

Any change in the input into the system that effectively increases the temperature generated at the friction surface can bring about marked changes in wear. On the one hand, those metals which do not readily form a protective oxide, e.g., stainless steel and nimonic alloys, have poor wear resistance at low temperatures, but improve as either the ambient or friction temperature rises. On the other hand, if annealing, recrystallization, or tempering occurs during wear, then it is likely that wear resistance will decrease because of a reduction in hardness. As far as materials

TABLE I

COEFFICIENT OF FRICTION AND WEAR RATE OF SOME METAL COUPLES

Metal combination	Coefficient of friction μ	Wear rate $(cm^3/cm \times 10^{-12})$
0.2% carbon steel on itself	0.62	157,000
60% Cu–37% Zn–3% Pb on steel	0.60	24,000
Stellite on carbon steel	0.60	320
Ferritic stainless steel on carbon steel	0.53	270
Tungsten carbide on itself	0.35	2

are concerned, in general it is beneficial to use those with a high thermal conductivity, which will conduct the heat away rapidly. Any environment that prevents surface oxidation during adhesive wear will result in a higher wear rate and a reduction in wear resistance. As far as surface preparation is concerned, roughness may cause excessive abrasive wear of the counterface, particularly if the rough surface is also much harder than the counterface. Generally speaking, it is preferable to confine wear to one of the components, i.e., the bearing, because the shaft operating in the bearing will be much more expensive to replace or repair. The formation of hard debris developed in situ will contribute to abrasive wear and cause machining of one or both surfaces if the debris is larger than the clearance and cannot be removed.

Wear is rarely catastrophic and for this reason it has not been as extensively researched as corrosion and fatigue. Wear usually leads to a reduction in operating efficiency and more frequent replacement or repair of worn components. Sometimes, however, a rare catastrophic form of wear failure enables research to be focused on a particular problem area. For example, the "wire wool" or "machining" failures that were frequently encounted during the 1950s and 1960s emphasized the critical importance of unacceptably high levels of metal contaminant particles in the lubricant used for rotor shafts and plain bearings of power generation plants. This caused the generation of hard "scabs" which became embedded in the white metal bearings, which subsequently rubbed against the shaft surface and "machined" long particles of metal, which then became compacted into "wire wool" (Archard and Hirst, 1956). It also became apparent that steels containing chromium were particularly susceptible to this type of failure. Chromium steels were introduced during this period for their increased oxidation and creep resistance in high-temperature applications.

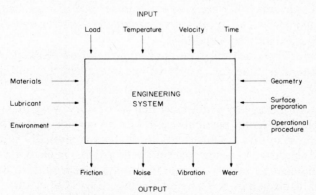

Fig. 1. Input into and output from engineering equipment.

These steels were notable for failures that usually occurred in the very early stages of running. Attention was later drawn to the poor running-in characteristics of steels containing chromium (Farrell and Eyre, 1975) and more recently this has been related to their oxidation characteristics.

In some situations the amount of wear that can be tolerated may be small, e.g., liquid or gas seals (piston rings), and in others the loss of a considerable amount of material may be tolerated, e.g., mineral processing equipment. A very large number of metals and alloys and an equally large number of surface treatments have been developed to obtain improvements in the life of engineering components. One of the most difficult problems, however, is to select the most appropriate material or treatment for a specific application; this will be discussed later.

The terminology used to describe wear of metals is still rather vague and imprecise, e.g., scratching, gouging, and galling are still widely used. Every effort should be made to adopt terminology that is clear and unambiguous. The OECD document on *Terms and Definitions* (OECD, 1969) should be used by all those concerned with selecting, applying, or researching materials to be used for tribological applications.

There are no standard tests for measuring wear resistance and because of the large number of variables there is still considerable controversy about the value of "wear testing" procedures. Wear testing tends to fall into three major groups.

1.   Laboratory techniques which are used under well-defined conditions so as to isolate a particular wear mechanism and to relate this to composition, structure, environmental variables, etc. The greatest value of this type of test is the understanding it provides of the fundamentals of friction and wear and the development of improved materials.

2.   Simulative rig testing, which is carried out in an attempt to reproduce those variables that are suspected in service to contribute to wear. The great danger in simulating some but not all of the variables is that misleading answers may be obtained.

3.   Service tests are the final answer but can be difficult to monitor, take considerable time for results to become available, and are usually very costly.

There is an increasing awareness that laboratory wear techniques will become more valuable, particularly if wear in service is studied with the same diligence so that the information obtained is mutually supportive. It is essential that the laboratory technique be applied in such a way that the wear mechanism known to operate in service is reproduced. These techniques are discussed in great detail elsewhere (ASTM, 1970).

II. Wear in Industry

Wear encountered in industry covers many quite different engineering situations and operating conditions, and it can be dangerous to overgeneralize about these. Surveys carried out, however, all produce similar results and show that the following wear types are most widely encountered:

Abrasive	50%	Fretting	8%
Adhesive	15%	Chemical	5%
Erosive	8%		

There are, however, many situations in which wear can change from one type to another. Adhesive wear may be responsible for generating hard debris which then contributes to abrasive wear. Two types of wear may operate simultaneously; e.g., abrasive wear and chemical wear operate together in marine diesel cylinder liners. It is therefore very important that a correct diagnosis of wear in a particular situation be obtained before solutions are sought. Attention should be drawn to the difficulty in applying fundamental information to real industrial situations, particularly where complex mechanisms interact.

A. Abrasive Wear

Abrasion involves the removal of material from a surface by the mechanical action of an abrasive which has an acicular profile and is harder than the surface being worn. The general surface characteristics of an abraded surface are shown in Fig. 2, from which it will be seen that damage takes the form of parallel grooves running in the rubbing direction. The number and size of the grooves and thus the volume of metal removed from the surface will vary considerably from one situation to another, and general descriptions such as scratching and gouging have been used to describe qualitatively the severity of damage. It must be emphasized, however, that in both cases an abrasive wear mechanism is responsible for the damage caused.

Abrasive wear usually takes place under two- or three-body conditions (Fig. 3). In the former, particles are transported along the surface with both a sliding and rolling action in such a way that a reduction in particle size does not normally occur. This is often referred to as low-stress abrasion because the stress is below the fracture stress of the abrasive. In three-body wear, abrasive materials are being deliberately broken down into a smaller size, and the stresses required are quite high. This is often

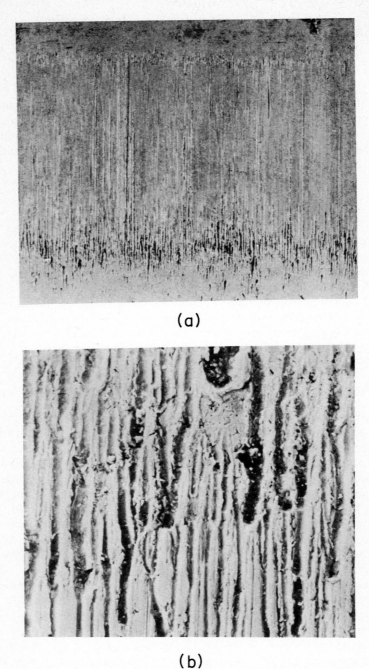

Fig. 2. Surface characteristics of abrasive wear as viewed in the scanning electron microscope. (a) ×40; (b) ×400.

described as high-stress abrasion. This situation may also arise in engineering bearings if a hard wear debris is produced in situ or if abrasives are accidentally introduced from an external source. Hard particles become embedded in the softer surface (the bearing) and then act as an abrasive lap to the harder surface (the shaft), causing a higher rate of wear.

Wear resistance of metals is directly related to their hardness, and the degree of penetration of the particle is given by

$$\frac{w}{H_V} = \frac{\text{load on the particle}}{\text{hardness of the surface}} \tag{1}$$

Wear volume (V) is given by

$$V = (w/H_V)AL \tag{2}$$

where A is the cross sectional area of the groove and L is the length traversed. Abrasive wear usually increases linearly with both the applied load and the sliding distance. If deviations occur, they are usually caused by either a reduction in particle size of the abrasive or clogging of the surface being abraded by the debris. Much of our knowledge of abrasive wear is the result of the work of Kruschov (1957), who obtained the results shown in Fig. 4. Relative wear resistance E is inversely proportional to hardness, and there is a constant of proportionality which holds good for pure metals. This constant, however, changes for more complex inhomogeneous alloys, and E is then related to the volume fraction of the phases present:

$$E = Be_1 + Ce_2 \tag{3}$$

where E is the relative wear resistance of the sample, equal to the wear of sample material divided by the wear of a standard under the same wear conditions; B the volume fraction of component 1; C the volume fraction of component 2; and e_1 and e_2 are the relative wear resistances of components 1 and 2, respectively.

It was concluded by Krushov (1957) and Boyes (1969) that the hardness value to be considered was not the annealed hardness but that of the worn surface, i.e., the fully work-hardened hardness. It was found that prior

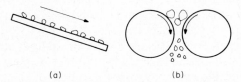

(a)   (b)

Fig. 3.   Abrasive wear conditions. (a) Two-body wear: particles sliding and rolling over a conveyor surface. (b) Three-body wear: roll crushing in mineral dressing.

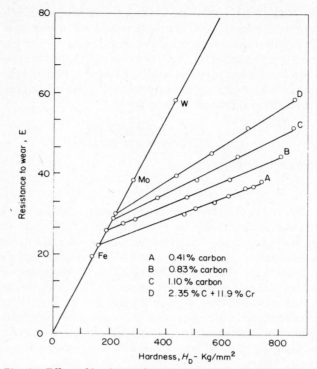

Fig. 4. Effect of hardness of metals on abrasive wear resistance.

work hardening did not increase the wear resistance because work hardening normally occurred during wear. Different microstructures of the same hardness behave differently in wear, and various attempts have been made to relate wear resistance to other properties, including Young's modulus and yield strength. The relationships obtained, however, have been no better than those related to hardness, and this is not surprising because all of these mechanical properties are related to one another. It is much more likely that wear resistance as obtained for a particular wear mechanism, for example, abrasive wear, is related to both the strength of the matrix and its fracture resistance as expressed either by ductility or fracture toughness.

Nonferrous metals, with the exception of stellites and tungsten carbides, are rarely used for abrasive applications. A great deal of information is available on the abrasive wear resistance of steels and cast irons. This enables a fairly rational approach to the selection of abrasion-resistant metals to be carried out. Serpik and Kantor (1965) have shown

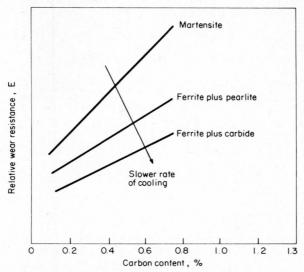

Fig. 5. Effect of carbon content and microstructure on abrasive wear resistance of steels.

that for steels E increases with the carbon content and also varies with the microstructure for a given carbon content. Considerable improvements in E are brought about by heat treatment to refine the pearlite, and this is attributed to the increased rate of work hardening as the interlamellar spacing of the ferrite is reduced (Fig. 5). For a given carbon content martensitic structures provide the greatest resistance to abrasion. It has also been shown that both the volume of the carbides present and their precise composition are important. Vanadium and niobium carbides are superior to both chromium and tungsten carbides for a given concentration of carbide (Fig. 6) (Serpik and Kantor, 1965). These differences are attributed to the hardness of the particular carbides (Table II). Kirk (1974) in a review of tool steels refers to the improved life of a high-speed steel containing 5% vanadium, which owes its outstanding "inbuilt" abrasion resistance to the high content of vanadium carbide present. Atterbury (1974), also commenting on experiences with tool steels, states that the advantage of the steel under discussion was attributable to the presence of vanadium instead of tungsten, which gives it slightly improved abrasion resistance.

The concentration of carbide is important and no further advantage is obtained above some optimum value. For white cast irons the volume of carbide was varied by increasing the carbide content within the range 1.0–3.25%, and no further improvement in abrasion resistance was ob-

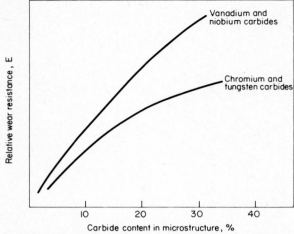

Fig. 6.   Effect of type and percent of carbides on abrasive wear resistance.

TABLE II

HARDNESS OF ABRASIVES AND SECOND PHASES[a]

| Mineral | Hardness | | Material or phase | Hardness | |
	Knoop	HV		Knoop	HV
Talc	20		Ferrite	235	70–200
Carbon	35		Pearlite, unalloyed		250–320
Gypsum	40	36	Pearlite, alloyed		300–460
Calcite	130	140	Austenite, 12% Mn	305	170–230
Fluorite	175	190	Austenite, low alloy		250–350
Apatite	335	540	Austenite, high Cr iron		300–600
Glass	455	500	Martensite	500–800	500–1010
Feldspar	550	600–750	Cementite	1025	840–1100
Magnetite	575		Chromium carbide (Fe, Cr)$_7$C$_3$	1735	1200–1600
Orthoclase	620		Molybdenum carbide Mo$_2$C	1800	1500
Flint	820	950	Tungsten carbide WC	1800	2400
Quartz	840	900–1280	Vanadium carbide VC	2660	2800
Topaz	1330	1430	Titanium carbide TiC	2470	3200
Garnet	1360		Boron carbide B$_4$C	2800	3700
Emery	1400				
Corundum	2020	1800			
Silicon carbide	2585	2600			
Diamond	7575	10 000			

[a] Courtesy of Climax Molybdenum.

tained above 40% by volume of carbide (Fig. 7) (Popov and Nagomyi, 1969). Above this level the presence of a brittle acicular hypereutectic carbide and the M_3C carbide prevented further improvements by causing fracture.

The carbide particle size in white cast irons was varied by adjusting the cooling rate from the solidification temperature, and the effect is shown in Fig. 8. These results reflect the whole range of carbide sizes that could be produced in this way, and it will be observed that considerable improvements are obtained as the carbide size is reduced. It can be concluded that as the carbide size increases it becomes easier for it to be dislodged from the matrix. A matrix of martensite was found to give the maximum wear resistance and is also the optimum structure for retaining the carbide. A metastable austenitic matrix may be more suitable under conditions of severe abrasion in which phase transformation or work hardening is possible and where martensite would be too brittle. Under less severe conditions, however, a pearlitic or martensitic matrix is preferred because phase transformation would not occur in the austenitic matrix. Surface treatments, particularly those produced by heat treatment, including carburizing of steel and the application of hard facing, are widely used for combating wear under abrasive conditions. In most cases "hard facing" is applied in situations in which considerable wear is expected and relatively thick layers are therefore laid down. Heat-treated surfaces are usually much thinner, and it is important, therefore, to realize that there is

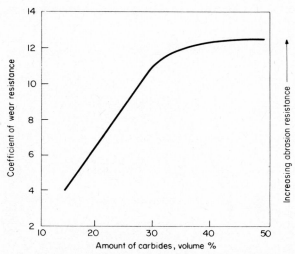

Fig. 7.   Abrasive wear resistance of cast iron containing carbides.

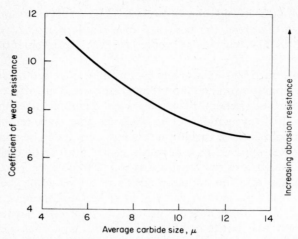

Fig. 8.   Effect of carbide particle size on abrasive wear resistance of white cast irons.

a complex interplay with factors other than those already discussed, particularly the thickness of the hardened layer and the strength of the substrate material used. It has been shown for boronizing that when it is applied to a higher carbon steel (0.35 as compared to 0.15% carbon), there is an improvement in the load-bearing capacity (Fig. 9) even though the hardness and thickness of the coating are identical in both cases (Eyre, 1975). It is extremely important, therefore, that these factors should also be considered at the design stage. In general, hard surfaces confer max-

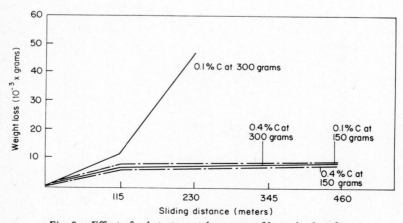

Fig. 9.   Effect of substrate metal on wear of boronized surface.

imum wear resistance when adequately supported by a strong substrate material.

Richardson (1968) has shown that the hardness of the abrasive particles must also be taken into account, and he refers to "soft" and "hard" abrasives. Figure 10 shows that the wear resistance of steel increases rapidly only when the hardness of the surfaces being abraded exceeds some critical value which is 0.5 of the hardness of the abrasive:

$$K_T = H_V \text{ of surface}/H_V \text{ of abrasive} > 0.5 \qquad (4)$$

It will be seen, therefore, that the nature of the abrasive itself must be considered before a decision about the minimum hardness of the metal surface is made. The hardness of some of the more common mineral abrasives and oxides and also the constituents encountered in ferrous metals are given in Table II.

In abrasive wear by loose particles, only a small fraction, usually less than 10% of the particles, results in wear, i.e., those that have an optimum attack angle; the remainder of the particles roll or slide across the surface, causing little wear (Sedrickas and Mulhearn, 1963). The attack angle is shown in Fig. 11, and when it is between approximately 50 and 120°, maximum cutting of the surface occurs, and thus wear resistance is reduced (Fig. 12). The abrasive wear rate is almost always directly proportional to load and is substantially independent of sliding speed, provided that con-

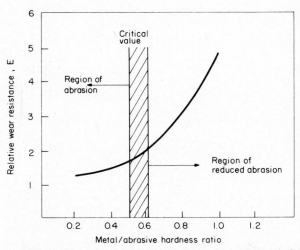

Fig. 10.   Effect of hardness of the abrasive on wear resistance of metals.

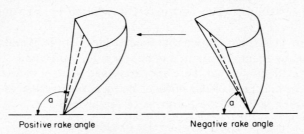

Fig. 11. Attack angle of abrasive particles. Angle *a* is referred to as the "attack angle." The arrow indicates the direction of sliding.

ditions do not favor a change in wear mechanism, which is only likely if thermal or corrosive effects become significant. Wear rate is also constant when the materials resisting abrasion and the abrasive itself remain fixed. If, however, the particle size of the abrasive varies, there is a critical value above which the wear rate remains constant, but there is a strong dependence of wear rate on particle size below this critical size (Fig. 13). Since this has been observed for different wear-resistant metals and also for both adhesive and erosive wear, it has been suggested that variation in particle geometry and the presence of cracks in the particles caused during manufacture may contribute to their breakdown during rubbing. Although the critical particle size has not been estimated with any great accuracy, it would appear to be less than 0.2 mm.

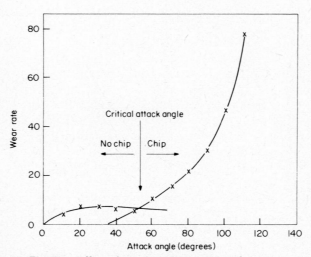

Fig. 12.   Effect of attack angle on wear resistance.

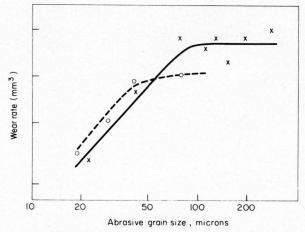

Fig. 13. Effect of abrasive particle size on wear of metals. × steel; ○ bronze.

1. EROSION, A SPECIAL FORM OF ABRASIVE WEAR

The word *erosion* is derived from the Greek word *erodore*, meaning to gnaw away, but generally refers to the phenomena of surface destruction and removal of material by essentially external mechanical forces. These forces usually arise because of multiple impacts produced by the dynamic impingement of a liquid on a solid surface or of a solid on another solid surface (Eyre, 1971). The occurrence of such multiple impacts between liquid droplets on a solid surface is a common natural phenomenon, e.g., raindrops impacting against glass windows. However, when rain drops strike against a stationary or slow-moving solid object, then the pressure generated by the impact is so low that little surface damage is observed. If the velocity of the solid object is very high, e.g., supersonic jet planes, rockets, or missiles, then rain drops can cause severe erosion damage after a relatively short length of time. This type of surface damage can also be caused by impacting solid particles such as dust or sand particles, in which case it is called "impact erosion." The occurrence of impact erosion is by no means associated only with natural phenomena such as rain, dust, or sand storms, but occurs frequently in man-made systems such as the erosion of steam turbines due to the impact of condensed liquid droplets, the sand erosion of water turbines, and other control devices. The general surface characteristics exhibited by impact erosion are shown in Fig. 14.

The second type of erosion damage, known as "cavitation erosion," can only be caused by liquids. This type of erosion is caused primarily by

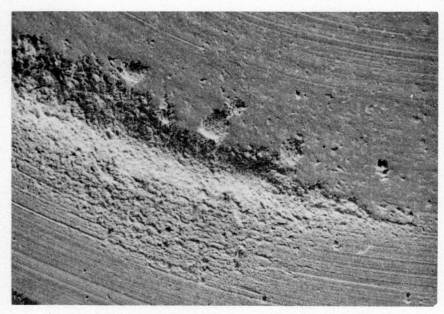

Fig. 14.   Surface characteristics of erosion as viewed in the SEM (×50).

flowing liquids or a solid object moving at a high velocity through a stationary or comparatively slow-moving liquid. Although erosion by
flowing liquids can occur naturally, e.g., erosion of river beds, especially
at rapids, it also causes major design problems in ship propellers, pumps,
valves, hydrofoils, bearings, diesel engine cylinder liners, etc. Cavitation
erosion will not be discussed further.

2.   IMPACT EROSION

The particle velocity and impact angle, combined with the size of the
abrasive, give a good measure of the kinetic energy of the erosive stream.
Volume wear is generally proportional to the (velocity)^3. When the impact
angle is small, generally less than 30°, cutting wear prevails and increasing
the hardness of the surface increases its wear resistance. At high angles
wear is caused primarily by deformation and impact, and a solution is
rather more complex (Fig. 15). A softer material with high elasticity and
deformability may be more suitable. Both natural and synthetic rubbers
produce good results at high impact angles because of their low moduli of
elasticity, and there would appear to be a good correlation between erosion resistance and ultimate resilience:

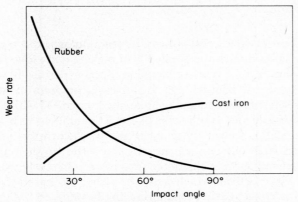

Fig. 15. Effect of impact angle on wear rate.

$$\text{ultimate resilience} = \tfrac{1}{2}(\text{tensile strength})^2/\text{elastic modulus} \qquad (5)$$

Equation (5) expresses the amount of energy that can be absorbed before deformation or cracking occurs. A solution to an impact erosion wear problem may therefore be obtained by changing the angle of attack, reducing the velocity of flow, or by choosing a more suitable material (Fig. 16).

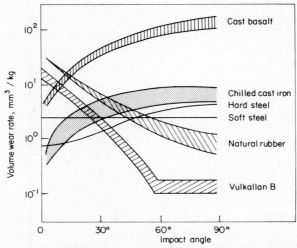

Fig. 16. Erosive wear resistance of some engineering materials.

B. Adhesive Wear

Adhesive wear involves contact and interaction between asperities on two surfaces, and although the applied load may be quite low, the contact stress is high because the real contact area of the asperities is very small. If the contact stress is high enough to cause plastic deformation and the surfaces are sufficiently clean and chemically compatible with each other, the formation of adhesive junctions is likely. Eventually the shear stresses developed by sliding cause these junctions to fracture and transfer of material occurs from one surface, usually the softer, to the harder counterface. The general surface characteristics of an adhesively worn surface are shown in Fig. 17, from which it will be seen that the most significant feature is transferred particles adhering to the surface. Microsections through these particles usually show that transfer has taken place in a number of operations, as indicated by the deformed and layered appearance. Because of frictional heating, it is likely that hardening or softening of the particles will also occur. The presence of hard adhered particles will undoubtedly cause further wear by abrasion.

Adhesive wear is promoted by a number of factors, the most important of which are the following:

1. The tendency for different counterface materials to form solid solutions or intermetallic compounds with one another is directly related to adhesive wear (Fig. 18). Lead, silver, and tin are very successful either as coatings or when alloyed with copper- or aluminum-based metals when sliding against ferrous metals. Titanium and nickel are very poor when sliding against copper and chromium because of their mutual solubility. Ideally, therefore, it can be concluded that friction couples should be chosen to have the minimum tendency to form solid solutions, and this is achieved by choosing (a) different crystal structures and (b) different chemical properties.

2. The cleaner the surfaces, the more likely it is that surface reactions, particularly primary bonding and diffusion, will occur to promote adhesive wear. Surfaces, therefore, should be contaminated with an oxide, solid lubricant, or EP additive. Friction is reduced and wear consequently decreases when oxygen enables a protective oxide layer to form in situ on one or both of the surfaces during sliding wear.

Those metals that do not easily form a thick protective oxide offer poor adhesive wear resistance, and both stainless steels and nickel-base alloys fall into this category. Figure 19 shows that at room temperature wear of nickel is high and only reduces to a much lower value when, above some critical temperature, a thick protective oxide forms. Available evidence

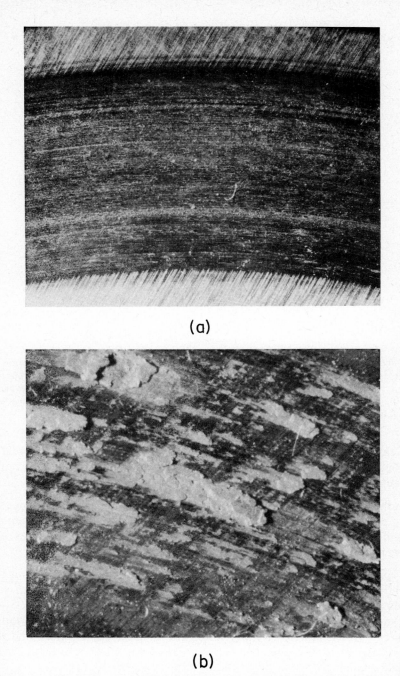

Fig. 17. Surface characteristics of adhesive wear (×8). (a) Nonadhesive (oxidative) wear. (b) Adhesive wear and metal transfer.

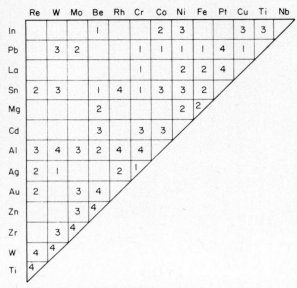

	Re	W	Mo	Be	Rh	Cr	Co	Ni	Fe	Pt	Cu	Ti	Nb
In				1			2	3			3	3	
Pb		3	2			1	1	1	1	4	1		
La						1		2	2	4			
Sn	2	3		1	4	1	3	3	2				
Mg				2				2	2				
Cd				3		3	3						
Al	3	4	3	2	4	4							
Ag	2	1			2	1							
Au	2		3	4									
Zn			3	4									
Zr		3	4										
W	4	4											
Ti	4												

Fig. 18. Tendency for different metals to adhere together. 1 denotes greatest resistance and therefore best combination for wear. 4 denotes least resistance and therefore worst for wear. 1—Two liquid phases, solid solution less than 0.1%; 2—Two liquid phases, solid solution greater than 0.1% or one liquid phase, solid solution less than 0.1%; 3—One liquid phase, solid solution between 0.1% and 1.0%; 4—One liquid phase, solid solution over 1%. Blank boxes indicate insufficient information.

indicates that the thickness of protective oxides needed to reduce wear is of the order of 1 to 15 μm, which is much thicker than the oxides normally present on metal surfaces.

Adhesive wear is more complex than abrasive wear, and the effects that may be attributed to oxidation and clean surfaces have already been referred to. When two metals slide together under adhesive conditions, the wear behavior can be described in one of three ways (Fig. 20):

Type A. This type of wear indicates a high wear rate throughout the duration of the test, which is usually fairly short. The counterface surfaces and the debris appear to be metallic.

Type B. This indicates a very low wear rate throughout the duration of the test, which is usually fairly long. The counterface surfaces and the debris appear to be oxide. The particle size of the debris is orders of magnitude smaller than that produced by type A wear.

Type C. Initially the wear rate is high and the surfaces and debris appear to be metallic. This situation changes fairly sharply, and the wear

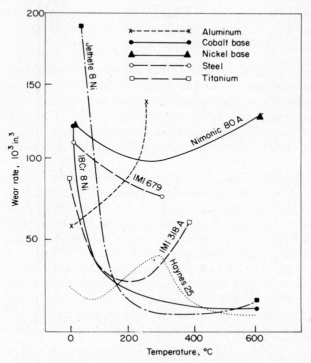

Fig. 19. Wear of metals with variation in temperature.

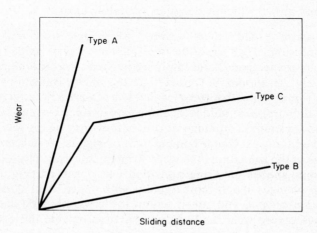

Fig. 20. Different types of wear behavior.

rate reduces to a much lower value for the remainder of the test with the production of oxide debris. The first part of the wear curve is similar to type A and the second part to type B.

It will be observed, therefore, that there are two forms of wear under adhesive conditions, one of which can be described as metallic with evidence of adhesion and the second described as oxidative with no evidence of adhesion. Sometimes components exhibit both forms of behavior: initially it is of a metallic adhesive type, and this region is often referred to as *running in,* which precedes the equilibrium or steady-state oxidative wear.

These different forms of wear can be obtained for the same material combination if the load and speed are varied over a sufficiently wide range and transitions from one type of wear A to B and B to A are observed. This behavior for both steels and cast iron has been widely reported (Welsh, 1965; Eyre *et al.,* 1969), but Welsh was the first to systematically study these effects. The pattern of wear for a 0.52% carbon steel is given in Fig. 21, which shows a change from a low value of 10^{-10}–10^{-9} cm^3/cm with both an oxidative wear track and debris to a much higher value of 10^{-5}–10^{-6} cm^3/cm, which then persists until a much higher load is reached, at which time the wear rate falls sharply to a much lower value. Welsh found it convenient to designate these wear rate transitions:

T_1: below which wear is oxidative and often referred to as "mild" wear. Above T_1, wear is metallic. This wear regime is often referred to as "severe" wear.

T_2: below which wear is metallic and is often referred to as "severe wear," and above which it becomes oxidative again.

These terms "mild" and "severe" are widely used and sometimes cause confusion. They describe in a qualitative way oxidative and metallic wear, respectively, and imply a low wear rate and high wear rate, respectively. It should be noted that the sharp transitions that exist between these respective wear regimes are markedly affected by both load and sliding speed. Mild wear rates may in some circumstances be too high to be acceptable, although of course they may be the lowest that can be achieved for a particular combination of metals. Welsh observed that the values of the transition load were strongly influenced by sliding speed, and at some speeds only one transition was observed (Fig. 22). The T_2 transition changes much more rapidly with speed than does the T_1. A combination of load and speed within the cross-hatched area will give continuous severe wear, and any combination outside this area will give mild wear.

By varying the carbon content in a range of steels this general pattern of

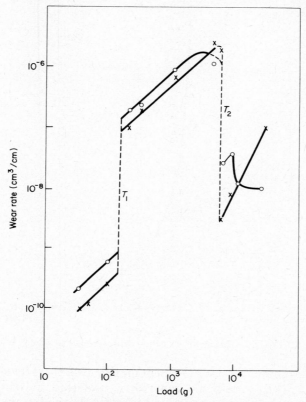

Fig. 21. Transitional wear behavior for a 0.52% carbon steel.

behavior was reproduced but with T_1 and T_2 displaced to different values. For the lowest carbon content (0.026%) severe wear was obtained over the whole load range with no T_1 and T_2 transitions. As the carbon content increases the T_1–T_2 range narrows, and above 0.75% carbon the severe wear range is eliminated entirely. A T_1 transition was induced in 0.026% carbon steel at very low loads and sliding speeds so this carbon content does not preclude the mild wear state. Figure 23 illustrates for two sliding speeds the way in which the range of severe wear is restricted as the carbon content increases. At 100 cm/sec T_1 and T_2 merge for carbon contents between 0.63 and 0.78%, and for higher carbon contents mild wear is the equilibrium form of wear at all loads.

For quenched and tempered 0.52% carbon steel at the higher values mild wear occurs over the whole load range. The first appreciable deviation occurs after tempering at 500°C when the hardness has reduced to 436

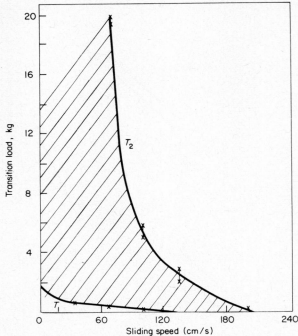

Fig. 22.   Change of transition load with sliding speed.

HV. This deviation increases in extent until at a hardness level of 286 HV the normal T_1–T_2 transitional behavior is produced. It was concluded therefore that the characteristics of soft steel changed to the characteristics of hard steel when a critical hardness value had been achieved which was within the range 348–436 HV.

Results from different investigations have been drawn together to compile Fig. 24, from which it will be concluded that the critical hardness is a function of the carbon content of the steel. Although these results were obtained by investigators who were using different wear equipment and test procedures, they complement each other very well and confirm the general patterns of behavior and also the effect of some metallurgical factors on these transitions.

During the course of these experiments Welsh observed that some low-alloy steels exhibited the same pattern of behavior, but that a 3% Cr/0.5% Mo steel produced results that were strikingly different in that the T_1–T_2 severe wear range was doubled in comparison with that of an equivalent carbon steel. This is an important feature which, taken together with the general experiences with "wire wool" failures (Archard

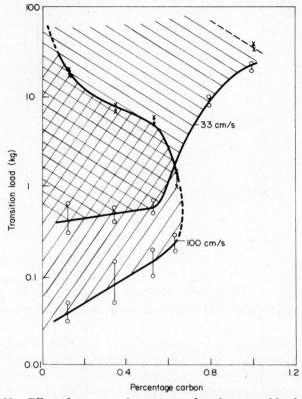

Fig. 23. Effect of percent carbon content of steels on transition loads.

and Hirst, 1956) and the poor running-in characteristics of chromium steels (Farrell and Eyre, 1970), should emphasize the need to explore these general patterns of wear for specific steels before introducing new alloys into engineering practice.

Graphitic cast irons were later investigated by Eyre and Maynard (1971) and Lea (1972), who showed that the general pattern of behavior reported by Welsh was also exhibited by these metals. However, before wear of cast irons is discussed it is necessary to draw attention to their particularly inhomogeneous nature. These alloys are produced by varying the carbon, silicon, and phosphorus content, or more specifically, the carbon equivalent value (CEV), which is given by

$$CEV = \% \text{ carbon} + \tfrac{1}{3}(\text{silicon} + \text{phosphorus})$$

and also the cooling rate, which is mainly a function of the section

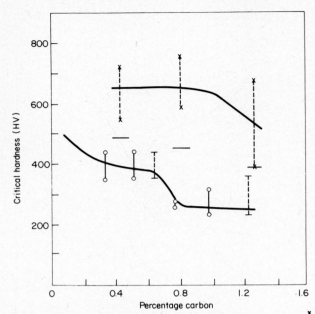

Fig. 24. Effect of percent carbon content of steels on the critical hardness. ⟨ in Hydrogen.
⟨ in Air.

thickness. The microstructure and also the mechanical properties produced are controlled by the CEV value and the section thickness. From
Fig. 25 it will be observed that a casting with different section thicknesses
will produce different microstructure and mechanical properties from a
melt with a constant composition.

The adverse effect of free ferrite in the microstructure is shown in Fig.
26, which shows a trend toward more metallic wear as ferrite increases
until the T_1 transition load is so reduced that metallic wear occurs for a
100% ferritic iron throughout the duration of the test (Eyre and Maynard,
1971). Tin, copper, and chromium are widely used as alloying additions to
promote the formation of pearlite in thicker sections. Care has to be exercised, however, in the use of chromium not to add too much or machinability may be adversely affected by the formation of excess carbides.

The advantageous effect of phosphide as a hard microstructural constituent in cast iron is shown in Fig. 27, from which it will be observed that
the T_1 transition is elevated to higher loads and also becomes less sharp
(Eyre and Williams, 1973). This is a consequence of the effect of the phosphide on two factors:

1. Phosphide eutectic increases the strength under compressive con-

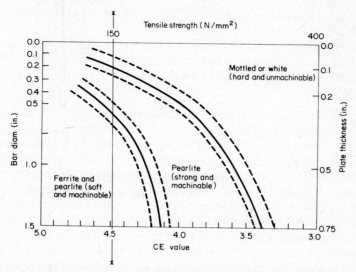

Fig. 25. Relationship between CEV, section thickness, and microstructure of gray cast iron.

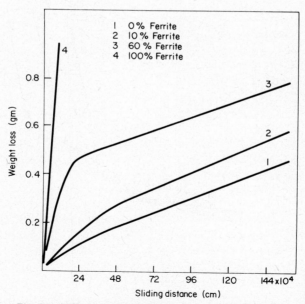

Fig. 26. Effect of free ferrite on wear of gray cast iron.

ditions and this increases the resistance to plastic flow and thus elevates the T_1 transition.

2.　Differential wear also occurs so that the phosphide stands proud of the matrix and thus carries a higher proportion of the applied load. Wear of the pearlitic matrix is then reduced because it carries less of the applied load.

There is, however, a limit to the amount of phosphorus that may be added because shrinkage porosity increases in the casting as phosphorus increases above about 0.8% and also the metal becomes even more brittle. For most applications, therefore, phosphorus is usually limited to a maximum of 0.7%. There are notable exceptions to this general rule. In the United Kingdom a cast iron containing 1.2% phosphorus has been used as a railway brake material for many years. More recently a 3.0% phosphorus iron has been shown to have significant improvements in wear resistance and these results have been confirmed in both laboratory tests and in field tests. This led to the adoption of this type of cast iron, particularly for medium-speed goods transportation. It should be noted that at higher sliding speeds there is no improvement because wear occurs predominantly by surface melting. In fact there is a slight reduc-

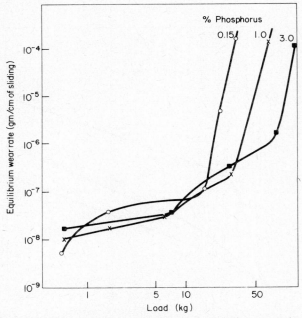

Fig. 27.　Effect of phosphorus content on wear of gray cast iron.

tion in wear resistance of these improved alloys because of the lower melting point (925°C) of the phosphide eutectic.

The transitional behavior of cast iron has been explored in some detail and expressed in a slightly different way to that adopted by Welsh. Figure 28 shows the variation of transition load with sliding speed, from which it will be observed that for nodular cast iron the T_1 and T_2 merge at approximately 600 cm/sec at a load of 4 kg. Below this load mild (oxidative) wear predominates and above this load the wear rate and wear mechanism will depend upon the interplay between load and sliding speed. Variation in heat treatment of this cast iron has a very marked effect, particularly in raising the load at which the T_1 and T_2 merge, and thus the range of mild wear is extended up to 8 kg (Eyre and Maynard, 1971).

It is quite clear that the wear mechanism encountered at particular loads and sliding speeds is affected by both the composition and microstructure of the material in its initial unworn condition. Transitions in behavior are also associated with the hardness of the base metal or alloy,

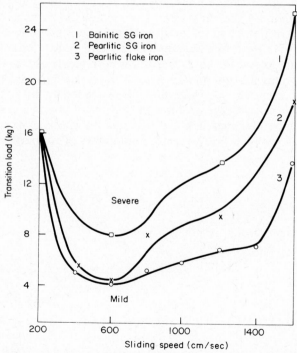

Fig. 28. Transitional wear behavior for nodular iron in different microstructural conditions.

and if a critical hardness is achieved, severe (metallic) wear may be suppressed. It is also clear that the adhesive wear concept is oversimplified and in general applies only to relatively soft pure ferrous or nonferrous metals. Under sliding conditions oxidation of one or both of the surfaces may take place to inhibit severe adhesive wear. Oxidation may take place during the early stage of the wear process itself and thus modify the subsequent wear behavior. With both steels and cast irons severe adhesive wear may only occur during the initial period of running in, and toward the end of that period an oxide builds up on the surface to prevent further metallic contact. It is therefore necessary to ensure that oxygen is available at the contacting surfaces so that the metal oxidizes sufficiently quickly to build up a film thick enough to prevent adhesion. In this respect steels containing chromium are particularly poor, as indicated by Welsh. Farrell and Eyre observed that the running-in characteristics of chromium steels were also markedly different. Figure 29 shows that at a given load and sliding speed it takes approximately three times as long to run in a 3% Cr steel as a 0.0% Cr steel of the same hardness and carbon content. This means that it is much more difficult to attain the equilibrium oxidative wear regime for the alloy steel, severe metallic wear continues for a longer period of time, and the total wear volume is subsequently greater.

Welsh (1965), Eyre and Baxter (1972), and others have noted that during running in and at the T_2 transition both steels and cast irons produce layers on the friction surface that are variously described as "hard," "white," or "nonetching." Attention has also been drawn to the association of white layer with the problem of scuffing of piston rings used in internal combustion engines. Figure 30 shows the white layer produced on gray cast iron within a few hundred centimeters of rubbing under condi-

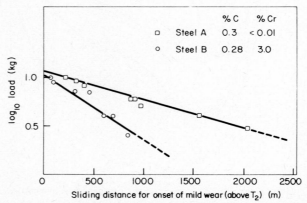

Fig. 29.   Running-in characteristics of steel containing chromium.

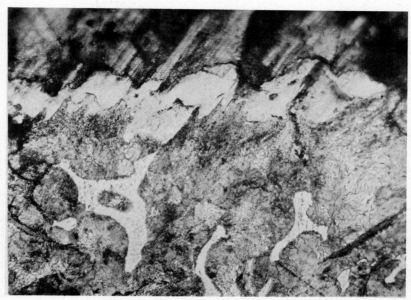

Fig. 30.   White layer on a gray cast iron piston ring (×200).

tions for which type C wear behavior is encountered, i.e., running in takes place prior to equilibrium mild wear. This surface is ''hard'' and values in the range 800–1200 HV have been obtained in comparison with the normal matrix hardness of approximately 250 HV. It is also nonetching in conventional etching reagents and its structure is difficult to resolve with the optical microscope. Using Mössbauer spectroscopy and electron microscopy, recent studies have shown that these surface layers may contain appreciable quantities of retained austenite and iron carbide, and have a submicron grain size. There is still some controversy over the conditions attained at the wear surfaces, but it is generally agreed that high flash temperatures are required, of the order of 700–1000°C. It has also been suggested that the temperatures are much lower and that adiabatic shear is responsible for the deformed microstructure.

In a study of scuffing behavior of gray cast iron, Fig. 31 was compiled, which indicates the load–speed transitional behavior, running-in wear, and also scuffing behavior assessed from metallographic observation of the presence of a white layer. Information compiled in this way gives an excellent guide to the wear characteristics of particular combinations of metals and shows the effect of load and speed likely to be encountered in service. The practical problem of scuffing resolves itself into the following issue. If lubrication is inefficient for any reason, metal-to-metal contact

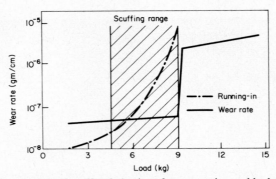

Fig. 31. Assessment of scuffing behavior of gray cast iron rubbed dry on steel.

occurs, which may produce "scuffed" surfaces. White layer transformation leads to the production of localized transverse fractures in the surface, which initiates very hard wear debris. Areas of "scuffing" damage are usually extremely localized, running as narrow bands in the rubbing direction (Fig. 32). When friction is high and scuffing develops, the areas in contact expand because of frictional heating and the load is carried even more locally than before; temperatures therefore rise even further. The piston ring seal is destroyed by the raised topography of the scuff band, gas escapes along each side of the scuff band, and further heating from the gas results. Scuffing will continue and eventually spread around

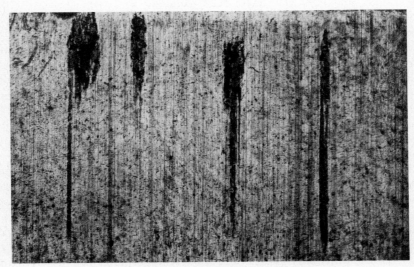

Fig. 32. Scuff bands on the rubbing face of a piston ring.

the piston ring surface unless some healing process is brought into play, e.g., oxidation or lubricant protection.

1. NONFERROUS METALS

Nonferrous metals, with the exception of stellite, and tungsten carbide are not used in abrasive wear applications although they are often subject to accidental abrasive wear caused by contaminant material getting into bearings. White metals, bronze, and aluminum alloys are widely used under sliding (adhesive) conditions, particularly where hydrodynamic lubrication is achieved for most of their operational cycle. For conditions in which some intermittent metallic contact occurs, particularly at the beginning and end of operation or in that part of the operation in which hydrodynamic lubrication cannot be achieved, attention must be paid to wear resistance and compatibility of the bearing and shaft. In most cases plain bearings operate against a plain carbon or low-alloy steel which is in either the normalized or heat-treated condition. A hardened shaft is recommended against the harder bronze bearings. The potential of a material for use in a bearing is particularly difficult to determine on the basis of its physical and mechanical properties. Bearing manufacturers have devised a number of standard bearing tests that are related to the type of application for which the bearing is intended. The final assessment is carried out through engine tests both in test beds and in the field. These tests and their relevance have been discussed by Pratt (1973), and there is general agreement that a plain bearing material requires the following properties, although it is not always easy to assess these characteristics directly.

(1) *High strength to resist load.* A high yield strength to resist plastic deformation and a high fatigue resistance under operating conditions. Increased hardness is a useful way to improve these properties.

(2) *Good conformability,* to adjust to deflections and imperfections in the shaft geometry. This requirement is best achieved with a low modulus and low yield strength, i.e., a material with a low hardness.

(3) *Good embeddability,* to absorb debris. This requirement will be achieved by the characteristics described in (2) above.

(4) *Good corrosion resistance.* Resistance to the special additives present in the lubricant to reduce wear and also against other environmental contaminants, particularly the ingress of sea water in a marine environment.

(5) *High hardness to resist erosion,* under conditions where small particles are carried in the lubricant.

(6) *Good boundary properties,* to resist seizure on contact with the shaft. In this respect lead and tin are essential alloying additions.

(7) *Cost.* It is essential that bearing alloys are cheap and easy to manufacture. This aspect is receiving considerable attention because of the high cost of traditional alloys including tin, indium, etc.

It is unlikely that any single bearing material will have all of these desirable characteristics, particularly since some of them are conflicting. The commercial development of bearing materials has taken place before the desirable characteristics of the materials have been theoretically understood, and therefore great credit is due to bearing manufacturers. Bearing materials are available for almost any known application because of the wide combination of materials and manufacturing processes available and the experience built up in their application. However, failures still occur, usually caused by selection of the incorrect material, lubricant, or counterface material.

2. Construction of Bearing Materials

The first metallic bearings were cast cylinders made of bronze or white metals which were fitted into a suitable housing. Such assemblies were comparatively weak and led to the early development of the practice of bonding the white metal to a supporting backing of steel or brass (bimetallic bearings) to improve strength. Bearings of this type, which were individually cast, were relatively expensive to make and have been largely superseded by mass production techniques in which a thin steel or bronze strip is coated with the appropriate bearing alloy by a continuous coating process. The coated strip is then cut to length and press formed to shape. These newer techniques permit the use of metal combinations and controlled microstructures combined with rapid manufacture that could not otherwise be achieved.

A further innovation is the three-component (trimetallic) bearing in which one of the harder bearing materials such as bronze, already bonded to a steel backing, is covered with a thin (20–200 μm) electrodeposited or cast overlay of white metal. The greatest developments in bearing practice have taken place in the manufacturing processes that are now available. These include static and continuous casting, sintering, heat treatment to produce more desirable microstructures, roll bonding, precision electroplating, use of diffusion barriers, and electrodeposition of composite materials.

3. Main Groups of Bearing Materials

a. White Metals. These are tin- or lead-base alloys often referred to as "Babbitt" metals after the inventor, who was awarded a British patent

TABLE III

WHITE METALS: BRITISH STANDARD 3332

Alloy %	Sn	Sb	Cu	Pb
BS 3332/1	90	7	3	0.35 max
BS 3332/2	87	9	4	0.35 max
BS 3332/3	81	10	5	4
BS 3332/4	75	12	3	10
BS 3332/5	75	7	3	15
BS 3332/6	60	10	3	27
BS 3332/7	12	13	1	74

in 1843 for his invention of a tin-base alloy for bearings. There are six standard tin-base alloys, the compositions of which are given in Table III.

These alloys have a tin-rich matrix containing some antimony and copper in solid solution. If the antimony content exceeds 8%, cuboids Sb−Sn will be present in the microstructure. Skeleton crystals or needle-like particles of copper−tin compound (Cu_6Sn_5) will also be observed. If the lead content exceeds 2% there will be some lead−tin eutectic around the grain boundaries, and with larger lead concentrations primary crystals of lead-rich phase will also be present.

Lead-base alloys have similar properties to tin-base alloys and are more widely used in the United States than in the United Kingdom. The corrosion resistance of lead-base alloys is somewhat inferior to the tin-base alloys. Microstructures are rather similar to the tin-base alloys. Both tin- and lead-base white metals have the best combination of properties for bearings, particularly their antiseizure resistance. Their fatigue resistance is, however, rather low for modern conditions in automotive engines. The fatigue resistance can, however, be extended by control of the overlay thickness (Table IV). White metals are still almost universally used for large bearings in generators, large engines, ships, and other relatively lightly loaded applications.

TABLE IV

EFFECT OF OVERLAY THICKNESS ON FATIGUE PROPERTIES

Lining thickness (mm)	Fatigue resistance (MN/m^2)
0.20	15
0.10	20
0.05	25

b. Copper–Lead Alloys. Copper and lead are immiscible in both the solid and liquid condition. In order to avoid complete segregation, chill casting is used, which produces a pronounced dendritic structure with crystals of copper and lead lying normal to the steel backing. An alternative method of manufacture involves production via a powder metallurgy route, which gives better control of lead distribution and an equiaxed structure. These different methods of manufacture give rise to marked differences in properties:

1. In the cast alloy the dendritic structure enhances the load-carrying ability and facilitates heat flow from the bearing surface.

2. Lead is the most susceptible microconstituent, subject to corrosion by the weak organic acids, which may accumulate in a well-used oil, and can therefore be removed selectively from the bearing. The long dendrites in cast alloys provide a ready path for corrosive fluid to penetrate, while the more evenly distributed lead in the sintered bearings offers greatest resistance.

These alloys generally contain between 20 and 30% lead with additions of up to 5% tin (Table V). Their strength is greater than that of white metals, particularly at engine operating temperatures. Their greater strength (or hardness) necessitates the use of hardened journals. These alloys cannot tolerate dirt to the same extent as the softer white metals. In order to overcome these disadvantages copper–lead bearings are generally "overlay plated" with a thin electrodeposit of lead plus 10% tin or lead plus 5% indium alloy.

c. Lead Bronzes. The high-temperature performance and load-carrying capacity of copper–lead alloys can be improved by tin additions. Up to 5% tin may be added, which goes into solution in the copper and thereby considerably strengthens the bearing.

Larger bearings often have a microthin white (tin-base) metal cast on them. This white metal layer should always be separated from the lead bronze by a nickel diffusion barrier to prevent interdiffusion, which could form a hard intermetallic layer.

d. Phosphor Bronze. The presence of phosphorus, usually up to 0.5%, has a marked effect on increasing the strength of the matrix, which increases the alloy's wear resistance under heavy loads.

e. Aluminum Alloy Bearings. Two alloys are in widespread use, one containing about 6% tin and the other 20% tin. The 6% tin alloy may be used in massive form or bonded to steel, but it requires hardened journals if wear of the journal is to be kept within acceptable limits. The 20% tin

alloy is a more recent development and its method of manufacture provides a good example of the advanced technology applied to bearing manufacture.

The optimum percentage of tin in the high-tin alloys was arrived at on
the basis of work by Hunsicker (1949) (Fig. 33), which shows that little
improvement in seizure resistance is obtained if the tin content exceeds
20%. The high cost of tin also encourages the lowest tin percentage commensurate with the required strength.

High-tin aluminum alloys in the as-cast condition have very poor mechanical properties because the tin forms a continuous network enclosing
the primary aluminum crystals. By cold working and low-temperature
heat treatment this continuous phase can be broken up to produce an interlocking network structure, the so-called reticular structure, which has
greatly improved mechanical properties. However, the alloy is still in
strip form and must be bonded to a thin steel strip backing. This is
achieved by a continuous pressure welding operation carried out between
rollers, bonding between the two strips being promoted by a very thin
sheet of pure aluminum.

This type of bearing has a higher load-carrying capacity than that of
copper–lead alloys and yet can be used in conjunction with soft journals.
The 6% tin alloy is often "overlay" plated with a lead–tin alloy, which

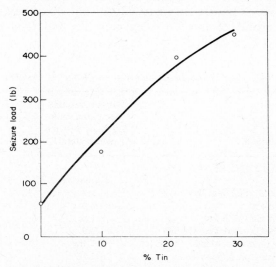

Fig. 33.   Effect of tin content on seizure resistance.

TABLE V

COPPER-BASED BEARING ALLOYS

Alloy BS 1400	Composition (%)				HV
	Cu	Pb	Sn	P	
LB5-C	75	20	5	—	45–70
LB2-C	80	10	10	—	65–90
LB4-C	85	10	5	—	45–70
PB 1C	89.5	—	10	0.5	70–150

improves the compatibility with the journal surface and allows the use of unhardened journals.

The aluminum–20% tin alloy has been one of the most important bearing developments in the past decade, and in the United Kingdom some 60% of automobile engines and 96% of diesel engines are fitted with bearings of this material. This alloy has the highest fatigue strength of the bearing materials discussed here (Table VI).

    f. Silver Bearings. Silver bearings, sometimes plain, sometimes with a thin overlay of lead–indium, have been adopted by the aircraft industry and are used on one well-known make of diesel engine. The silver is electrodeposited on a steel backing and is about 0.5-mm thick. Electrodeposited silver is much harder than cast silver, and silver bearings are unequaled with respect to load-carrying capacity and fatigue resistance.

TABLE VI

FATIGUE AND WEAR PERFORMANCE OF BEARING MATERIALS

Material	Fatigue strength compared with 0.012-in. Sn-base white metal	Journal wear rate compared with Sn-base white metal
Tin-base white metal (7% Sb, 3% Cu), thickness 0.012 in.	1 (lowest)	1 (lowest)
Tin-base white metal, thickness 0.005 in.	1.3	1
Lead-base white metal	0.95	1
Copper–lead (70% Cu–30% Pb)	1.8	2.2
Copper–lead (74% Cu–22% Pb–4% Sn)	2.9	—
Lead–tin overlay on copper–lead	2.16	0.9
Lead–indium overlay on copper–lead	2.16	0.9
Tin–aluminum (20% Sn–1% Cu)	2.4 (highest)	1.6 (highest)

Unfortunately, they are prone to seizure and are very sensitive to the nature of the lubricant and to certain lubricating-oil additives.

4. GENERAL COMMENTS ON SELECTION OF BEARING MATERIALS

The performance characteristics of these various groups of metallic bearing materials are summarized qualitatively in Table VII, and Table VIII gives more specific recommendations on the maximum permissible bearing loads. It is generally recommended that if the bearing pressure is greater than 4500 lb/in.^{2}, then the journal surface should be hardened. The ratings of these established alloys agree reasonably well with their known performance characteristics under engine conditions. Of the materials given in Table VI the bottom three alloys are normally overlay plated because of their poor seizure resistance.

In a recent investigation Stott (1976) has suggested that in the selection of bearing materials for operation under lubricated conditions it is also wise to consider their dry wear characteristics. If lubricant failure occurs, it is necessary that the bearing metallurgy be sufficiently good to wear at an acceptably low rate with the production of a minimum of wear debris. From a study of wear of the different bearing alloys against a 1.0% carbon–1% chromium steel heat treated to 750 HV, he compiled wear curves over a wide range of load and speed. This covered the whole spectrum from low to very high wear rates and in addition he obtained the resultant metallographic structure of the wear surface. This approach enabled him to compile a wear/metallography atlas for each alloy that should enable others to select the most suitable bearing alloy. It also allows a good correlation with failure in service to be studied. Figure 34 shows a plot of wear rates compared over the whole range of load and speed, and the results confirm that phosphor bronze is considerably superior to aluminum bronze as a bearing alloy. He concluded that the oxidation resistance of the aluminum bronze made it more difficult for a protective oxide to form, and thus it was prone to adhesive wear.

5. SINTERED METALS

Sintered metals have been widely used in wear applications, particularly tungsten carbide, for both cutting tools and wear insets for wire and tube making dies. Sintered metals are also being increasingly used in more general applications for wear, particularly where lubrication is used. Sintered iron piston rings are now widely used in both gasoline and diesel

TABLE VII

COMPARATIVE PROPERTIES OF BEARING ALLOYS

Bearing alloy	Brinell hardness at room temperature	Fatigue strength[a]	Conformability and embeddability[a]	Maximum operating temperature (°C)	Seizure resistance[a]	Corrosion resistance[a] Weak organic acids	Corrosion resistance[a] Strong mineral acids
Lead-base white metal (lead babbitt)	15–25	5	1	130	1	3	3
Tin-base white metal (tin babbitt)	20–30	5	1	130	1	1	2
Sintered copper–lead	20–35	3	2	150	2	5 (without overlay) / 3 (with overlay)	3
Cast copper–lead	25–40	2	2	160	2	5 (without overlay) / 3 (with overlay)	3
Lead bronze, cast or sintered	40–70	1	3	170	2	5 (without overlay) / 3 (with overlay)	3
Aluminum, low tin	50–80	1	4	170	4	2	3
Aluminum, high tin	30–40	2	3	160	3	2	3
Aluminum babbitt	25–40	2	2	160	3	3	3
Aluminum–silicon (overlay plated)	—	—	—	—	—	—	—
Silver	50–80	1	4	200	4	3	2
Phosphor bronze	60–110	1	5	220	3	3	4
Silicon bronze	70–120	1	5	220	4	3	3
Leaded bronzes (steel backed)	40–70	2	3	180	3	4	2
Porous bronze	30–50	3	2	130	1	3	4
Porous iron	40–60	2	3	130	2	3	5
Bronze–graphite composites	—	—	2	180	1	3	4
Plastic and filler in metal matrix	—	—	1	160	1	2	2

[a] Arbitrary scale with 1 being the best material and 5 the worst.

TABLE VIII

PERMISSIBLE LOADS FOR RECIPROCATING PLAIN BEARINGS

Bearing alloy	Maximum bearing pressure (lb/in.²)	Bearing alloy	Maximum bearing pressure (lb/in.²)
White metal	<2500	Low-tin lead–bronze	4000–6000
Copper–lead	2500–4000	High-tin lead–bronze	6000–10,000
Aluminum–tin	2500–5000	Phosphor–bronze	10,000–15,000

engines in competition with cast iron (Hewitt *et al.*, 1975). Their advantages are the following:

1. Porosity can be of advantage in acting as a reservoir for lubricants and is generally in the range 15–25%. The greater the porosity, the lower the strength, and as always a compromise is necessary.

2. It is possible to produce alloys and mixtures of materials that would be difficult or impossible to produce in any other way.

Sintered iron exhibits transitional wear behavior in the same way as do wrought and cast metals (Eyre and Walker, 1976). The influence of pressing and sintering variables on the reduction in mild wear rate and the increase in the transition load (Fig. 35) are attributed to their general effect on increasing the mechanical strength of the sintered compacts. It was observed that in comparison with pure iron in the wrought condition the sintered iron gave a lower rate, a higher transition wear load, and less tendency to gross adhesion, in spite of the fact that its mechanical strength was considerably lower. A sintered iron wear pin was impreg-

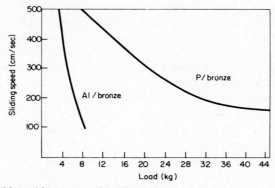

Fig. 34.   Load/speed/constant volume loss diagram for bronze versus steel (EN 31) (for wear of 10^{-7} cm³/cm).

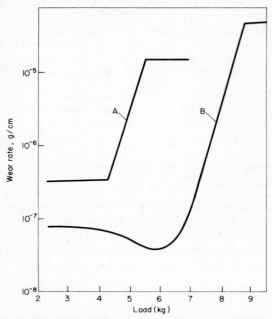

Fig. 35. Wear behavior of sintered iron. A—compact pressure 309 N/mm², sintered at 1050°C; B—compact pressure 464 N/mm², sintered at 1200°C.

nated with a commercial oil which included a colloidal dispersion of graphite, the outside of the pin was then wiped dry, and comparative wear tests were carried out at 6.5 kg and 300 cm/sec sliding speed. Figure 36 shows that wear of the impregnated pin occurred at a very low rate for a considerable time. During this period, the oil could be observed to ooze out of the pin and collect at the sliding interface. At the end of the test when the oil supply ceased, a high wear rate immediately occurred.

Phosphorus is widely used in gray cast iron to improve wear resistance, particularly under sliding conditions, and it was therefore decided to investigate the possibility of transferring this technology to sintered iron–phosphorus alloys. Commercial powders containing phosphorus were already available, but it was felt that even higher concentrations would be of particular interest. Alloys containing up to 5% phosphorus were produced and wear tests carried out over a wide load range. Considerable improvements were obtained, particularly at the higher loads; e.g., at a wear rate of 10^{-6} gm/cm the load bearing capacity was increased from 4 to 22 kg for the 3.0% phosphorus alloy.

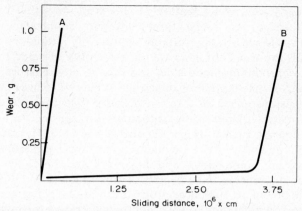

Fig. 36. Effect of impregnated lubricant on wear of sintered iron. A—not lubricated; B—lubricated. Conditions of test: 6.5 kg, 300 cm/sec.

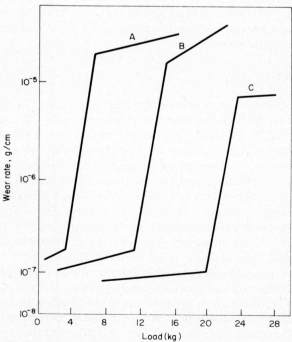

Fig. 37. Effect of counterface metal on wear of sintered iron. A—steel disk; B—chromium disk; C—gray cast-iron disk.

A particularly interesting feature of this work was that experiments were also carried out using three different counterface materials. Although the mild wear rates were not appreciably altered by varying the disk material, the transition load was considerably affected by a change from steel to chromium-plated steel and then to gray cast iron (Fig. 37). This emphasizes once more how dangerous it may be to base the selection of a wear-resistant metal on a test procedure that does not include the counterface material likely to be encountered in service. In a simulative study of sintered piston rings Hewitt *et al.* (1975) observed that chromium-plated surfaces (i.e., cylinder bores) were particularly susceptible to damage by some alloys, and they attributed this to the softness of the compact. Increasing the strength of the compact by alloying eliminated this problem. It appeared from these studies that the particular problem could be attributed to wear debris penetrating the microcracks in the chromium surface, which ultimately caused brittle fracture of particles of chromium. This hard debris itself caused problems by becoming embedded in the soft piston ring and increasing the level of abrasive wear. This further demonstrates that one wear mechanism may change to another or that two may operate simultaneously. This causes particular problems in any attempt to translate information from research investigations into industrial practice.

6. ALUMINUM ALLOYS

Aluminum–silicon alloys are widely used in all types of internal combustion engines for cylinder blocks, cylinder heads, and pistons (Shivanath *et al.*, 1977). However, aluminum–silicon alloy cylinder bores are prone to scuffing under conditions of poor lubrication such as those which exist during starting or warming up of engines. For this reason, aluminum alloy cylinder blocks usually have cast-iron liners or are electroplated with harder materials. A few years ago, an important breakthrough in the automobile industry in the United States allowed the designers to dispense with the usual cast-iron liners in a production engine. They used a die-cast hypereutectic aluminum–silicon alloy of composition similar to B.S. LM30 alloy as a cylinder block material. The cylinder bore surface was etched by an electrochemical technique, which removed aluminum, leaving the primary silicon particles standing proud of the matrix. The silicon particles were found to possess excellent scuffing resistance, especially when they rubbed against a hard iron-plated aluminum alloy piston. Recently a European manufacturer has also adopted this technology to produce a high-speed series of engines. A Japanese attempt to develop a more machinable hypoeutectic aluminum–silicon alloy of B.S. LM26 type which could be used as a linerless cylinder block material has also been described.

The long established "Y" alloys (Al–Cu–Ni–Mg or Mn) are still preferred for casting heavy-duty pistons for diesel engines. However, it appears that B.S. LM13-type "Lo-Ex" aluminum–silicon alloy is specified universally for automotive pistons because of its lower cost, superior castability, good bearing qualities, lower expansion characteristics, and superior corrosion resistance.

In view of the wide use of aluminum–silicon alloys in applications requiring good sliding wear resistance, there appears to be rather limited published information on wear of these alloys. Some data indicate that the wear rate of aluminum–silicon alloys decreases linearly with an increase of silicon content from the eutectic composition. Similar conclusions were obtained by wear testing three complex hypereutectic aluminum–silicon alloys with silicon content varying from 14.5 to 25%, sliding against a rotating and reciprocating iron disk. A fine dispersion of silicon particles also proved beneficial. Silicon content appears to have no significant effect on wear of aluminum alloys until it exceeds 20%. The composition of the sliding surface appears to be very important in determining the wear rate of aluminum–silicon alloys; e.g., flake graphite cast iron causes less wear and seizure than does carbon steel. Al–21.6% Si alloy showed mild wear when it slid against the same material, but its wear rate increased as the silicon content of the mating surface decreased. All other combinations of aluminum–silicon alloys produced severe wear. Other investigators found no discernible effect due to the composition of disk materials when they were either cast iron or steel. Experimental results also showed that wear of the hypoeutectic alloy (11% Si) was lower than that of a hypereutectic alloy (22% Si).

These differences in reported observations may be caused by several factors, which include differences in experimental techniques and in alloy composition. Most of the above-mentioned investigations were carried out on commercial alloys that contain alloying elements other than silicon. These alloying additions may modify the wear characteristics of aluminum–silicon alloys by solid solution strengthening and precipitation hardening.

Shivanath et al. (1977) have recently reported results obtained on binary alloys of aluminum and silicon. The main objectives of their investigation were to determine the effect of silicon content and the effect of particle size as varied by sand and chill casting and the use of sodium modification of the eutectic alloy. The effect of a surface etching treatment was explored, using sodium hydroxide on the modified and unmodified eutectic alloy. It was considered that there was little value in rubbing the aluminum alloys against themselves, and a steel counterface material was considered to be more appropriate and comparable with industrial practice.

Aluminum–silicon alloys exhibit two forms of wear under dry sliding conditions (Fig. 38). These are oxidative and metallic wear and they have characteristic metallographic features. Oxidative wear occurs when an oxide or mixed oxide/metal surface covers the wear pin and a similar layer is produced on the steel counterface. There is some evidence of localized deformation of the substrate and breaking up of the silicon particles. The metal–oxide interface is much rougher than the oxide–disk interface, which indicates that initially some roughening occurred, which then became smoother by oxide filling the rough valleys. Because of this, the oxide thickness varies considerably between about 10 and 80 μm. Wear of the alloy occurs by a process of spalling of the oxide layer fairly

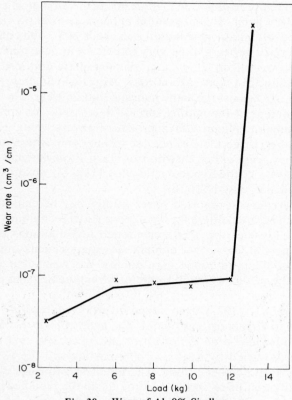

Fig. 38.   Wear of Al–8% Si alloy.

irregularly over the surface, and at any one time only part of the surface is in contact with the disk counterface. The aluminum and steel surfaces are always separated by this patchy oxide layer. Wear rates are low because the amount of metal removed is confined to the thickness of the oxide. The amount of wear of the steel was negligible although some transfer of iron to the aluminum was detected, and this observation confirms the work of Beesley and Eyre (1976). The exact nature of this complex surface, however, was not fully explored in this investigation.

Metallic wear occurs when deformation of the surface occurs on a fairly massive scale because the yield strength has been exceeded. Fracture occurs with the separation of large fragments of metal, some of which become adhered to the steel counterface. The degree of plastic deformation and the rate of metal removal prevent the production of an oxide layer during metallic wear.

There is no clear correlation between the mechanical properties of these alloys and their wear characteristics, and it would appear to be more appropriate to look for a correlation with their properties under compression and also at elevated temperatures. The brittle hypereutectic alloys behave in a relatively ductile manner during severe wear, unlike their behavior under tensile conditions. The transition load gives a good indication of the change from oxidative to metallic wear, which will result in marked changes in the wear rate and therefore express a measure of the load bearing capacity of the alloys (Fig. 39).

The two wear regimes can therefore be recognized by their distinctive wear rates:

$$\begin{array}{lll} \text{oxidative wear rate} & 10^{-8}-10^{-7} & \text{cm}^3/\text{cm} \\ \text{metallic wear rate} & 10^{-5}-10^{-4} & \text{cm}^3/\text{cm} \end{array}$$

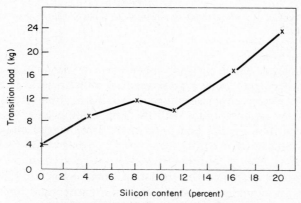

Fig. 39. Effect of silicon on transition load.

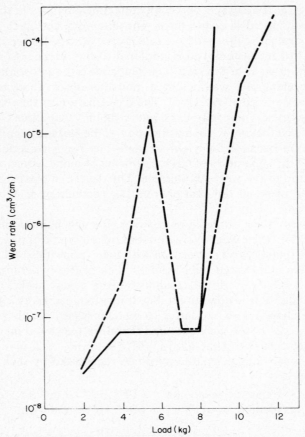

Fig. 40. Effect of surface etching on wear of Al–11% Si alloy. ————Etched alloy; —·—unetched alloy.

Surface etching, which has only been used on the 11% Si unmodified alloy, completely suppresses an intermediate metallic wear behavior (Fig. 40). Adhesive wear is suppressed because at the beginning of the test only silicon particles are in contact with the steel counterface. Running in of these surfaces then occurs and only in the later stages does aluminum come into contact with the disk because of the removal of the silicon particles. However, the conditioning which has occurred during running in inhibits adhesive wear, and the oxidative wear regime continues.

These results complement those reported earlier and open up many possibilities for exploiting surface etching techniques to produce more compatible tribological surfaces, and these may even have greater implications in lubricated wear situations.

C. *Fretting*

Fretting is a wear phenomenon that occurs between two relatively tightly fitted surfaces having an oscillatory relative motion of small amplitude. The general view is that fretting is usually met with in situations in which the amplitude is generally less than 25 μm and certainly not greater than 130 μm. All solids fret, although the nature of the fretting is unique to the particular combination of solids in contact and their environment (Waterhouse, 1972).

A classic case of fretting is shown in Fig. 41 in which a steel pin has operated against a gray cast-iron bush. This assembly is part of a conveyor system, and the components are subject to vibration. The surfaces are partially covered with a tenacious red/brown film around the edges of which can be observed clean smooth surfaces which have been polished in situ. A large volume of red/brown dust accumulated in the lubricating oil, which was gradually changed in consistency to that of a paste. This

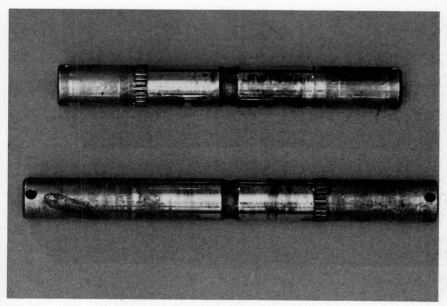

Fig. 41.   Fretting of a carburized steel shaft.

paste eventually extruded out through the oil holes provided in the cast-iron bush. X-ray analysis showed that the red/brown powder was α-Fe_2O_3, which has a hardness of Mohs 6, approximately 600 HV. This abrasive dust, which is an impure form of "jewelers' rouge," was responsible for the production of the smooth areas. The presence of the red/brown oxide prevented the ingress of the lubricant and the situation rapidly worsened. Filling up of the normal clearances between the pin and bush gradually increased the frictional resistance, the noise level increased, and in some cases seizure occurred. It was difficult to provide an increased flow of lubricant or to find an improved lubricant. It was therefore decided that surface treatments applied to the existing design offered the best chance of success, particularly in the short term. After a few controlled laboratory experiments and field trials, manganese phosphate treatment proved to be a satisfactory solution.

Fretting may occur on such parts as splines, bearings and shafts, flexible couplings, spindles, seals, joints, bolts, and shackles. Fretting situations can be divided into two main groups:

(1)   Those for which the contacting surfaces are not designed to move relative to each other, e.g., shrink fits, press fits, bolted joints, and keyways. One of the main dangers is that ultimately fretting damage may initiate fatigue failure.

(2)   Those for which relative movement takes place for part of the operating time, e.g., bearings and couplings. The main danger is the presence of the debris between the moving surfaces gradually leading to increased friction and noise, and in some cases seizure.

1.   FRETTING MECHANISMS

During fretting there appear to be three basic processes operating although there is no universal agreement on this interpretation (Waterhouse, 1972):

(1)   Mechanical action, disrupting oxide films on the surface, exposing clean and possibly strained metal which would be reactive and which, in the presence of the atmosphere, oxidizes rapidly during the half cycle after the disruption, to be redisrupted on the return half cycle.

(2)   The removal of metal particles from the surface in a finely divided form by a mechanical grinding action or by the formation of welds at points of contact at a surface other than the original interface, either by direct shearing or by a local fatiguing action. The atmosphere would play no part in this process except where fatiguing was involved, when it could introduce an element of corrosion fatigue.

(3)   Oxide debris resulting directly from (1) or as a result of the oxidation of the metal particles in (2) acting as an abrasive powder which continues to damage the surfaces.

2.   EFFECT OF SOME METALLURGICAL VARIABLES ON FRETTING

Fretting is worse when the friction coupled materials are similar in composition and mechanical properties, and this evidence favors the suggestion that adhesion plays an important part in the initiation of the debris. Hardness has been the main variable investigated and it appears to influence fretting in two possible ways. A higher hardness implies a higher yield, tensile, and fatigue strength. For the following reasons a decrease in damage is to be expected if the hardness is increased:

(1)   The higher the yield strength, the smaller the real area of contact will be under load and therefore less damage will result.

(2)   An increase in yield strength may exceed the load at which adhesion would normally occur and this should have the effect of eliminating fretting completely.

(3)   An increase in hardness will increase the fatigue strength and this should reduce or eliminate fretting provided that fatigue resistance is an important parameter. In the case of fretting producing red/brown oxide, which ultimately leads to failure by an increase in friction, noise, and ultimately seizure, fatigue is not thought to be important. Adhesion and the role of surface treatments are thought to dominate.

Wright (1957) found that increasing the amount of soft ferrite in gray cast iron increased the rate of damage, and increasing the phosphorus content decreased the rate of damage. Both of these results are consistent with the widely held view of adhesive (metallic) wear of this material.

Rolfe (1952) found that a 1% C–1% Cr steel of 208 HV produced more damage on a case-hardened steel with a hardness of 860 HV than it did on a mild steel of hardness less than 200 HV. Graham (1963) on the other hand obtained a good correlation between hardness and fretting damage in the range 150–600 HV, obtained by keeping the steel constant and obtaining the hardness range by suitable heat treatment.

The evidence is therefore somewhat confusing, and it would appear that if adhesion is the main problem, then hardness as such is likely to be insignificant. If, however, fatigue is the predominant problem, then using a harder material will increase the fretting resistance.

The abrasive action of the debris is also thought to be of some significance because of its high intrinsic hardness, and therefore the harder surfaces will be less damaged by the abrasive. Wright found that a steel of 190 HV suffered only twice the damage of that of a steel of 800 HV under the same conditions, and therefore it was suggested that hardness was of minor importance. It is important, however, to bear in mind that the 190-HV steel is likely to have work hardened considerably while the 800-HV steel would remain at roughly the same hardness. The actual work-hardened hardness of the two steels may have been in the ratio of their fretting resistance if significant abrasion by the debris occurred.

3. PREVENTION OF FRETTING

a. Mechanical Methods. Slip can be prevented between the surfaces by eliminating vibration from the mechanism or by separating the surfaces so that there is no metal-to-metal contact. The insertion of a rubber or a polymer seal may eliminate the problem, although attention has been drawn to fretting of metals by polymers (Stott *et al.*, 1977).

b. Metallurgical Methods. A wide variety of surface treatments has been tried, and in general it has been found that those that prevent adhesion are most successful; in this respect phosphating, tufftriding, and Sulphinuz have all had some success.

c. Lubrication Methods. Lubrication must separate the points of contact and must wash away any debris inadvertently generated, particularly early in the design life. Roughening the surface to provide microscopic valleys and the use of an oil of low viscosity so that flow is readily facilitated and debris is washed away has been successful. Use of an oil with an oxidation inhibitor such as phenyl-α-naphthylamine is recommended. Use of solid film lubricants can give protection until such time as the surfaces have run in and optimized themselves so that fretting is subsequently prevented.

III. Surface Treatments Used to Reduce Wear

A very wide range of treatments has been devised to improve the wear resistance of both ferrous and nonferrous metals (Table IX) (James *et al.*, 1975), and it is increasingly difficult to keep abreast of new developments taking place, which include ion implantation, ion plating, etc. The selection of the most appropriate treatments has been largely based on experience and these treatments have usually been used to rescue a component which has worn too rapidly, thus causing early failure or uneconomical operation. Surface treatments are now being increasingly considered at

TABLE IX

CLASSIFICATION OF SURFACE TREATMENTS

Type 1 Material added to surface	Type 2 Surface chemistry altered	Type 3 Surface microstructure altered
A. Welding	H. Interstitial hardening	K. Mechanical treatment
A1 Gas	H1 Carburizing	K1 Peening
A2 Arc	H2 Nitriding	K2 Rolling
A3 Plasma	H3 Carbonizing	K3 Machining
B. Thermal spraying	H4 Cyaniding	L. Thermal treatment
B1 Flame	H5 Sulfonitriding	L1 Flame hardening
B2 Arc	H5 Boriding	L2 Induction hardening
B3 Plasma	I. Diffusion treatment	L3 Chill casting
B4 Detonation	I1 Siliconizing	M. Thermomechanical treatment
C. Cladding	I2 Aluminizing	M1 Martensitic work hardening
C1 Brazing	I3 Chromizing	
C2 Explosive bonding	I4 Sulfurizing	
C3 Diffusion bonding	J. Chemical treatment	
D. Miscellaneous	J1 Etching	
D1 Spark hardening	J2 Oxidation	
D2 Powder coating	D. Miscellaneous	
D3 Organic finishes	D6 Impacting abrasives	
D4 Painting		
Multiple processes[a]		
E. Electrodeposition		
E1 Electrolysis (aqueous)		
E2 Metallizing		
E3 Anodizing		
E4 Electrophoresis		
F. Vapor deposition		
F1 Physical		
F2 Chemical		
G. Chemical deposition		
G1 Chemical plating		
G2 Phosphating		
G3 Chromating		
C. Cladding		
C3 Diffusion bonding		
D. Miscellaneous		
D5 Hot dipping		

[a] E.g., plating followed by diffusion treatment.

the design stage and it is necessary, therefore, to offer some guidelines to their selection and application. Halling (1976) has recently suggested that the development of surface technology will be one of the most important growth areas over the next decade, and this emphasizes its importance in the control of wear. It has been thought prudent, therefore, to consider surface treatments separately rather than discuss them individually in the previous sections on wear mechanisms. This argument is reinforced by the knowledge that some treatments are appropriate to the solution of wear problems caused by different wear mechanisms.

A. Reasons for Surface Treatments

Surface treatments are particularly important for the following reasons:

1.   It is often possible to manufacture engineering components from cheap, readily available metals that are relatively easy to produce with minimum energy consumption. Surface treatments can then be used to confer the desired fatigue, corrosion, and wear characteristics.

2.   Under hostile conditions many metals have reached their limits of endurance and as conditions become even more arduous, satisfactory operation will only be achieved by selection and use of the optimum surface process and material to protect that surface.

3.   It is quite likely that even greater demands will be made on science and technology to bring about further improvements, particularly in extending the life of engineering components and also in reducing the energy consumed in manufacturing processes. It will be essential, therefore, to understand fully the role of surface properties so as to appreciate the scientific principles which need to be explored to achieve these improvements.

4.   It would be possible to change the properties locally, e.g., in the area of a bearing.

5.   It is desirable to obtain wear properties in large components, e.g., a blast furnace hopper bell, which would otherwise present considerable manufacturing problems.

6.   It will be possible to repair worn components either on site or after removal to a workshop, and thus to facilitate minimum interruption to production.

The very wide range of treatments currently available may be divided into three main groups (Table X).

Type 1.   Those treatments involving the addition of extra material, usually of a composition different from the surface of the base metal. The

TABLE X

Surface deposition (Type 1)	Conversion coatings (Type 2)	Surface hardening (Type 3)
Electrodeposition	Phosphating	Carburizing
Hard facing	Sulphinuz	Nitriding
Plasma spraying	Tufftriding	Induction hardening

design has to take into account the recess or space into which the material will be deposited. Machining subsequently to the surface finish and the geometry required by the designer may be important. One of the main problems is likely to be the adhesion of the coating to the substrate, and considerable emphasis should be placed on adequate surface preparation.

Type 2. Those treatments involving no addition of extra material which would cause significant growth or change of surface dimension. All of the changes occur within the original surface of the base metal.

Type 3. Those treatments involving a change in composition which enables an increase in surface hardness to be achieved through heat treatment.

Some of these treatments may fall into more than one group, but since this is a first-stage rationalization only, a more detailed consideration later will emphasize other considerations.

B. Metallurgical Features of Surface Treatments

The main metallurgical effects of those treatments that subsequently modify friction and wear properties are

(1) change of surface composition, e.g., by carburizing, nitriding, ion implantation, etc.;

(2) change of surface hardness either by electroplating, weld deposition, or heat treatment;

(3) change of surface metallographic structure, e.g., by boronizing, carburizing, either directly by diffusion or indirectly through subsequent heat treatment;

(4) change of surface topography. Most engineering surfaces have a directional topography produced by conventional machining processes. A random surface can, however, be produced either by etching, phosphating, or shot blasting, and this type of surface may well be the best to enable an efficient supply of lubricant to reach asperity contact areas.

In general, those treatments used to increase surface hardness are applied where abrasion resistance is the main requirement, and those that change the surface chemistry are used to improve wear resistance under adhesive conditions. Under abrasive conditions the selection of the appropriate surface thickness is relatively simple, but this is more complex under adhesive conditions. In general, engineers have faith in those processes which produce relatively thick, hard coatings. For adhesive conditions there is no simple relationship between wear resistance and thickness, particularly since many of the treatments available have been devised to aid running in. Improved wear resistance often persists long after the surface treatment has been worn off because the treatment has allowed the surface topography, hardness, and microstructure to optimize themselves so that a better "run-in" surface is produced. An awareness of this feature is very important, particularly since some of the newer processes, including ion implantation and chemical vapor deposition, produce very thin coatings. Suh (1977) has concluded that for adhesive wear resistance the surface treatment should produce very thin coatings that are also soft and ductile to offer maximum resistance to work hardening so that dislocations run through the coating. The reluctance of engineers to specify thin, soft coatings has not yet been completely overcome.

Some of the more important surface treatments will now be discussed in the light of their main advantages and industrial applications.

C. Type I: Electrodeposition

The object of electroplating is to graft a metal with specialized physical and mechanical properties onto the surface of a component manufactured from a cheaper base metal. Two particular points that distinguish electroplating for wear applications from those processes concerned with decorative appeal should be emphasized:

1. Much more rigid control of the electroplating conditions is required.
2. The deposit thickness lies between 0.05 and 5 mm, in comparison with only microns for decorative plating.

These processes are therefore rather specialized, and engineers should rely on the services of a manufacturer who understands the basic requirements more fully than does the decorative plater. The properties of the more widely used electroplated metals are given in Table XI. Copper, tin, and silver are usually applied in thicknesses of between 0.05 and 0.25 mm for bedding in or running in of components such as pistons, piston rings, and bearings.

TABLE XI

Properties of Electroplated Metals

	Chromium	Nickel	Iron	Tin	Copper	Cast iron
Hardness (Brinell)	1100	220	200	4	50	220
Recovery hardness (annealed at 600°C)	600	75	140	—	—	—
Tensile strength (kg/mm)	—	32	60	1.5	35	18
Melting point (°C)	1800	1455	1530	232	1083	1250
Coefficient of friction	0.12	—	0.30	—	—	—
Coefficient of heat conductivity (cal/cm sec °C)	0.16	0.20	0.19	0.15	0.9	0.11
Corrosion resistance	good	excellent	bad	excellent	good except in acids	bad
Wear resistance	excellent	moderate	moderate	bad	bad	good
Oil spreadability	excellent	good	good	good	good	good
Young's modulus (kg/mm² × 10³)	10	11	20	15	10	9

Nickel and iron have different mechanical properties when electro-deposited than they have in the cast or forged condition. Forged nickel and iron are never used in wear applications because they scuff so easily. However, electrodeposited iron shows wear resistance comparable to that of gray cast iron. This is thought to be associated with the effect of internal stresses on increasing the hardness and also on the orientation of the crystal structure, which is modified. Both effects result in a higher surface shear strength. Nickel is used where corrosion is severe and iron would be unsuitable as well as where the mechanical and/or thermal stresses are high. Both nickel and iron are widely used to build up worn components, for example, worn crosshead pins, that are under size. As much as 3 mm on the diameter can be recovered with 2.5 mm of nickel followed by 0.5 mm of chromium. Iron is used to build up large bore diesel engine cylinder liners, which can require a total layer thickness on the diameter of up to 10 mm after reboring concentric and parallel. They are built up mainly by electrolytic iron with a final wear-resistant layer of porous chromium plating.

Chromium is the most widely used electrodeposited coating for wear resistance; in fact it has no engineering application in its pure state in either the cast or wrought condition. In the electrodeposited condition the structure is much finer and the crystal structure contains many faults and built-in complexes of the oxide and hydroxide type. A high hardness is developed, coupled with good mechanical strength, good corrosion resistance, and a low coefficient of friction, which gives chromium plate a very high resistance to mechanical wear. Under extreme loading conditions the risk of cold welding between chromium plate and other metals is less than that between any other metal couples, and therefore there is less chance of seizure. One major disadvantage, however, is its nonwettability, so that it will not readily retain an oil film; this presents problems under marginally lubricated conditions. A porous chromium plating technique was developed in 1935 whereby porosity is etched into the liner by reversing the current flow after deposition to meet dimensional requirements. This process produces two basic types of porosity (Fenton and Oolbekink, 1973).

1. CHANNEL-TYPE POROSITY

The porous surface consists of a network of comparatively large open channels that are all interconnecting. This structure is shown in Fig. 42 and is reported to promote very rapid distribution of the oil over the surface of diesel engine liners by capillary action. The oil in the channels is also mechanically spread across the surface of the liner. It has been

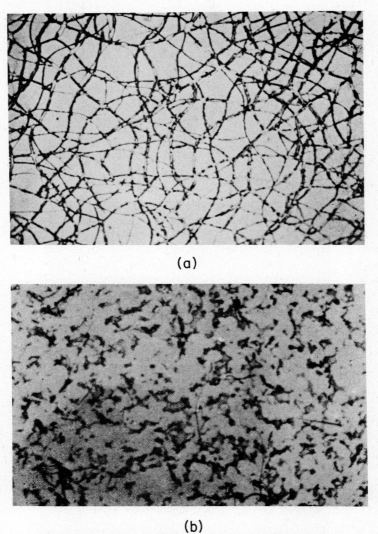

(a)

(b)

Fig. 42.   Surface structure of chromium plating. (a) Channel type; (b) intermediate type.

shown, however, that under poor lubrication conditions rapid
plating may occur. Figure 43 shows microcracks d
channeling; these cracks gradually become more exte
til particles of chromium break away to form an abra
care is required, therefore, in selection of any wear-resis
sure that it is used under the optimum conditions
more catastrophic form of wear from initiating in defec

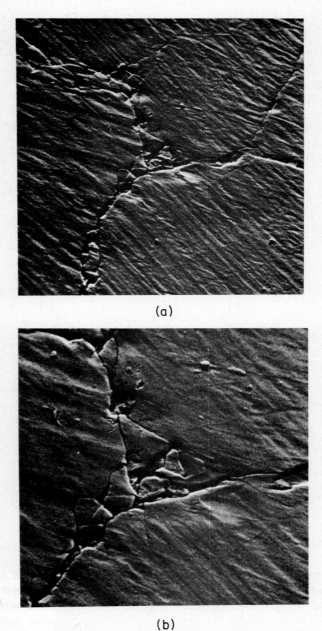

(a)

(b)

Fig. 43.   Surface of Cr plated disk. (a) Fragmentation of Cr at channel cracks; (b) micro-
cks penetrating into the Cr matrix (×375).

2. INTERMEDIATE POROSITY

This type of surface is intermediate between channel-type and pin-point-type porosity. Intermediate porosity is produced by breaking down the original channels by finish honing to give a series of fairly wide opened-out pits that are not interconnected (Fig. 42). The same good oil retention properties of the channel type are reproduced, but the distribution effect is lessened by breaking up the interconnecting channels. The overall percentage of porosity as well as porosity distribution across wear surfaces can be controlled so that better oil distribution can be achieved than with plain surfaces.

The intermediate-type finish is to be preferred for cylinder liner applications in which the use of the channel type would lead to excessive oil consumption because of the leakage of oil beneath the compression piston ring into the combustion zone.

Chromium plating has many useful applications in diesel engine components, crankshaft main journals, gudgeon pins, packing and plain bearings, and pump shafts. There is considerable evidence that hard chromium plating is superior to gray cast iron as a top ring material in reciprocating engines and also as liners of large marine diesel engines. Figure 44 shows average wear on a chromium liner of 0.073 mm per 1000 h compared to 0.167 mm per 1000 h for a cast-iron liner.

3. HARD FACING

Surface coating involving the deposition of metal or ceramics either by oxyacetylene or arc processes has been widely used to build up wear-resistant layers. These techniques were developed primarily to enable components to be repaired from the ravages of wear (and corrosion). It is possible to carry out extremely localized rebuilding of components involving either small or large areas either on site or in the repair shop. These coatings are used mainly in wear applications involving abrasion, and in more recent years the techniques available have been more widely included at the design stage. This is particularly so in the aircraft industry, where the high melting-point oxidation- and abrasion-resistant alloys that have been developed are the only means of combating wear problems at operating temperatures up to 600°C.

Surface coating techniques involving the deposition of material by welding and spraying processes are enumerated in Tables XII and XIII and illustrate the very wide range of metals and alloys that can be deposited by these techniques. The main processes used for deposition are those using an oxyacetylene flame with deposition of metal from either an

Clan MacGregor			Sulzer RD 76	
Cyl no.		Wear mm	No. of hours	Wear mm/1000 h
1	Chrome	0.79	8447	0.09
2	Cast iron	1.55	8480	0.18
3	Chrome	0.50	8447	0.06
4	Cast iron	1.23	8276	0.15
5	Cast iron	1.22	8480	0.14
6	Chrome	0.57	8480	0.07

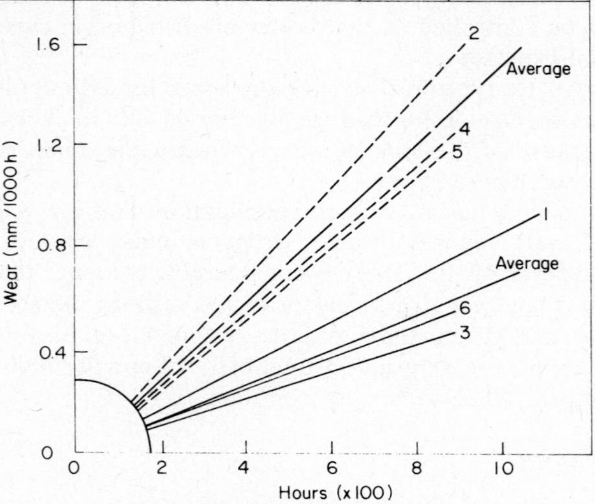

Fig. 44. Wear of diesel engine liners.

alloy rod or powder in a flame and then projection of the particles onto the surface.

Within recent years the plasma gun has become more widely used, particularly for the higher-melting-point metals and ceramics. The main objective in deposition is to raise the temperature of the surfacing rod or powder and the substrate surface to the correct bonding temperature range. A large number of these alloys melt at temperatures between 1200 and 1500°C, but they will not adhere well unless the surfaces are cleaned and therefore fluxes are required for this purpose. Fluxes also improve wetting so that a satisfactory metallurgical bond is developed. Self-fluxing alloys were developed during the 1940s (e.g., by the Sprayweld process), and these are fused in a second operation carried out after spraying. Fusing may be carried out by heating with an oxyacetylene torch, by induction heating, or by placing the alloy in a furnace with either a partially or completely protective atmosphere.

TABLE XII

PROCESS FOR HARD SURFACE FACING

Arc welding processes	Gas welding processes	Spraying processes (with or without fusing)
Manual metal-arc (stick electrode)	Oxyacetylene Gas welding Stick filler rod	Oxyacetylene spraying using (i) pure or metal alloy wire consumable (ii) ceramic rod consumable
Submerged arc (using wire or strip electrode with flux)		Oxyacetylene spraying using (i) pure metal or alloy[a] powder consumables (ii) refractory powder consumables
MIG (metal–inert gas consumable wire electrode)		Plasma arc using powder consumables (not transferred arc)
TIG (tungsten–inert gas or argon arc)		Oxyacetylene powder fusing
Electroslag		
Atomic hydrogen		

[a] A number of alloys of the hard-facing variety may be fused after power spraying by heating the component in an oxyacetylene flame or in a suitable furnace, e.g., in vacuum.

Moore (1976) has drawn attention to the microstructure of the hard facing as the critical factor in determining its suitability for a particular application. Although the composition of the hard facing consumable may be constant, the deposition rate, cooling rate, etc. all have a marked effect on the structure produced. It is the structure-dependent properties that control its wear behavior. A new classification specifies hard facing consumables according to the general microstructural features of typical deposition by the use of a four-digit representation (Fig. 45). The first of the digits distinguishes among steels, white irons, tungsten carbide alloys, cobalt alloys, and nickel alloys. The second completes a qualitative subdivision into a total of 25 metallurgical groups. The specification of these groups is the major factor in describing the type of wear to which the alloy is particularly resistant. Table XIV summarizes the system and shows its relation to tribological properties. A secondary division of the consumables within a group is achieved by using the hardness (Rockwell C) of the weld metal as the final two digits. These latter figures are indicative of the relative brittleness of the various alloys within a group. They are not useful for comparisons among different groups, however. An alphanumeric suffix is also used to specify the deposition technique and consumable form.

Examples of the economics of hard facing were quoted recently (Gregory, 1977) from examples in the steel industry. The cost per ton of

TABLE XIII

ALLOYS USED FOR HARD FACING

Group	Subgroup	Alloy system	Prime features	Typical macro-hardness (R_c)	Process suitability			Finishing method	
					Electrodes	Cored wire semiauto	Fully auto wire	Turn	Grind
Steels (up to 1½% C)	Ferritic	Carbon and C/Mn	Tough crack-resistant base for hard facing	up to 25	✓	✓	✓	✓	✓
		Alloy martensitic	Tough impact resistant—easily used—reasonable cost—good hardening	25–65	✓	✓	✓	✓	✓
		High speed complex martensitic	Extreme hardening by thermal treatment	25–65	✓	✓	✓		✓
	Austenitic	Chromium/manganese (15% C–15% Mn)	Work hardening corrosion resistance	up to 65	✓	✓	✓	✓	
		Manganese steel (10–20% Mn)	Work hardening	up to 65	✓	✓	✓		✓
		Chromium nickel steel	Stainless tough corrosion resistance at high temperatures	up to 30	✓	✓	✓		✓
Iron (above 1½% C)	Ferritic	Martensitic iron up to 15% alloy G/Mi Cr/Mn Cr/N	Carbide particles in hard compression strength erosion and abrasion resistance	50–65	✓	✓	✓		✓
		Chromium iron 25/35% and M for Mn	Carbide particles in hard matrix erosion, abrasion, oxidation resistant at high temperatures	50–65	✓	✓	✓		✓

TABLE XIII (*cont.*)

Group	Subgroup	Alloy system	Prime features	Typical macro-hardness (R_c)	Process suitability			Finishing method	
					Electrodes	Cored wire semiauto	Fully auto wire	Turn	Grind
	Austenitic	10/20% chromium + M for Mn	Fair resistance to abrasion—good resistance to oxidation at high temperature—resistance to cracking better than martensitic iron	45–46	✓	✓	✓	✓	✓
Copper alloys		Copper + 5/10% tin Copper + 2/5% iron	Antiseizing resistance to friction		✓	×	×	✓	
Nickel alloys		Nickel + 10/20% chrome Boron + 2/5% Nickel + 15/25% Chromium and 10/15% Mn and 0/5% tungsten	Hot-hardness, corrosion, erosion resistance at high temperatures	30°C 45–50 500°C 40–45	✓	✓	✓	✓	
Cobalt alloys		Cobalt + 20/50% Chromium and tungsten Carbon 1.0–3.0%	Hot-strength abrasion and erosion resistance at elevated temperatures. High temperature creep strength	30°C 35–45 500°C 25–30	✓	✓	✓	✓	
		Tungsten carbide particles in soft matrix	Extreme abrasion resistance	60–65	✓	×	×		✓

427

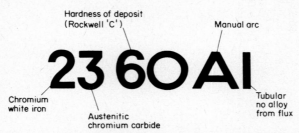

Fig. 45.   A new classification for hard facing consumables.

wear of sinter breaker bars was reduced by a factor of 4 by the application of tungsten carbide (Table XV). Moore also discusses this aspect in some detail and emphasizes that great care is required in analyzing the wear problem, particularly with respect to how much impact loading is involved (Moore, 1976).

D.   Type II: Chemical Conversion Coatings

Chemical conversion coatings are produced by the transformation of the metal surface into a new nonmetallic form with properties different from those of the original surface. The major examples of conversion coatings are phosphating, tufftriding, and sulfidizing.

These conversion coatings are relatively thin and are therefore intended principally to aid running in under nonabrasive conditions rather than to improve the long-term behavior of the sliding surface. Some claims are made that they improve the long-term as well as the short-term wear characteristics even after they have been completely worn away.

The use of conversion coatings enables engineers to:

(1)   permit the substitution of a less expensive, more easily fabricated material, without sacrificing performance;
(2)   substantially upgrade the performance potential of an existing design, particularly by reducing wear damage during running in; and
(3)   extend the life of a component.

1.   PHOSPHATING

The excellent adhesion properties of phosphate coatings, which are attributed to epitaxial growth, led to their development and use as surface treatments for painting and as coatings to which lubricants adhere more readily.

During the phosphating process,

$$Fe^{2+} + 2H_2PO_4^- \rightarrow Fe(H_2PO_4) \downarrow$$

the surfaces become black or gray and take on an etched appearance, giving rise to interconnecting surface channels surrounding plateaus which become covered with phosphate crystals (Fig. 46) [see Perry and Eyre (1977)]. This process can be applied to a wide range of ferrous and nonferrous metals and does not adversely affect the properties of the metal being treated, i.e., there is little heating during processing (40–100°C) that could cause tempering or softening of the metal being treated. Zinc, iron, and manganese phosphate treatments have all been used, but manganese phosphate is preferred for wear applications.

These coatings have been found to improve the wear characteristics of steel surfaces, and Midgley (1957) attributes this to the following:

1.   During running in, the presence of the phosphate crystals helps to prevent metal-to-metal contact and thus prevents adhesive wear. An oxide film soon forms as sliding continues and the mild region of wear is reached without significant damage.

2.   During the wear cycle, after running in has been completed the phosphate coating does not in itself increase wear resistance, but the run-in surfaces have a greater chance of survival.

2.   TUFFTRIDING

This is a liquid salt bath treatment carried out at approximately 570°C. Cyanide compounds liberate carbon and nitrogen in the presence of ferrous metals. Nitrogen is more soluble than carbon at the process operating temperature and diffuses into the surface, while the carbon forms iron carbide particles at or near the surface. These particles act as nuclei, precipitating some of the diffused nitrogen to form a tough compound zone of carbon-bearing epsilon iron nitride. None of the brittle iron nitride is formed. The compound zone will usually be about 0.0006 in. deep, but the nitrogen diffusion zone may be up to 0.030–0.040 in., which is achieved after 1 to 3 h.

Although there is a considerable increase in hardness, the improved adhesive wear properties are attributed to the formation of the white nitride-rich compound zone which reduces the weldability of the surface.

Figure 47 shows that below 250 psi no running in of gray cast iron occurs, i.e., mild wear is initiated almost immediately; above this load untreated gray cast iron exhibits increased weight and dimensional loss as

TABLE XIV

Prime use	Type[b]	Description
Service at high loading[c]: medium temperature[d]		
Steels		
Metal-to-metal	11xx	Pearlitic steels[e]
Impact: heavy	12xx	Austenitic Mn steels
—⌡	13xx	Austenitic stainless steels[e]
light	14xx	Low carbon martensitic steels
Cutting edges	15xx	Tool steels
	16xx	Martensitic stainless steels
Abrasion	17xx	High carbon austenitic steel
	18xx	High carbon martensitic steels
Service at medium loads: medium temperature: abrasive conditions		
Chromium white irons		
Impact: heavy	21xx	Austenitic irons[e]
light	22xx	Martensitic irons
Abrasion: coarse	23xx	Austenitic chromium carbide irons
	24xx	Complex chromium carbide irons[e]
fine	25xx	Martensitic chromium carbide irons
Service at low loads: medium temperature: abrasive conditions		
Tungsten carbides (Minimum 45% tungsten carbide)[h]		
Rock cutting and drilling	31xx	Carbide particles + 5 mesh, brazing alloy filler metal
Abrasion: coarse	32xx	Mixed carbide, av. particles + 20 mesh
—[i]	33xx	Mixed carbide, av. particles − 20 + 40 mesh
fine	34xx	Mixed carbide, av. particles − 40 + 100 mesh
	35xx	Mixed carbide, av. particles − 100 mesh
Service at medium loads: high temperature: especially involving metallic friction and/or corrosive liquids and gases		
Cobalt alloys		
Impact: heavy	41xx	Complex superalloy Co–Cr–Mo type
light	42xx	Hypoeutectic Co–Cr–W type
Abrasion	43xx	Hypereutectic Co–Cr–W type
Service at medium loads: high temperature: expecially involving abrasion and/or oxidation		
Nickel alloys		
Impact	51xx	Complex superalloy
Abrasion	52xx	Modified Ni–B eutectic
	53xx	Tungsten carbide (−45 wt. %) nickel alloy matrix
Corrosion	54xx	Solid solution

[a] From Moore (1976).
[b] xx indicate digits representing Rockwell C hardness.
[c] High load >100 Mn/m²; low load <30 MN/m².
[d] High temperature >500°C. Medium temperature 200–500°C.
[e] This alloy is used for general rebuilding and as hard-facing underlays.

significant running in occurs, while tufftriding completely inhibits running in up to the mild/severe transitional load (Eyre and Dutta, 1975).

Examination of the two surfaces shows marked differences in both surface topography and subsurface microstructure (Fig. 48). The untreated cast iron has a very rough torn surface appearance with evidence of phase transformation at the tips of the asperities as a result of the high flash temperatures developed. The tufftrided surface, however, is much smoother,

AWRA Hard Facing Alloy Classification and Application[a]

Features	Examples
Strong, inexpensive	Butter layers, spindles, shafts, rollers, track links, sprockets, tractor idler wheels
Tough, work hardening	Crusher jaws, rolls, mantles, ball mill liners
Tough	Tramway rails, crossings, bearings at medium temperatures, tractor track grousers
Strong	Clutch parts, railway points and crossings
Strong, secondary hardening	Machine tools, lathe tools, shears, guillotine blades, cutting knives
Hard, corrosion/heat resistant	Punches, dies, metal forming tools
Hard, work hardening	Crushing rolls, hammers, tractor grousers
Hard	Post-hole augers, earth scoops, conveyor screws, drag line brackets, pump housings
Strong, provide good support to brittle overlays	Crushing equipment (e.g., jaws, rolls, hammers, mantles), pump casings, impellors
Hard	Agricultural plow shares, tines, mill scraper blades, wear bars, bucket lips
Strong ⎫ large primary chromium carbides	Screen butt straps, quarry screen plates, chutes, grizzy bars, dragline teeth
Hard ⎭	Sizing screens, ball mill liner plates, pump impellors, crusher jaws, harrows
	Wet applications in mining and crushing industries, e.g., ball mill liners
Massive carbides useful as individual cutting edges	Rock drills, oil drills, sand mixer blades, oil well fishing tools
Strong deposit	Bucket teeth, ripper points, oil drill collars, auger blades and teeth, oil well drills, bulldozer end tips
	Rock drills, ditcher teeth, dry cement pump screws, suction dredge blades
Hard deposit	Plow tines, ditcher teeth, concrete pump screws, churn drills, tool joints
Tough, creep resistant	Hot and cold shear blades, valve seats
Strong	Exhaust valves in diesel engines
Hard	Scrapers, feeders screws, etc., in chemical mining, cement industries
Tough, creep resistant	Blast furnace bells, hoppers, forging dies and hammers, hot trimming dies
Hard, fine borides	Hot forging dies, parts subject to hot erosion in chemical plants
Hard, carbides and fine borides	
Inert	Valve bodies and parts subject to oxidation

[f] >0.5% surface strain heavy impact, >0.2% surface strain light impact.
[g] Includes types containing up to 45% tungsten carbide.
[h] Mesh size in US standard.
[i] Arc welding usually produces lower toughness and lower abrasion resistance than gas welding or metal powder spraying.

with no evidence of phase transformation products, which indicates that lower flash temperatures have been reached. These surface topographical and microstructural differences indicate that tufftriding operates by reducing the tendency for both adhesion and metal flow in the first few minutes of contact. It is during this period that surface roughening associated with high contact stresses generates high "flash" temperatures which result in "scuffing."

The advantages obtained by tufftriding a wide range of steels and cast

TABLE XV

ECONOMICS OF THE USE OF HARD FACING

Materials used	Production cost (£)	Fitting cost (£)	Total cost (£)	Service life		Cost per Ton of sinter (p)
				Tons of sinter	Ratio	
Standard mild steel bar	31.50	8.89	40.39	81,000	1	0.05
Hard faced with tungsten carbide	53.75	8.89	62.64	467,000	5.8	0.01

irons are illustrated in Table XVI, which is based upon a Falex lubricant seizure test. It will be observed that the seizure load for AISI 1025 steel increased from 400 to 1500 lb after tufftriding for 90 min. These effects have been confirmed by the many successful industrial applications of this process for components such as gears, cam shafts, slideways, piston rings, and die blocks. (Gregory, 1970).

3. SULPHINUZ

This is an accelerated nitriding bath where the nitriding potential is increased by the addition of a sulfur compound which acts as an accelerator. By this means sulfur, as well as nitrogen and carbon, is incorporated in the surface of the parts treated. Sulphinuz baths are operated at temperatures between 540 and 600°C and for times up to 3 h. A light etching

(a)   (b)

Fig. 46. Surface structure of phosphating. (a) ×800; (b) ×1600.

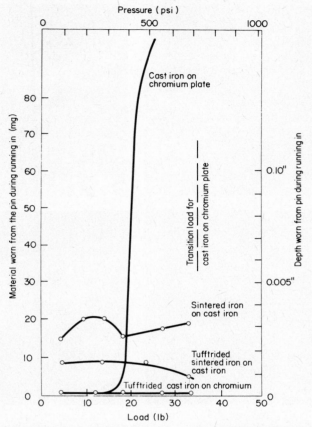

Fig. 47. Running-in wear of gray cast iron and sintered iron.

zone rich in sulfur (5%), nitrogen (2%), and carbon (2%), and approximately 0.0005 in. deep, is produced and a further zone rich in nitrides extends to a depth of about 0.015 in.

The effect of the Sulphinuz process is rather similar to that obtained by the tufftriding process, and it is not easy to make a rational decision concerning which to use. However, the improvements obtained are best illustrated for stainless steel, which is particularly prone to adhesion during running in (Fig. 49).

Both the tufftriding and Sulphinuz processes have the disadvantage that the process operating temperatures are sufficiently high to temper or anneal many ferrous metals, and under high load bearing applications this may be a disadvantage, particularly if fatigue is also involved. Some slight

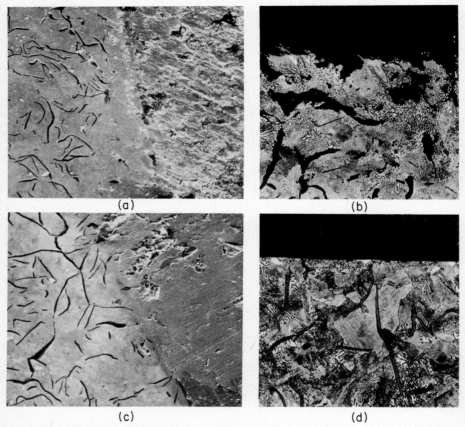

(a)   (b)

(c)   (d)

Fig. 48.   Effect of tufftriding on wear during running in after 80,000 cm. (a), (b) not tuff-trided; (c), (d) tufftrided.

distortion is also possible, and because these processes are the last in the production line, any distortion may be difficult to detect and expensive to rectify. The distortion, however, is less than one would expect with carburizing.

The surface conversion coatings referred to above form the main bulk of the more important industrial processes that are now widely used (Gregory, 1970). They are not designed to improve abrasion resistance because some of them do not bring about marked increases in hardness, and in any case the coatings are relatively thin. Conversion coatings are primarily designed as aids to running in and as such markedly improve the rather poor properties of carbon and alloy steels, austenitic stainless

TABLE XVI

ADVANTAGES OBTAINED BY TUFFTRIDING A WIDE RANGE OF STEELS AND CAST
IRONS; BASED UPON A FALEX LUBRICANT SEIZURE TEST[a]

Material	As cast (lb)	(sec)	Gaseous nitrided (lb)	(sec)	Tufftride treated 90 min/570°C (lb)	(sec)	Tufftride treated 180 min/570°C (lb)	(sec)
AISI 1025	400	9	1350	49	1500	60	1500	57
AISI 1045	450	14	1350	47	1450	55	1600	60
80-60-03 nodular iron	700	26	1350	48	1800	75	1950	75
60-45-15 nodular iron	600	19	850	31	1750	74	2150	89
Meehanite type GD	450	13	450	12	500	13	1000	61
Meehanite type GA	550	16	600	20	850	13	1000	54
Meehanite type GM	500	18	650	24	1100	43	1400	41

[a] Given in pounds of pressure and time in seconds at pressure to seizing.

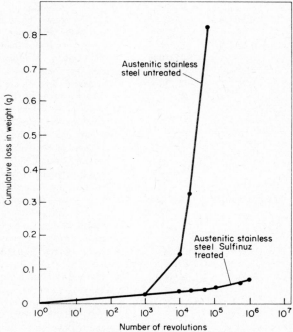

Fig. 49. Effect of Sulphinuz on wear of stainless steel.

steels, and cast irons under metal-to-metal rubbing conditions in which lubrication is poor or nonexistent.

E. *Surface Hardening Techniques*

Metals may be surface hardened by a wide range of techniques, the efficiency of which are affected by the correct selection of alloy composition and the technique applied to produce the optimum depth of hardening. Metals and alloys that may be treated by the various techniques are summarized in Table XVII.

The use of surface hardening techniques is widely practiced to improve

TABLE XVII

SURFACE HEAT TREATMENTS

Process	Case depth (mm)	Case hardness[a] (HV)	Remarks and applications
Carburizing	to 0.5	S 736 C 230+	Can be used on low-C, med-C, C–Mn, C–Mo, Cr–Mo, Ni–Cr–Mo, and Ni–Mo steels. Light case depths for high wear resistance and low loads, e.g., water pump shafts, sockets, and small gears. Moderate case depths used for high wear resistance and moderate to heavy service loads, e.g., valve rocker arms, bushes, and gears. Heavy case depths for high wear resistance to sliding, rolling, or abrasive action and for high resistance to crushing or bending loads, e.g., transmission gears, king pins. Extra heavy case for maximum wear and shock resistance, e.g., camshafts
	0.5–1.0	S736 C 230+	
	1.0–1.5	S 736 C 230+	
	1.5+	S 736 C 230+	
Carbo nitriding	0.075–0.5	S 800–890 C 320–350	Can be used on low-C, med-C, Cr, Cr–Mo, Ni–Mo, and Ni–Cr–Mo steels. Produces a hard wear-resistant clean surface. Typical uses include ratchets, small gears, hydraulic cylinders
Nitriding	to 0.75	S 900+	Can be used on any steel containing Al, Cr, Mo, V, or W, including 13% Cr and austenitic stainless steels. Applications include valves, shafts, and anywhere involving sliding or rolling friction

TABLE XVII *(cont.)*

Process	Case depth (mm)	Case hardness[a] (HV)	Remarks and applications
Tufftriding	to 1.0	S 450	Can be used on low-C, med-C, C–Mn, Cr, Cr–Mo, Cr–Mo–V, and stainless steels. The surface is not brittle. Wear resistance and fatigue properties are improved. Also improved is corrosion resistance, except on stainless steels. Typical uses include levers, gears, spindles, pistons, and pumps
Flame hardening	0.75–3.2	S 370–630 C 240	Can be used on med-C, C–Mn, Cr–Mo, and Ni–Cr–Mo steels. Produces moderately high surface hardness with unaffected core. Typical uses include sprocket and gear teeth, track rails and runner wheels, lathe beds
Induction hardening	0.75–3.2	S 736	Can be used on med-C, C–Mn, Cr–Mo, and Ni–Cr–Mo steels. Produces high surface hardness and unaffected core. Applications include cam shafts, sprockets and gears, shear blades, and bearing surfaces

[a] S, Surface; C, Core; +, and higher.

abrasive wear resistance, and the prime object is to achieve maximum hardness to the depth dictated by the amount of wear that will take place during the design life. The use of alloy steels for carburizing, particularly those containing chromium, causes some difficulty because of a tendency to overcarburize with a consequent buildup of brittle surface carbides which are liable to cause premature spalling under impact loads. The precise effect of surface hardening on improving wear resistance under sliding conditions is quite complex, but generally speaking, increasing the hardness is not advisable unless the loads are high and fatigue stresses are also involved. It is likely that those treatments that may also be classified as surface hardening techniques, e.g., tufftriding, would be more successful under metal-to-metal sliding conditions.

F. Developments in Coating Techniques

Developments in the semiconductor and atomic energy fields are now being translated into hardware for wear resistance, and within recent years a number of new processes, particularly chemical vapor deposition (CVD), ion plating, and ion implantation have received considerable attention.

1. CHEMICAL VAPOR DEPOSITION†

The CVD technique involves the exposure of a heated substrate to a mixture of gaseous reactions so that they interact at the surface to form a solid deposit. The technique is most widely used to produce hard coatings consisting of carbides, silicides, and borides of the transition metals. If the optimum conditions are chosen, a coating of uniform thickness can be obtained over an object of complex shape because the CVD reaction only takes place at the substrate surface. It is therefore possible to coat hundreds or thousands of small components simultaneously with excellent control of coating thickness. The application of titanium carbide coatings to tungsten carbide tool inserts is the best commercial example of the CVD process.

The improvements in flank wear, crater depth, and, most important of all, increase in cutting speed shown in Fig. 50 are attributed to the success of TiC in limiting diffusion and adhesion. Titanium carbide reduces the coefficient of friction between tool and chip with consequent reduction of cutting forces and temperatures. It also forms oxide layers close to the tip of the tool, and these are transported into the crater zone. Diffusion of tungsten from the tool matrix into the chip is prevented and adhesion over the rake face is diminished. Tools that have to be reground would require recoating, so the technique is limited to single-use edges. If the work piece is oxidized or abrasive particle inclusions are present, then the thin TiC coating (0.0002–0.0003 in.) can be penetrated and the advantages of the coating are wasted.

One of the limitations of the process is the reaction temperature, which has commonly been about 1000°C and which can have a serious softening effect on steel substrates. It is necessary, therefore, to select the base metal very carefully so that either the temperature does not temper the structure or that air hardening can be facilitated after the completion of the CVD process. Lower reaction temperature processes are now becoming available, which will remove this limitation.

2. ION NITRIDING

This process is sometimes also referred to as glow discharge or plasma nitriding. Edenhofer (1976) has recently reviewed the subject, which had

† From James (1977).

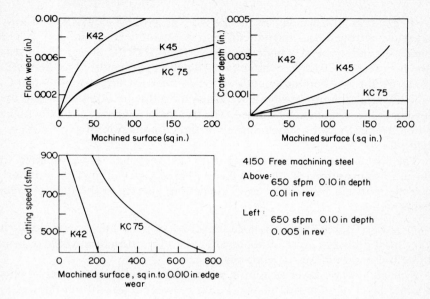

4150 Free machining steel

Above:
650 sfpm 0.10 in depth
0.01 in rev

Left :
650 sfpm 0.10 in depth
0.005 in rev

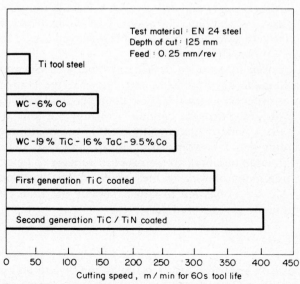

Fig. 50. Effect of TiC coatings on improving machining characteristics. Improvement in cutting tool performance resulting from developments in tool materials.

its beginnings in Germany in the 1930s. The process consists of a low-pressure nitrogen gas introduced into a vacuum (1–10 mbars) where under the influence of a high dc voltage (500–1000 V) the gas is ionized and becomes electrically conductive. The energy transferred to the workpiece by the ion bombardment is used to heat the target to 400–600°C, which is the optimum temperature range for diffusion of nitrogen into the matrix of the surface to be treated. The treatment time will normally range from 20 min to 20 h, depending upon the combination of properties required.

The mechanical and wear properties are reported to be superior to those of conventionally nitrided cases. Three types of surface structure can be achieved by this process, depending also upon the type of steel being treated.

Type I. Tool steels with more than 10% of alloy elements produce a very hard diffusion layer of about 1100 HV. This structure would appear to be the optimum for abrasion resistance.

Type II. A two-layer structure, which consists of a diffusion layer plus a white layer of γ' nitride phase (Fe_4N) between 5 and 10 μm thickness, can be produced on lower alloy steels and nitriding steels. The γ' nitride layer is extremely ductile and wear resistant and is therefore not ground off and plays an important part in extending the life of plastic processing machine parts.

Type III. Carbon steels and cast irons are usually ion nitrided to produce a monophased ϵ-white layer of about 10 to 15 μm. The subcase material will be relatively soft (300–500 HV). The ϵ layer is said to have good wear resistance under sliding conditions.

Ion nitriding appears to have developed rapidly in Germany, France, Austria, and Japan and is now being introduced in the United Kingdom. Components as different as the 0.030-in.-diameter steel balls for ball-point pens and a 23-ft-long extrusion cylinder weighing 3.8 tons are currently being processed. A wide range of components used in automobile manufacture, including gears, rocker arms, valves, and cam shafts, are now being treated in large numbers so the process appears to be economically viable. The treatment of forging dies made in hot working tool steel (Hl3-X40 Cr Mo V5 1) has led to longer service life than that obtained from conventionally gas nitrided parts.

3. ION IMPLANTATION

The composition and structure of surfaces can be altered by implanting selected ions that are introduced by an electrical potential at around 100 kV. The process was originally developed for making silicon semiconductor devices, but has now been applied to the general improvement in tribological properties, particularly reduction in friction and an increase in wear resistance.

There are a number of important advantages, which include the following:

(1) It is possible to implant selectively a range of both metallic and nonmetallic ions either singly or in combination.

(2) The process is carried out cold so there is no degradation in size or surface finish, and therefore the process can be applied to finished articles.

(3) It is possible to implant both ferrous and nonferrous base metals.

(4) Selected area implantation is possible.

The ions B^+, N^+, and Mo^+ implanted into steel to doses between 10^{16} and 10^{18} ions/cm^2 reduce wear by more than a factor of 10. The improvements are greatest for N^+ and Mo^+ and least for B^+. Dearnaley (1974) reports an interesting observation that the combination $Mo + S$, with the sulfur implanted after the molybdenum ions, produced a larger reduction in friction than that produced by Mo or S alone. Although he was unable to detect the presence of MoS_2 in the worn surfaces, it is interesting to speculate on the chemical interactions of lubricant constituents and the metal bearing elements. The literature contains repeated references to the advantages of having molybdenum in the metal and sulfur in the oil, because under these conditions maximum reduction in friction and increase in wear life are achieved. The work of Rounds (1972) and Matveevsky *et al.* (1973) supports this hypothesis.

It was also observed that the effects of ion implantation were still detected long after the implanted layer had been worn away and it was concluded that a better "run-in" surface had been achieved, which thus enhanced the subsequent wear behavior of the surface.

References

Atterbury, T. D. (1974). *Met. Mater. Tech.* **6**, 551–558.

Archard, J. F., and Hirst, W. (1956). *Proc. R. Soc. London Ser. A* 236–397.

ASTM (1970). Publ. 474.

Beesley, C., and Eyre, T. S. (1976). *Tribol. Int.* **9**, 63.

Boyes, J. W. (1969). *J. Iron Steel Inst. London* **4211**, 57–63.

Czichos, H. (1974). *Tribology* **8**, 14–20.

Dearnaley, G. (1974). *Rev. Mater. Sci.* **4**, 93–123.

Edenhofer, B. (1976). *Metallurgist* **8**, 421–426.

Eyre, T. S. (1971). Ph.D. Thesis, Brunel Univ., Uxbridge, England.

Eyre, T. S. (1975). *Wear* **34**, 383–397.

Eyre, T. S., and Baxter, A. (1972). *Met. Mater.* **6**, 435–439.
Eyre, T. S., and Dutta, K. (1975). *Int. Mech. Eng. Conf. Piston Ring Scuffing* Paper C74/75.
Eyre, T. S., and Maynard, D. (1971). *Wear* **18**, 301–310.
Eyre, T. S., and Walker, R. K. (1976). *Powder Metall.* **1**, 22–30.
Eyre, T. S., and Williams, P. (1973). *Wear* **24**, 337–349.
Eyre, T. S., Iles, R. F., and Glasson, D. W. (1969). *Wear* **13**, 229–245.
Farrell, R. M., and Eyre, T. S. (1970). *Wear* **15**, 359–372.
Fenton, P. A. D., and Dolbekink, A. (1973). Diesel Engineers and Users Assoc., Preprint 356.
Graham, W. A. (1963). M.Sc. Thesis, Univ. of Oklahoma, Norman, Oklahoma.
Gregory, E. N. (1977). *Met. Constr. Br. Weld. J.* **1**, 14–15.
Gregory, J. C. (1970). *Tribol. Int.* **3.3**, 73–80.
Halling, J. (1976). *Prod. Eng.* **1**, 19.
Hewitt, J. C., Clayton, A., and Cox G. J. (1975). *Powder Metall.* **18**, 35, 168.
Hunsicker, A. (1949). A.S.M. Sleave Bear. Mats.
Hurst, W. (1965). *Metall. Rev.* **10**, 38, 145–172.
James, D. M., Smart, R. F., and Reynolds, J. A. (1975). *Wear* **34**, 375–382.
James, M. R. (1977). *Met. Mater. Tech.* **9**, 483–486.
Kirk, F. A. (1974). *Met. Mater. Tech.* **6**, 543–547.
Kruschov, M. N. (1957). *Inst. Mech. Eng. Conf. Lub. Wear* 655.
Lea, M. (1972). Ph.D. Thesis, Brunel Univ., Uxbridge, England.
Matveevsky, R. M., Markov, A. A., and Bujanovsky, I. A. (1973). *Inst. Mech. Eng. Eur. Conf. Tri., 1st* Paper C268/73.
Midgley, J. W. (1957). *J. Iron Steel Inst. London* **193**, 215.
Moore, A. J. W. (1976). *Tribol. Int.* **9**, 265–270.
O.E.C.D. (1969). Friction, Wear & Lubrication, Terms & Definitions, Publication 26,587, Paris.
Perry, J., and Eyre, T. S. (1977). *Wear* **43**, 185–197.
Popov, V. S., and Nagornyi, P. L. (1969). *Russ. Cast. Prod.* **8**, 337.
Pratt, G. C. (1973). *Int. Metall. Rev.* Rev. Paper No. 174, 18, 62.
Richardson, R. C. (1968). *Wear* **11**, 245.
Rolfe, R. T. (1952). *Allen Eng. Rev.* No. 29, 6.
Rounds, F. G. (1972). *ASLE Trans.* **15**, 54–66.
Sedrickas, A. J., and Mulhearn, T. O. (1963). *Wear* **6**, 457.
Serpik, N. M., and Kantor M. M. (1965). *Frict. Wear Mach.* **19**, 28.
Shivanath, R., Sengupta, P. K., and Eyre, T. S. (1977). *Br. Foundryman* **70**, 12, 349–356.
Stott, F. H., Bethume, B., and Higham, P. A. (1977). *Tribol. Int.* **10**, 211–215.
Stott, G. (1976). Ph.D. Thesis, Brunel Univ., Uxbridge, England.
Suh, N. P. (1977). *Wear* 44, 1–16.
Waterhouse, R. B. (1972). "Fretting Corrosion." Pergamon, Oxford.
Weish, N. C. (1965). *Phil. Trans. R. Soc.* **11**, 435–439.
Wright, K. H. R. (1957). *Proc. Conf. Lub. Wear Inst. Mech. Eng.* p. 628.

Wear of Metal-Cutting Tools

E. M. TRENT

Department of Industrial Metallurgy
The University of Birmingham
Birmingham, England

I. Introduction

The wear of tools used for cutting metals and alloys is a subject of great importance in relation to the economics of the engineering industry. It was estimated in 1971 (Zlatin, 1971) that over $40 billion was spent in the

United States on the machining of metal parts. The wear of tools influences these costs in two ways. When a tool is worn out it must be reground or replaced, and the cost of the tools used is part of the total machining cost. For example, in the United Kingdom nearly 20 million carbide cutting tools are used per year at a cost of the order of £ 50 million. However, this is not the major way in which machining costs are influenced by tool wear. Costs are lowered by increasing the rate of metal removal, which means increasing the cutting speed or the feed, or, in some instances, the depth of cut. In practice many factors may limit the rate of machining but, when cutting iron, steel, and other high-melting point metals, the rate of wear of tools increases as cutting speed is raised, and the ultimate limit to the speed or feed is determined by the ability of the tool to resist wear.

It is the search for new tool materials to make possible increased rates of metal removal and thus reduce costs which has been responsible for the development in the past 80 years of two major classes of tool material—high-speed steels and cemented carbides. That these have been very effective is exemplified by the increase in typical cutting speeds for machining steel from 5 m min^{-1} (16 ft min^{-1}) using carbon steel tools to 150 m min^{-1} (500 ft min^{-1}) with steel-cutting grades of cemented carbide. High-speed steel and cemented carbide tools are now used almost universally. For specialized applications still higher rates of metal removal can be obtained using other tool materials, e.g., alumina, diamond, or cubic boron nitrides.

For these reasons, the aspects of cutting tool wear that have been the subject of research are related to the cutting of the higher-melting-point metals, especially steel and cast iron since far more of these are machined than of all the other metals and alloys combined. Tool wear has been studied mainly in relation to high-speed steel and cemented carbide tools. Cutting tool wear will therefore be discussed here mainly in this context, although other work materials will be mentioned, and the wear of more specialized tool materials will be considered briefly.

II.   Metal-Cutting Operations

The conditions to which metal cutting tools are subjected are extremely variable, depending on the type of machining operation, the cutting conditions, and the material being cut. The operations of turning, facing, planing, milling, drilling, tapping, sawing, and many others impose very different conditions on the tools, but all have certain features in common. In all metal-cutting operations a thin layer of metal is removed by a

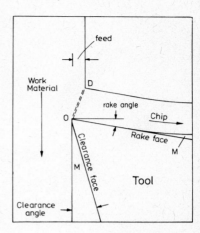

Fig. 1. Features of metal cutting.

wedge-shaped tool which is driven asymmetrically into the work material as shown diagrammatically in Fig. 1. Because of the asymmetrical movement of the wedge tool, the metal removed, called a "chip," moves across one face of the tool, known as the "rake face," either in the form of a continuous ribbon or as small fragments. The new metal surface produced by the cutting operation moves away from the tool edge and is prevented from making extensive contact with the "clearance face" of the tool by the "clearance angle." The thickness of the layer of metal removed, termed the "feed," can vary greatly from less than 0.025 mm (0.001 in.) to over 2.5 mm (0.1 in.) but it is commonly between 0.075 and 1 mm (0.003–0.040 in.). The rate of metal removal is controlled largely by varying the feed and the "cutting speed," which is the velocity of the work material past the tool edge, and may vary from a few inches per minute to 600 m min^{-1} (2000 ft min^{-1}) or even more in exceptional cases.

The simplified two-dimensional diagram (Fig. 1) is essential when considering the basic wear processes on tools, but, because it is a *two-dimensional* model, it is an oversimplification. It does not represent the conditions that exist in the third dimension, along the cutting edge, which are often critical in relation to tool life. The shape of the tool edge and the length of edge engaged in cutting at any time are very variable. This is clear if operations as different as drilling, forming, sawing, and turning are considered. Figure 2 shows a variety of tool shapes. Critical positions for wear may be at a nose radius or at the position where the surface of the work material intersects the cutting edge.

A very important parameter is the time of cutting, i.e., the time any part of the edge is engaged in cutting, and whether this is continuous or inter-

Fig. 2. Some basic metal-cutting tools.

mittent. The time of continuous cutting can be very variable. Frequently the tool is cutting continuously for several seconds to a minute. In more exceptional cases cutting may be continuous for an hour or more, as when turning a long shaft. In many milling and other operations, where rotating multitoothed cutters are employed, the cutting time for each tooth may be much less than 1 sec, and the cutting cycle may be repeated as many as five or ten times per second.

Cutting is carried out either in air or with the use of a "coolant" or lubricant, which may be a water emulsion or a mineral oil with or without additives such as fatty acids, chlorine, and sulfur compounds. These may make an appreciable difference to the character and rate of wear on the tools.

III.   Descriptive Treatment of Tool Wear Phenomena

To cut efficiently, most tools must have a very precise shape, and tool life is terminated when a very small change in the shape of the tool edge has occurred. Complete tool failure happens when the whole edge of the tool is carried away, but this is generally avoided by replacing or regrinding the tool before catastrophic failure begins. Tool life is considered to be finished when signs of approaching failure are detected. These depend on observations by the machine operator and may consist of a change in sound during cutting, change in surface appearance of the chip or the workpiece, change in dimensions of the workpiece, increased heat generation, or observation of wear on the tool.

These symptoms of impending failure are the result of gradual changes in the shape of the tool edge, and have been frequently described in the literature. The characteristic forms of wear and the descriptive terms used are summarized in Fig. 3.

A. Flank Wear

Flank wear is the most commonly observed wear phenomenon, occurring eventually in most cutting operations. The depth of the wear land (V_B in Fig. 3) can often be measured with reasonable accuracy and is the most frequently used quantitative parameter of tool wear. A maximum permissible amount of flank wear before replacing the tool is sometimes the cri-

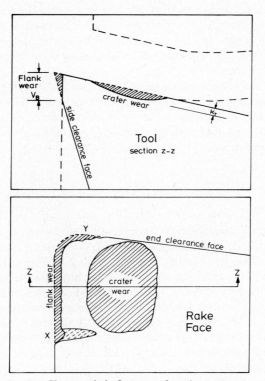

Fig. 3. Characteristic features of cutting tool wear.

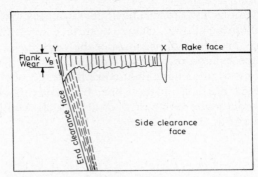

Fig. 4. Flank wear.

terion of tool life, and values between 0.4 and 1.6 mm (0.015 and 0.060 in.) have been specified for different operations. Flank wear is sometimes very even, but more usually it varies along the length of the cutting edge, as shown in Fig. 4. Frequently there is accelerated wear at the position where the initial work surface intersects the cutting edge—X in Figs. 3 and 4. Increased wear is also often observed at the nose of the tool or at other sharp changes in profile of the cutting edge—Y in Figs. 3 and 4. This is sometimes referred to as "nose wear" and may extend along the end clearance face (Fig. 3).

The rate of flank wear increases as the cutting speed is raised under most but not all conditions. In many laboratory tool tests the rate of flank wear as a function of cutting speed under standardized cutting conditions is used as a measure of the performance of tool materials or the machinability of work materials.

B. Cratering

Cratering is the term used to describe a form of wear frequently observed on the rake face of the tool where the chip moves over the tool surface (Figs. 1 and 3). A groove or "crater" is formed, usually starting at some distance from the cutting edge—e.g., 0.2–0.5 mm. It is most commonly observed when cutting steel and other high-melting-point metals at relatively high cutting speeds. The crater gradually becomes deeper with time and may terminate tool life by weakening the cutting edge. An example of a crater on a WC–Co cemented carbide tool is shown in Fig. 5. The depth of the crater, K_T in Fig. 3, is sometimes measured in laboratory tool testing together with flank wear and is used as an index of the efficiency of the tool.

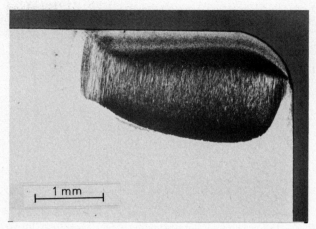

Fig. 5. Crater on rake face of WC–Co alloy tool used to cut steel at high speed.

Both flank wear and cratering are often referred to in the literature as if they were distinct forms of wear, each with its own mechanism. However, as will be shown, there are a number of wear mechanisms by which both the wear land on the flank and the crater on the rake face can be formed. The terms *flank wear* and *cratering* should, therefore, be regarded only as descriptive, not as characterizing a wear mechanism.

C. Deformation

The shape of the tool may be changed by plastic deformation resulting from the high stress and temperature near the edge during cutting (Fig. 6). Since this does not necessarily involve removal of any material from the tool, it is not strictly a "wear" process. However, the deformation may result in much accelerated wear on the tool flank or nose with sudden collapse of the tool. The upper limit to the rate of metal removal may occur when the tool is no longer able to resist the stress and temperature near the edge. Tool deformation may, therefore, initiate the wear processes which terminate tool life (Loladze, 1958b; Trent, 1968a).

D. Fracture

Because tool materials, being of high hardness, usually have low ductility and because cutting tools are wedge shaped with relatively sharp edges, fracture is one of the causes of tool failure. Cutting tools are not

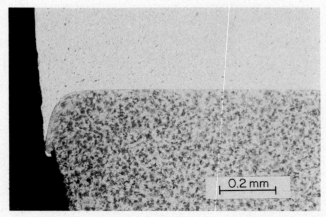

Fig. 6. Deformation of tool edge under action of compressive stress.

normally designed with a knowledge of the stresses imposed on them. Practical experience in tool design and intuitive understanding of the behavior of tool materials have resulted in tool shapes for particular applications that minimize the incidence of fracture while maintaining cutting efficiency. Adaptation of tool design has made possible the use of cemented carbides in a wide range of cutting operations, but there are still many cutting conditions in which cemented carbides fail by fracture and the tougher tool steels must be used. So far the use of such tool materials as alumina and diamond, which are harder but less tough than cemented carbide, has been very limited because of their susceptibility to fracture. Fracture, as opposed to wear, will not be further considered here, except to say that no hard and fast line can be drawn between wear and fracture. Some wear processes take place by the carrying away of small fragments of tool material that separate from the tool by a process of fracture. Also, repeated thermal and mechanical stresses on tools used where cutting is frequently interrupted result in a fatigue type of cracking which may be the first stages of fracture (Braiden and Dugdale, 1970; Jonsson, 1975).

IV. Conditions at the Tool/Work Interface

An appreciation of the conditions existing at the tool/work interface is essential in order to understand the wear processes that change the shape of the tool. Two of the most important factors determining conditions at the interface are stress and temperature.

A. Stress

Unfortunately research has provided far too little reliable information concerning the stresses acting on cutting tools to enable accurate predictions to be made of the stresses on real tools in industrial cutting operations or to make more than provisional generalizations concerning the variations of stress as a function of work material properties, cutting speed, feed rate, and tool geometry. Many measurements of the *forces* acting on cutting tools have been reported, and there is good agreement among different observers as to the force values over a wide range of cutting conditions (Eggleston *et al.,* 1959; Williams *et al.,* 1970). There are two difficulties in determining *stresses* acting at the tool/work interface: (1) the area of contact on the tool upon which the force acts is not clearly defined and (2) the stress is not uniformly distributed over the contact area and distribution of stress is difficult to determine.

For certain simple cutting conditions estimates have been made of the stress and stress distribution on the rake face of cutting tools. The methods employed have included photoelastic studies (Amini, 1968; Usui and Takayema, 1960), the use of split tools equipped with strain gauges (Kato *et al.,* 1972; Loladze 1958a), and calculation from force measurements combined with studies of strain and strain rate in the work material during the cutting operation (Roth and Oxley, 1972). The stress has been analyzed into a compressive stress normal to the rake face and a shear stress parallel to this face, resulting from the chip drag. Stress distribution has been considered for a plane section normal to the cutting edge, as in Fig. 1, for the simple case of a continuous machining operation where the situation is not made more difficult by complex tool shape or by the formation of a built-up edge on the tool (a phenomenon that will be

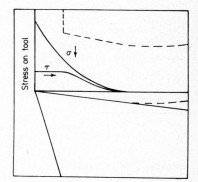

Fig. 7. Stress distribution on cutting tool. σ, compressive stress normal to rake face; τ, shear stress parallel to rake face. [After Zorev (1963).]

considered later). No attempt has yet been made to estimate stress distribution in three dimensions during cutting.

In relation to this simplified situation, there is reasonably good agreement that the stress distribution has the general characteristics proposed by Zorev (1963) and shown diagrammatically in Fig. 7. The compressive stress is a maximum at or near the cutting edge and decreases rapidly with distance from the edge to zero where the chip breaks contact with the tool. The shear stress is more uniform near the edge until, at some distance from the edge, it also decreases to zero. The few reported estimates for the values of the stresses when cutting different metals and alloys vary considerably. Stress on the tool is clearly dependent upon tool geometry. In particular, the compressive stress decreases with increase in rake angle (Loladze, 1958a). For tools of the same geometry the compressive stress on the tool is related to the yield stress of the work material. Table I shows some of the values which have been reported for compressive stress near the cutting edge.

Thus normal stress may be very high close to the cutting edge. When cutting materials with very high yield strength—for example, heat-treated alloy steels or nickel-based alloys—the stress may be high enough to cause tool failure by plastic deformation even at low cutting speed where the tool temperature is low.

B. Temperature

The temperature at the tool/work interface is of even greater significance than stress in relation to tool wear. As with stress, the distribution of temperature at the interface is important and experimental determination of temperature is difficult. Several methods have been employed, including thermocouples inserted in the tool, tool/work thermocouples, and measurement of radiant heat from the tool surface (Lenz, 1971; Pesante,

TABLE I

COMPRESSIVE STRESS NEAR THE CUTTING EDGE

Work material	Compressive stress	
	$(N\ mm^{-2})$	$(tons\ in.^{-2})$
Pb	30	2
Al	90	6
Brass	900	60
Steel (0.5% C, normalized)	1200	80

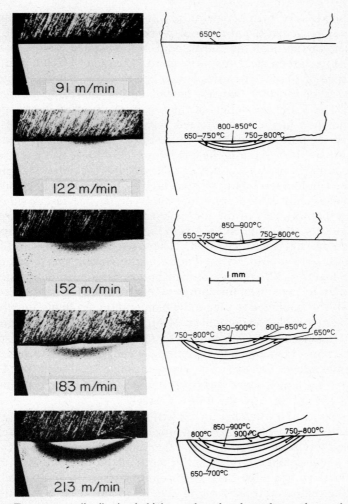

Fig. 8. Temperature distribution in high-speed steel tools used to cut low-carbon steel at a feed rate of 0.25 mm (0.010 in.) per revolution for a cutting time of 30 sec at the speeds shown.

1967), but probably the most certain estimates of interface temperatures have come from measurement of structural and hardness changes in steel tools in those regions of the tool heated above the tempering temperature during the cutting operation (Smart and Trent, 1975; Wright and Trent, 1973). This method has demonstrated that when cutting metals and alloys of relatively high melting point (e.g., steel, iron, nickel, and titanium and their alloys), very high temperatures and very steep temperature gradients may be generated at the tool/work interface. For example, Fig. 8 shows sections through the cutting edge of high-speed steel tools used to cut very-low-carbon steel at speeds up to 213 m min⁻¹ (700 ft min⁻¹) for a cutting time of 30 sec. The tools are etched to show structural changes in those parts of the tool that are heated during cutting to temperatures above 650°C. From the details of the structural changes the temperature at any position in the heat-affected region can be estimated to $\pm 25°C$ where the temperature lies in the range 650–900°C. Figure 8 shows that when cutting this steel under the conditions used, there was a temperature maximum on the rake face at a distance of approximately 1.5 mm from the edge. The temperature may be as high as 1000°C in this region, which is completely concealed from visual observation. It is not obvious to the observer that any part of the tool is even red hot. The temperature gradients are very steep, and even at very high cutting speed the temperature near the edge did not reach 650°C. There was a steady rise in the maximum tool/work interface temperature as the cutting speed was raised.

These features of temperature distribution have been demonstrated to be common to tools used for cutting steels at relatively high speeds. The steel that was machined to obtain the temperature gradients shown in Fig. 8 had a very low carbon content (0.04%). When the carbon content of the steel is increased, the *distribution* of temperature in the tool remains much the same, but a much lower cutting speed is required to generate the same temperature. For example, when cutting a steel with 0.42% C, the same interface temperature was generated when cutting at 55 m min⁻¹ that required 183 m min⁻¹ for the 0.04% C steel.

This characteristic temperature distribution in tools used to machine steel at moderate to high speeds is not, however, a feature observed in tools used to cut all other metals. When cutting titanium and its alloys, there is a high-temperature region in the tools on the rake face, but it is much closer to the edge than when cutting steel. In tools used to cut nickel, the temperature at the tool edge is found to be very little lower than further along the tool rake face. During the cutting of alloys of the lower-melting-point metals aluminum and copper, much lower temperatures are generated at the tool/work interface, and much higher cutting speeds can be used. In all cases as the cutting speed is increased, the tool temperature is raised.

The upper limit to cutting speed when machining with tool steels is determined by the ability of the tools to withstand the compressive stress near the edge at the temperatures close to the edge. When cutting steel, the temperature distribution is favorable to high-speed machining because the part of the tool near the edge subjected to maximum compressive stress remains relatively cool. However, heat is conducted back from the high-temperature region toward the edge and, above a limiting cutting speed, the tool deforms at the edge. This eliminates the clearance angle, and contact with the freshly cut metal surface extends down the tool flank. A new heat source is formed at this position and, heated from two surfaces, the collapse of the tool is rapid and catastrophic (Fig. 9) (Smart and Trent, 1975).

High yield stress in compression at elevated temperatures is thus a basic requirement of tools in order to achieve maximum rates of cutting the higher melting point metals. Table II (Trent, 1968b) shows the temperatures at which different classes of tool material will resist a compressive stress of 750 N mm^{-2} (50 tons in.$^{-2}$).

C. Seizure at the Tool/Work Interface

Apart from high stress and high temperature, two other important conditions at the tool/work interface are (1) the flow of work material past the tool surface is rapid and unidirectional and (2) the tool is always cutting into clean metal, i.e., it is generating surfaces that have not been exposed to the atmosphere before making contact with the tool. In moving away from the cutting edge, both the chip and the newly machined surface tend to sweep away with them the air and other gases or liquids in contact with the tool. The higher the cutting speed, the more effective is the sweeping action in removing gases which could react with freshly cut metal surfaces to inhibit metal-to-metal contact. There is evidence that in high-speed cut-

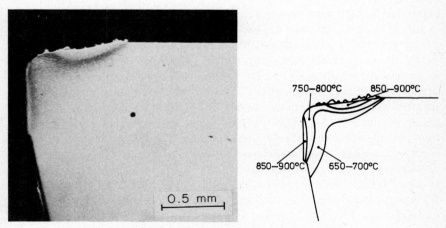

Fig. 9. Temperature distribution in tool used to failure.

TABLE II

TEMPERATURE FOR 5% STRAIN AT COMPRESSIVE STRESS
OF 750 N mm^{-2} (50 tons in.$^{-2}$)

Tool material	Temperature (°C)
Carbon steel	450
High-speed steel	750
Cemented carbide WC + 6% Co	1000
WC + 15% TiC + 8% Co	1100

ting [e.g., over 30 m min^{-1} (100 ft min^{-1})] the new metal surfaces react with oxygen from the air and effectively remove it from small pockets in the small angle clefts between tool and work material on the clearance and rake faces of the tool, i.e., in the positions marked M in Fig. 1 (Trent, 1967b).

Tools when they begin to cut have oxide films on the surface and usually oil and other contaminants. There is very little evidence to indicate how rapidly these are removed by the unidirectional flow of work material across the surface. This must be dependent upon many details of particular cutting operations. In continuous cutting, once such layers are removed, the unidirectional flow of work material will tend to inhibit the formation of new surface films. When cutting is interrupted, new oxide films are formed on the tool unless, as frequently happens, the tool surface remains covered with a layer of work material or a protective lubricant. In frequently interrupted cutting, e.g., in milling, the formation and removal of oxide films may be important in relation to wear.

All of the conditions at the tool/work interface that have been described tend to promote metallic bonding and seizure between the surfaces. In many metal-cutting operations, the relative speeds of the surfaces and the compressive stress forcing them together are considerably higher than in the process of friction welding. For example, steel may be friction welded at peripheral speeds of 16–50 m min^{-1} under a stress of 45–75 N mm^{-2} (3–5 tons in.$^{-2}$). It is therefore extremely difficult to prevent seizure at the interface in metal cutting. *There is good evidence to indicate that under most conditions obtaining in industrial metal-cutting operations, seizure is the normal condition over most of the tool/work interface* (Trent, 1967a; Wright and Trent, 1974). This is the most essential feature distinguishing the conditions that determine the wear mechanisms of tools used for metal cutting as compared with tools used for metal-forming operations or for cutting other materials. It is responsible for the

character of the specialized tool materials that have been developed for
metal cutting over the past 50 years.

Because of its importance in relation to wear, the character and extent
of seizure will be considered further before discussing the wear processes. Seizure is not normally considered as one of the possible conditions under which one solid surface can move over another. The conditions normally considered are shown diagrammatically in Fig. 10a–c.
Under hydrodynamic lubrication (Fig. 10a) the two surfaces are separated
by a layer of fluid that is thicker than any irregularities in the surfaces so
that there is no contact between the moving solid surfaces. Wear is possible by abrasive action of entrapped hard particles. In Fig. 10b, c any
layer of lubricant or other fluid is commensurate in thickness or thinner
than the irregularities of the solid surfaces so that these make contact at
their asperities. Wear may take place by abrasive action of hard particles

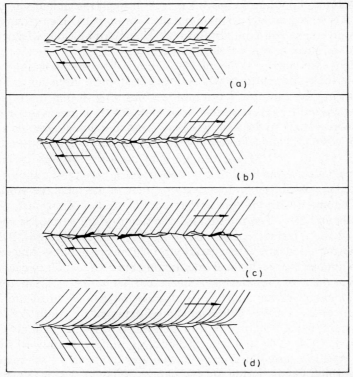

Fig. 10.   Diagrams showing features of sliding and seizure. (a) Hydrodynamic lubrication; (b) boundary lubrication—light wear; (c) galling or scuffing conditions; (d) seizure conditions.

or by formation of short-lived bonds where asperities come in contact. Transfer of metal from one surface to the other and reaction with the surrounding fluids take place. Wear may be light as in Fig. 10b, but with high loads and ineffective lubricants it may be very severe and rapid (Fig. 10c). Under the latter condition localized seizure becomes more extensive and surfaces are severely damaged ("scuffed" or "galled"). Relatively little attention has been devoted to this latter condition because failure has already been initiated and the precise mechanism of the final stages of failure is regarded as relatively unimportant.

The most extreme condition (Fig. 10d), where seizure is complete—i.e., relative movement at the whole interface ceases—is generally not considered at all because this is usually accompanied by failure in some other part of the system as when a bearing fails or a piston seizes in a cylinder. Metal cutting can continue under the condition of complete seizure because the area of the seized surface is relatively small, sufficient power is available to continue to move the tool through the metal, and the tool material is strong enough to maintain the tool shape. Under seizure conditions, movement can continue only by shear through the weaker of the two bodies close to but not at the interface.

Evidence for seizure at the interface comes from metallographic examination of tools after cutting. When cutting is stopped in the normal way by withdrawing the tool, the worn surfaces are often found to be partly or wholly covered by layers of the work material. In some cases the whole chip remains stuck to the tool. Carefully prepared metallographic sections through worn tool and adherent work material preserve the interface. Figures 11–15, 30, and 31 are examples of sections through carbide and high-speed steel tools used to cut steel under a variety of conditions. On the rake face and often on the clearance face of the tools the work material is in contact with the tool not just on the "hills" of the surface but in the "valleys" also. The interlocked character of the surface is such that sliding at the interface would not be possible. The adhesion of the surfaces during all stages of preparation of the microsections suggests that there is not merely mechanical interlocking but also metallurgical bonding at least at parts of the interface (Smart and Trent, 1975; Trent, 1967a).

The character of the interface is of great importance in relation to the wear processes. Further evidence can be obtained by studies of the cutting process, using a method in which cutting is stopped suddenly (Wright and Trent, 1974). The tool is propelled very rapidly downward from the cutting position shown in Fig. 1. In this way evidence of the metal flow pattern which exists during cutting can be retained. Using a quick-stop method in which the tool is propelled from its cutting position by an explosive charge, separation of the tool from the work and/or chip occurs in one of several ways. If metallurgical bonding is weak or absent, separation occurs mainly at the interface, exposing the worn surface of the tool, with perhaps a few very small fragments of work material adhering. If metallurgical bonding is strong separation may occur by fracture of the work material through the shear plane (O–D in Fig. 1), leaving the whole chip adhering to the tool as seen in Fig. 11. Separation may also occur by fracture through the chip close to the interface, leaving a layer of the chip adhering to part or all of the surface of contact (Fig. 12). The surface of the work material left on the tool often exhibits a ductile tensile fracture, demonstrating that the strength of the metallurgical bond between tool and work material was higher than the tensile strength of the flowing chip.

Absence of adhering metal on a tool after a quick stop does not necessarily mean that there was no atomic bonding, but that the tensile strength of the interface was lower than that of the chip. Further evidence is provided by metallographic sections through the forming chips and the tools after quick-stop tests. Figure 13 is characteristic of the flow pattern observed in the work material near the center of the cutting edge when cutting many metals and alloys, particularly at relatively high speeds. This pattern is consistent with seizure at the interface, resulting from either mechanical interlocking or metallurgical bonding. Movement of the chip over the tool surface takes place by intense shear in a thin layer of the work material, usually between 10 and 75 μm in thickness (Fig. 14), the under surface of which was anchored to the tool while the upper part was

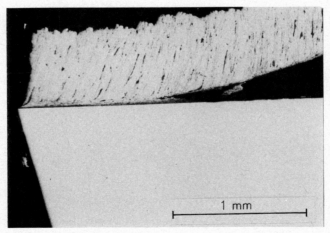

Fig. 11. Quick stop after cutting very-low-carbon steel. Metallographic section through chip bonded to tool.

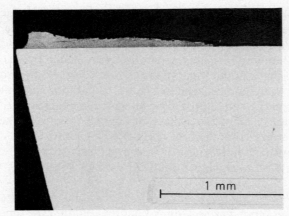

Fig. 12.   Quick stop after cutting steel. Part of chip bonded to tool.

moving away at the speed of the chip. Tool wear under these conditions must be considered not in terms of the velocity *at the interface,* but in terms of a very steep *velocity gradient, approaching zero velocity at the interface.* This thin layer extends along the rake face from the cutting edge to the position where the chip leaves the tool surface, and it some-times extends from the cutting edge for some distance down the clearance face also (Fig. 15). In this thin layer, which will be called here the "flow zone," both the amount and the rate of strain are extremely intense, so that the normally observed structures of the work material are completely

Fig. 13.   Section through quick stop when cutting copper showing pattern of flow around cutting edge.

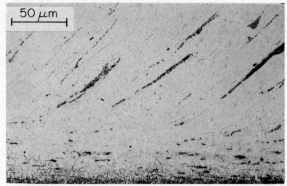

Fig. 14. Flow zone at interface between tool and chip—detail of Fig. 11.

Fig. 15. Flow of work material along rake and clearance faces of carbide tool used to cut steel.

destroyed. Current research demonstrates that dynamic recovery and possibly dynamic recrystallization occur in the flow zone, and wear of the tool must be considered as taking place by interaction with the work material in this condition. The behavior of the work material in the flow zone is more like that of a very viscous fluid than that of a plastic metal as normally conceived.

The flow zone is also the main heat source increasing the temperature of the tool. It is the energy required to shear the work material in the flow zone that is responsible for the rise in tool temperature and not "friction" at the interface. The tool is heated by conduction from the flow zone and the tool temperatures previously discussed can therefore be regarded as characteristic of the behavior of the work material strained under the conditions of the flow zone.

D. The Built-Up Edge

The seizure/flow-zone model represents the conditions at the tool/work interface when cutting many metals over a wide range of cutting conditions (also when cutting many alloys, particularly at high cutting speed). When cutting many alloys that contain more than one phase (e.g.,

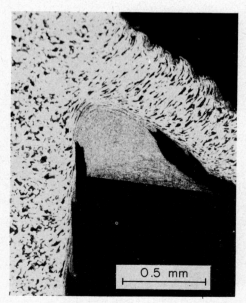

Fig. 16.   Section through built-up edge formed when cutting steel.

carbon steel, cast iron, alpha–beta brass at moderately low speeds, seizure occurs at the interface, but the first adhering layers of work material are so strengthened by the shear strain which they undergo that yield then occurs in less strained layers further from the tool surface. Continuous repetition of this process results in a relatively large agglomerate of very hard work material accumulated to form a built-up edge as shown in Fig. 16. If the cutting speed is raised, the temperature increases and a limit is reached at which the built-up edge can no longer support the stress of cutting so that it is then replaced by a flow zone.

When a built-up edge is formed, it reaches an equilibrium size for any set of cutting conditions and a dynamic equilibrium is established with fragments being constantly added and torn away (Heginbotham and Gogia, 1961; Williams and Rollason, 1970). The cutting action then takes place at a considerable distance from the edge of the tool, and it is possible for cutting to proceed without the chip or the new work surface making contact with the tool for considerable periods of time. More frequently there is intermittent contact with the tool as fragments of the built-up edge break away. If cutting is interrupted and started again, the whole built-up edge may be broken away. For operations in which cutting is interrupted frequently this action can have a large adverse influence on tool life.

E. Sliding at the Tool/Work Interface

It is rare for seizure conditions to exist over the whole area of contact between tool and work material, even when cutting at high speed. There are two factors that tend to inhibit seizure in the peripheral regions of the contact area. The compressive stress forcing the two surfaces together diminishes and, of more importance, atmospheric oxygen and lubricants

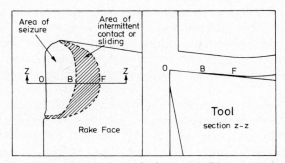

Fig. 17.   Characteristic regions of seizure and sliding on cutting tool.

have access to the interface. There is evidence that in these regions seizure is prevented or is only local and intermittent (Trent, 1969; Trent, 1976). Figure 17 shows, for a turning tool, those areas where seizure normally occurs and the peripheral regions. Movement of the chip across the rake surface of the tool in these regions can often be shown to involve alternate sticking and sliding. For example, Fig. 18a shows a longitudinal section through a piece of steel chip near the outer edge, where the intermittent stick–slip character is evident, while Fig. 18b is a section through the center of the same piece of chip, showing a flow zone characteristic of seizure. Thus both seizure and sliding conditions can exist at the same time on different parts of the tool/work interface. Sliding conditions are promoted by low cutting speeds and the presence of an effective lubricant.

V. Cutting Tool Wear

The mechanisms of tool wear under conditions of seizure and sliding will be considered separately. Since the condition of seizure is dominant, especially under those conditions of cutting that are of industrial importance, wear under seizure conditions will be considered first.

A. Diffusion wear

The term *diffusion wear* is used to describe a wear mechanism in which the tool shape is changed by diffusion into the work material of atoms from one or more of the phases in the tool material. In other words, the tool surface is dissolved away by the work material flowing over it. Three basic requirements for diffusion wear are (a) metallurgical bonding of the two surfaces so that atoms can move freely across the interface, (b) temperature high enough to make rapid diffusion possible, and (c) some solubility of the tool material phases in the work material. This is therefore a wear mechanism that becomes important when cutting relatively high-melting-point metals and alloys (e.g., iron, steel, and nickel-based alloys) at high cutting speed, when high interface temperatures are generated.

Evidence for the diffusion-wear mechanism is circumstantial since direct observation of the process is not possible. Its significance was first demonstrated in relation to the wear of cemented carbide tools when cutting steel (Opitz and König, 1967; Trent, 1963). It offers an explanation for the great reduction in rate of wear which results from the introduction into

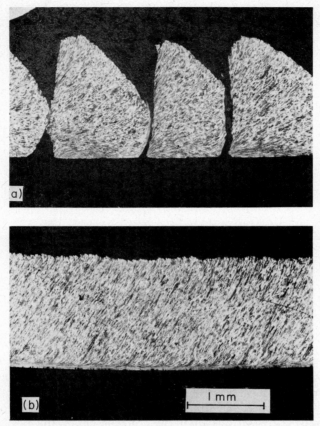

Fig. 18. Sections through steel chip (a) near outer edge in region of sliding; (b) near center in region of seizure.

tungsten carbide–cobalt (WC–Co) alloy tools of even a small percentage of titanium or tantalum carbide (TiC or TaC) when cutting steel at high speed. The introduction into the cemented carbide based on WC of as little as 10–20% of either TiC or TaC will often make possible a cutting speed three times as high for the same tool life.

When cutting most steels with WC–Co alloy tools, cratering wear (Figs. 3 and 5) occurs under conditions which depend upon the cutting speed and feed rate. In Fig. 19 the occurrence of cratering wear when cutting a medium-carbon steel with a WC + 6% Co tool is plotted as a function of cutting speed and feed rate, using tools of a constant geometry (Trent, 1959). Cratering wear is not observed below the lower line and be-

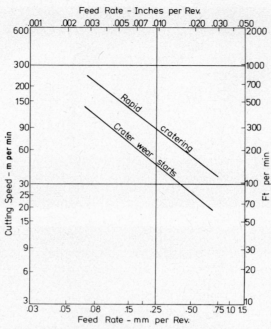

Fig. 19. Machining charts for 0.4% C steel cut with WC–Co alloy tool.

comes so severe that it destroys the tool very rapidly above the upper line. A crater such as that shown in Fig. 3 occurs always on that part of the tool rake face where the temperature is at a maximum when cutting steel, as was demonstrated in the tests using high-speed steel tools (Fig. 8). Examining the worn surface in the crater in detail, the carbide grains, almost without exception, are seen to be smoothly worn through, as shown in Fig. 20, rather than fractured. On most parts of the crater surface there are usually ridges and grooves parallel to the direction of chip flow. Often these are seen to be ridges starting from a large WC grain (Fig. 21). On the almost flat surface between the crater and the cutting edge, this flow pattern is absent but the carbides in the tool may be worn to give a surface that has an etched appearance (Fig. 22). The surface of the carbide grains in this region shows markings related to the crystallographic orientation of the carbide grains rather than to the flow direction of the chip over the tool.

The characteristics of these worn surfaces can be explained in terms of the temperature distribution at the interface and the flow pattern of the work material over the surface as pictured diagrammatically in Fig. 23. The maximum temperature at any part of the tool surface is in the crater

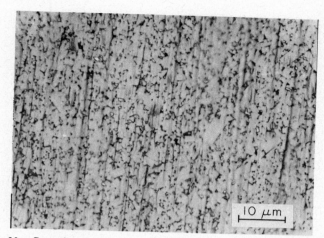

Fig. 20.   Smooth wear in crater of WC–Co alloy tool after cutting steel.

and the velocity gradient very close to the tool surface is very steep in this region so that atoms of W, C, and Co diffusing into the work material are immediately swept away. Large grains of WC that are unfavorably oriented for dissolution into the steel (WC is a very anisotropic compound) are attacked less rapidly than the others. They remain projecting from the surface, modifying the flow of work material and initiating ridges on the worn crater surface. Over the flat area between the crater and the cutting edge (Fig. 23), the work material immediately adjacent to the tool

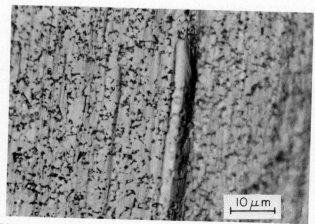

Fig. 21.   Ridge starting from large WC grain in crater of WC–Co alloy tool after cutting steel.

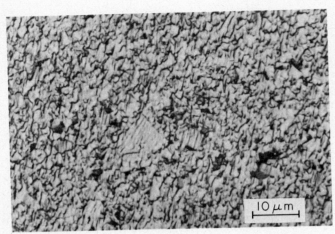

Fig. 22. Etched structure on rake face of WC–Co alloy tool near cutting edge after cutting steel.

surface tends to be stagnant or almost stagnant—a "dead metal" zone equivalent to that observed in extrusion. Also the temperature is lower in this region (Fig. 8) so that the rate of wear by diffusion is lower. The W, C, and Co atoms diffusing into the chip are not immediately swept away. The configuration of the surface depends on the rate of loss of atoms from each minute area of the surface, being affected by crystallographic orientation of WC grains and the presence in these grains of slip planes and other defects energetically favorable for solution. In effect the tool surface in this region is being etched in hot steel and has an etched appearance when adhering work material is cleaned away by acid treatment.

These differences in surface characteristics, dependent on flow pattern, are very similar to those observed in electrochemical machining. Where there is rapid flow of electrolyte in the narrow gap between the electrodes, the surface generated on the anode is highly polished, but at the periphery

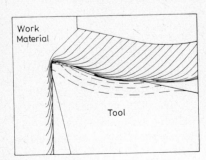

Fig. 23. Characteristic flow pattern in chip and temperature pattern in tool after cutting steel.

of the gap, where electrolyte flow is less vigorous, the anode surface is etched to leave structural features such as grain boundaries and second phase particles at different levels.

These observations point to two major parameters which influence the rate of diffusion wear: temperature and the rate of flow immediately adjacent to the surface. Diffusion rates are strongly dependent on temperature and typically in metals the rate of diffusion is doubled for temperature increments of the order of 20°C. Attempts have been made to investigate the diffusion-wear process by heating together stationary couples of the tool and work materials in vacuo, followed by measurement of changes in composition and observation of structural changes at the interface. Some useful qualitative information on the behavior of different tool materials can be obtained in this way although observations have to be interpreted with care since they may be misleading. The method is not useful for predicting quantitatively the rates of diffusion wear for two reasons: First, the condition of the work material in a stationary couple is entirely different from that at the tool/work interface. In the flow zone the extreme strain and strain rate must be accompanied by very high rates of dislocation formation and annihilation, with dynamic recovery and possibly recrystallization and very large grain boundary area. In such disturbed structures the rate of diffusion at any temperature will be much higher than in stationary crystal structures. Second, with stationary couples, the initially high rate of loss of atoms from the tool surface is rapidly reduced as the concentration of tool metal atoms builds up in the layers of work material adjacent to the interface. During cutting, there is no possibility of a high concentration of solute atoms building up in the flow zone. Both of these factors must lead to a much higher rate of wear than would be predicted from alffusion rates measured in experiments with stationary couples.

When cutting steel with WC–Co alloy tools, cratering occurs because WC has appreciable solubility in steel at elevated temperatures. When cutting steel at high speed, temperatures of 1000°C or more occur on the tool rake face at the position of the crater. The rate of crater wear increases with cutting speed since this increases both the temperature and the rate of flow in the work material, and it was found in the early years of use of tungsten carbide tools that their usefulness for steel cutting was limited because of this cratering wear.

Observation of flank wear on WC–Co tools used to cut steel at high speed suggests that wear on the flank under these conditions is also mostly caused by a diffusion process. In many cases, seizure at the tool flank can be demonstrated (Fig. 15) and the carbide grains are smoothly worn through as in the crater. There are two main differences between interface conditions in the crater and on the flank wear: (1) the temperature on the flank is much lower than in the crater when cutting steel at high speed, and the temperature difference may be as much as 300°C (Fig. 8); and (2) the rate of flow of work material on the flank very close to the interface is much higher than in the crater. This is because the velocity of the work material relative to the tool is higher than the chip velocity, usually by a factor of 2–4, and also because the flow zone on the flank is very thin, giving an extremely steep velocity gradient near the tool surface. In spite of the lower temperature at the flank, wear by a diffusion mechanism takes place at an appreciable rate in the same cutting speed range as that for cratering on the rake face. Within this speed range the rate of flank wear increases rapidly as cutting speed is raised.

The problem of machining steel at high speeds with cemented carbide tools was solved in the late 1930s by incorporation in the tool material of one or more of the cubic carbides TiC, TaC, and NbC. By the end of the Second World War all the major manufacturers of cemented carbides were producing steel cutting grades of good quality containing between 5 and 25% by weight of the cubic carbides, the balance being WC and Co.

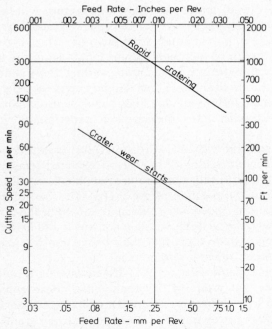

Fig. 24. Machining chart for 0.4% C steel cut with WC–TiC–Co alloy tool.

The advantage achieved in terms of cutting efficiency can be seen by comparing the chart of Fig. 24, showing the occurrence of cratering wear when cutting a medium-carbon steel with a tool containing 15% TiC, 77% WC, 8% Co, with that of Fig. 19 for a 94% WC, 6% Co alloy tool. Although cratering wear starts at about the same speed with the tool containing TiC, the wear rate increases only very slowly with cutting speed, and a speed three times as high can be used. The development of these alloys was the result of empirical tool testing of new compositions; the explanation for the improved performance was achieved in later laboratory investigations.

The cubic carbides are present in the tool alloys in the form of particles a few microns in diameter which are a solid solution containing all the TiC and TaC and a large percentage of WC. These solid solution cubic carbide ("mixed crystal") grains are dissolved much less readily than WC in the hot steel moving over the tool surface, so that the wear rate by diffusion is greatly reduced. Most of the steel-cutting grades of cemented carbide contain both cubic carbide grains and WC grains, and examination of worn crater surfaces clearly shows the preferential attack of the hot steel on the WC grains. Figure 25 shows the rake surface of a tool containing TiC and TaC after cutting steel for five times as long under the same conditions as for the WC–Co alloy tool shown in Fig. 5. Figure 26 is a scanning electron microscope (SEM) photograph of the worn tool surface after removing adhering metal in acid. The cubic carbide grains remain almost unworn while the WC grains have been dissolved from between them to a depth of 2–4 μm. During cutting, the work material was in contact with the whole of this surface, both on the hills and in the valleys, and diffusion appears to be the only mechanism that can explain a worn surface of this

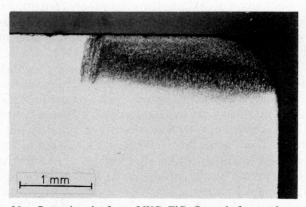

1 mm

Fig. 25.   Crater in rake face of WC–TiC–Co tool after cutting steel.

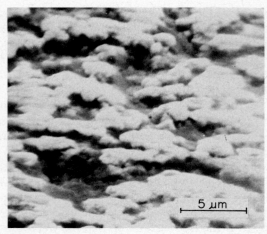

Fig. 26. SEM photograph of crater in rake surface of WC–TiC–Co alloy tool used to cut steel.

character. The cubic carbides, much more slowly dissolved in the work material, are left projecting from the surface in the stream of the flow zone. If sufficiently undermined, the projecting grains can be broken away and removed with the chip, and diffusion wear can be accelerated by this mechanical action.

Flank wear is also greatly reduced by the addition of these cubic carbides, where this is controlled by a diffusion process. Very great economic advantages to the engineering industry result from these changes in composition. As the proportion of added TiC, TaC, and/or NbC is increased, the permissible speeds for cutting steel are increased. This suggests the elimination of WC and the use of tools consisting basically of TiC or TaC. Tools based on TiC are commercially available and they can be used for cutting steel or cast iron at speeds higher than those employed with WC-based cemented carbides (Mayer *et al.*, 1970). They have not replaced WC-based tools on a large scale because their toughness is inadequate. In the past seven years the problem of how to make use of the diffusion wear resistance of TiC has been solved, at least for "throw-away" tool tips that are not reground, by coating tools with thin layers of TiC (or sometimes TiN). Layers of a thickness between 5 and 10 μm are deposited by chemical vapor deposition (CVD) on the surface of standard tool tips of WC-based cemented carbide. For many steel-cutting applications these coated tips permit higher cutting speeds or give much longer tool life at the same cutting speed (Fig. 27) while retaining the essential toughness of the WC-based substrate (Schintlmeister and Pacher, 1975).

A diffusion wear process is highly specific to the materials used. The grades of cemented carbide discussed and the coated carbides have been developed specifically for the machining of steel. Thus vanadium carbide (V_4C_3), although isomorphous with (and forming a continuous series of solid solutions with) TiC, is not used because it is no more resistant to solution in hot steel than is WC. Likewise the steel cutting grades of cemented carbide confer no advantage if used to cut titanium and its alloys. For this purpose WC–Co alloys are used because the cubic carbides based on TiC are more rapidly dissolved in hot titanium than is WC (Freeman, 1974).

High-speed steel tools are also worn by a diffusion process when cutting steel and other metals at high speed. In the case of steel tools it is basically the atoms from the matrix that diffuse across the interface and are carried away in the work material. Figure 28 shows a section across an interface worn by a diffusion process. The grain boundaries in the tool are undeformed up to the interface and the carbide particles, less rapidly dissolved than the matrix, project from the tool surface and are eventually mechanically detached. Thus, although the carbide particles in the steel tools are more resistant to diffusion wear than the matrix, they afford little extra wear resistance because they are isolated and can be undermined. To make use of the "diffusion resistance" of carbides, these must constitute the majority of the structure, as in the cemented carbides, so that they support each other and cannot be undermined.

Wear by diffusion takes place at an appreciable rate at 700°C on the rake face of steel tools and probably at a considerably lower temperature

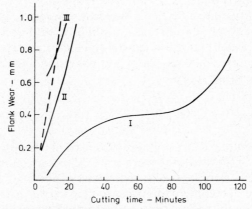

Fig. 27. Wear rates of coated and uncoated tools used to cut steel. - - -, 70 m/min, uncoated carbide; ———, I, 70, II, 100, and III, 125 m/min, TiC-coated. Feed, 0.76 mm/rev. Work material, Ni–Cr–Mo steel 360 HV.

Fig. 28. Section through rake face of high-speed steel tool showing diffusion wear after cutting steel.

on the tool flank. Wear rate increases as cutting speed is raised, and it is probably wear by diffusion that is responsible for the Taylor relationship (Taylor, 1907) between tool life, based on flank wear rate, and cutting speed,

$$Vt^n = C$$

where V is the cutting speed, t the tool life, and n and C are constants.

The relationship for a number of steel work materials is shown in Fig. 29 (A.S.M. Metals Handbook, 1948). At low speeds tool wear rates by diffusion become insignificant because interface temperatures are too low. If tool wear were entirely by diffusion, tool life should become effectively infinite at low speeds, but these curves cannot be extrapolated back because other wear mechanisms come into operation and may cause short tool life at low cutting speed.

B. Attrition Wear

The detachment of particles of microscopic size from the tool surface to be carried away in the stream of work material under conditions of seizure has already been described as a secondary process in diffusion wear. Such mechanical detachment of microscopic particles is called here *attrition wear*. It is one of the wear processes controlling tool life under conditions of seizure and is important mainly at relatively low cutting speeds.

Worn surfaces produced by attrition are characteristically rough com-

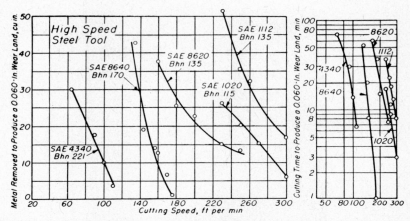

Fig. 29. Taylor tool-life curves. Metal removed and cutting time to produce 1.5-mm (0.06-in) flank wear on a high-speed steel tool for five annealed steels. Feed 0.23 mm (0.009 in.) per revolution. (From A.S.M. Metals Handbook Supplement, 1948.)

pared with the smooth surfaces resulting from diffusion. Figures 30 and 31 show sections through carbide and high-speed steel tool surfaces worn by attrition to which the work material remains adhered. Carbide grains are detached individually from cemented carbides, or are fractured and fragments of microscopic size are removed. This form of wear is promoted by conditions that cause an uneven flow of work material over the tool surface or intermittent contact with the tool surface. Flow adjacent to the interface tends to become even and laminar at high speeds, and it is often found that the rate of wear by attrition decreases as the speed is raised.

The occurrence of a built-up edge is one of the features of the cutting of steel, cast iron, and other multiphased alloys at relatively low cutting speed. As has already been described, the built-up edge is a rather unstable structure to which work material is constantly being added and broken away. Tool life in the presence of a built-up edge can be very variable. Under good conditions, where the built-up edge is relatively stable, it can act as a protective cover, almost entirely preventing contact between tool and work material for long periods of cutting. This occurs under some conditions when cutting cast iron with WC–Co cemented carbide tools. For this reason it is common practice to use this class of carbide tools for cutting cast iron, and to cut under conditions at which a built-up edge occurs. If higher speeds are used, the built-up edge disappears, diffusion wear occurs, and the use of steel-cutting grades of carbide is advantageous when cutting cast iron.

When cutting steel, a built-up edge of such stability and form as to pre-

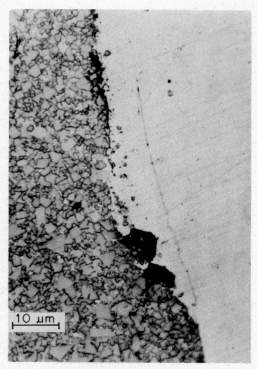

Fig. 30. Section through cemented carbide tool (WC–Co alloy) showing attrition wear after cutting steel.

vent wear seldom occurs. Sometimes the whole built-up edge, or large fragments of it, is torn away periodically when cutting is interrupted, bringing with it particles of the tool. Attrition wear caused in this way is often rapid and may be much more severe on cemented carbide tools than on high-speed steel tools. High-speed steel tools often give a longer tool life and are preferred for cutting steel in the lower speed range in which a built-up edge occurs. The reason for the high rate of attrition wear of carbide tools under these conditions is probably that, when part of the built-up edge is separated from the tool, it imposes local tensile stresses. The relative weakness and lack of ductility of carbides in tension leads to localized fracture through carbide grains or along grain boundaries. With steel tools, separation after seizure bonding may involve fracture in the tool (Fig. 31), but is more likely to occur at the interface or within the work material.

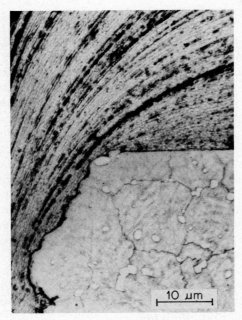

Fig. 31. Section through high-speed steel tool after cutting steel showing attrition wear.

Any condition that tends to promote uneven flow of the work material over the surface can lead to attrition wear—for example, vibration and lack of rigidity in the machine tool—and attrition wear is not confined to conditions under which a built-up edge occurs. Attrition wear and diffusion wear may occur simultaneously. When examining the flank wear on cemented carbide tools, it is common to observe both the smoothly worn carbide grains characteristic of diffusion wear and the fractured carbide grains attributed to attrition. The resistance of carbide tools to attrition wear depends both on their composition and on their structure. The WC–Co alloy tools are usually more resistant to attrition wear than are those containing the cubic carbides. Where carbide tools are used to cut steel in the lower speed range, it is usual to use WC–Co alloys rather than the "steel-cutting grades." Under these conditions, the rate of wear is very dependent on the carbide grain size rather than on the hardness. Fine-grained carbides with grain size of the order of 1 μm or less are much more resistant to attrition wear than are those with coarse grain—2 μm or coarser.

C. High-Temperature Shear

When high-speed steel tools are used to cut steel, nickel, and titanium and their alloys at high speed, a mechanism of wear may come into operation which is much more rapid than diffusion. In moving over the tool under conditions of seizure, the work material exerts a shear stress, and at those parts of the tool surface that reach the highest temperature, the stress may be high enough to shear away layers of the tool material. This can be observed in sections through the tool after machining (Wright and Trent, 1974). Figure 32 shows a section through the crater in the rake face of a high-speed steel tool which had been used to cut a very low carbon steel at 183 m min⁻¹ (600 ft min⁻¹) for 30 sec. In the crater the interface temperature had reached approximately 1000°C and at the rear of the crater are layers of the tool material sheared away by the moving chip. It is unexpected that an almost pure iron can be responsible for shearing the surface of high-speed steel, an alloy developed especially for its high strength. This occurs because (1) the tool steel is heated to a temperature much higher than that at which it is designed to have high strength and (2) the stress imposed on the tool is that required to shear the work material at an extremely high strain rate, and this stress is high enough to shear the tool material at a lower strain rate.

Craters worn by this mechanism have been observed in tools used to cut many steels and alloys of titanium and nickel. When cutting work materials with higher amounts of alloying elements, the shear stress exerted on the tool is increased and shearing of the tool surface occurs at

Fig. 32. Section through crater in high-speed steel tool showing wear by shear at elevated temperature.

lower cutting speeds and at lower interface temperatures, e.g., 750–800°C. This is a rapid wear mechanism and tools exhibiting this form of wear usually fail after a very short cutting time. Therefore, it is a wear mechanism that determines the upper limit to the rate of metal removal with high-speed steel tools.

A wear mechanism of this character is responsible also for the last stages of wear on the flank of steel tools when such a large wear land has been generated that very high temperatures occur at this position (see Fig. 9). It is the mechanism of collapse of the tool edge.

This wear mechanism has not been observed when cutting with cemented carbide tools although some destruction of the tool structure by shear stress was observed in cutting some highly creep-resistant nickel alloys. The upper limit to the ability of cemented carbide tools to remove metal is more usually determined by the ability of the tool to resist the compressive stress at elevated temperature or by resistance to diffusion wear.

D. Abrasion

Abrasion is one of the most common wear mechanisms between moving metal surfaces. Abrasion involves the removal of material from a softer surface by a harder one by processes sometimes referred to as repeated plowing or the formation of "microchips." It can be caused by the entrapment of hard particles between two softer surfaces or by hard particles carried in a fluid flowing over the surface. It is generally assumed that the abrading substance must be considerably harder than the surface that is worn. In metal cutting the tool is always considerably harder than the material being cut in order that it may function as a cutting tool. However, many work materials contain particles of phases which have a hardness much higher than that of the workpiece as a whole. Many of these second-phase particles are harder than the matrix in high-speed steel tools and retain this hardness to higher temperature. For example, in cast iron and steel there are particles a few microns across, consisting of carbides of iron, chromium, titanium, vanadium, etc., and also oxides, in particular Al_2O_3 but also silica and some silicates. Not many of these particles are as hard as or harder than the carbides used as the basis for cemented carbide tools.

Such particles in the work material may cause abrasion, especially on steel tools, under conditions of seizure as well as under sliding conditions, and in most published work on cutting tool wear the majority of flank wear, in particular, is attributed to abrasion. However, very little direct

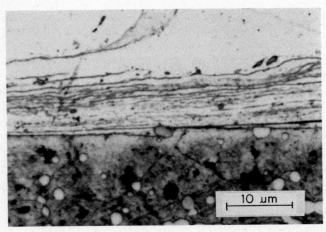

Fig. 33. Section through rake face in high-speed steel tool used to cut austenitic stainless steel showing abrasive action by Ti(C,N) particle.

evidence of abrasion as a mechanism of tool wear has been published. Wright (Wright and Trent, 1974) shows photomicrographs demonstrating the abrasive action of Ti(C, N) particles in austenitic stainless steel, plowing into the rake surface of high-speed steel tools (Fig. 33). The particles in this case are of the order of 5 μm in diameter. There is some doubt as to the effectiveness of such particles in causing tool wear. They strike the surface at a very oblique angle and are likely to become embedded in the tool surface, where they would act as a constituent of the tool around which the work material flows, like any other carbide particle at the tool surface.

There is good evidence that steel deoxidized with aluminum causes a higher rate of tool wear than does that treated with certain other deoxidants. This has been attributed to the abrasive action of Al_2O_3 particles formed in the steel as a deoxidation product. It is unlikely that direct observation of abrasive action by these particles will be possible because of their small size. An alternative explanation for the higher rate of tool wear when cutting Al deoxidized steels is that in these steels certain other plastic nonmetallic inclusions are absent or are modified and that these other inclusions act to reduce the rate of wear in steel treated with other deoxidants. An example of this action will be described later.

Although the evidence is not strong, it seems probable that well-dispersed hard particles of small size (e.g., less than ~5 μm) have little influence on the rate of tool wear, but that large hard particles or concentrated aggregates of small particles may greatly increase the rate of wear.

One piece of evidence in this respect comes from the machining of aluminum–silicon alloys. With silicon contents of 13% or less the silicon is present in these alloys as small particles only a few microns thick and, although there are a very large number of these, the rate of wear both of high-speed steel and of carbide tools is low and high cutting speeds can be used [e.g., 500 m min^{-1} (1500 ft min^{-1}) with WC–Co tools]. If the silicon content is increased, e.g., to 19–23%, the structures contain, in addition to the eutectic, large cubic crystals of primary silicon up to 100 μm across. These result in a very severe increase in the rate of wear so that tool life is short with WC–Co tools, even at 100 m min^{-1} (300 ft min^{-1}). In this case it can be demonstrated that the large silicon particles are forced into direct contact with the cutting edge, and the severely worn surfaces are observed to be coated with silicon (Trent, 1970). The finer silicon particles apparently have little effect because they are able to flow past the tool edge, making little direct contact with the tool.

If large densely packed aggregates of hard particles are present in the work material, these also can cause rapid tool wear by abrasion. In practice this may occur on the surface of castings, causing rapid wear on parts of the tool edge that cut through the casting surface during machining. Considerable increase in tool life can be obtained by treatments that clean the casting surface and remove entrapped mold material.

E. Sliding Wear

So far the wear processes considered have been mainly those that operate under seizure conditions, although abrasion may occur on those parts of the tool in which the work material slides over the tool surface. Figure 17 indicates those parts of the tool surface for which sliding or stick–slip action is observed (Fig. 18a). These are positions in which atmospheric oxygen has at least limited access to the tool/work interface and may prevent complete seizure. The rate of tool wear in these regions is often found to be much higher than at those parts of the tool surface where seizure occurs. Figure 34 shows a deep groove on the flank of a tool used for cutting steel at a constant depth of cut. The groove is at the position where the surface of the bar crossed the cutting edge of the tool. Rapid wear at this position is often observed both with cemented carbide and with high-speed steel tools. Figure 35 shows the rake face of a carbide tool used to cut steel and here also there is a groove where the outer edge of the chip passed over the tool. This accelerated wear often occurs also at the nose radius and down the end clearance face, and may result in the formation of a worn surface with a series of grooves spaced apart by the same distance as the feed and are known as feed marks (Fig. 36).

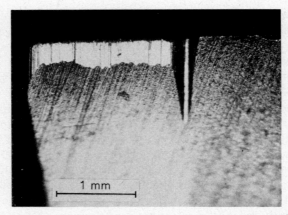

Fig. 34.   Sliding wear on flank face of WC–Co alloy tool after cutting steel.

The mechanism of wear at these locations has not been studied in detail but is probably akin to wear under other metal working operations in which sliding conditions occur. It seems probable that the wear process involves the transfer of submicroscopic particles of the tool material to the work surface. It may involve oxidation of the tool surface, or other chemical interaction with the surrounding atmosphere, followed by mechanical removal of the product of the reaction. The worn surfaces at these positions of accelerated wear are different in character from those in regions of seizure. On carbide tools all the carbide grains are smoothly worn through and show fine scratches in the direction of metal flow. In

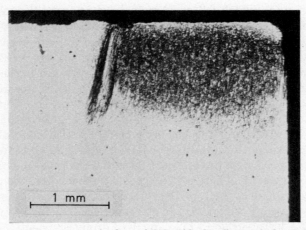

Fig. 35.   Sliding wear on rake face of WC–TiC–Co alloy tool after cutting steel.

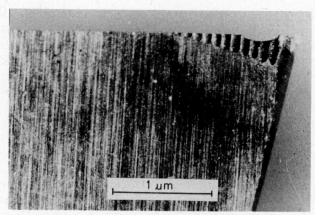

Fig. 36. "Feed marks"—a form of wear on end clearance face of tool used to cut steel.

these regions both types of carbide grain in the steel cutting grades of cemented carbide are equally worn when cutting steel.

The rate of wear in these regions is sensitive to the surrounding atmosphere, particularly at low cutting speed (e.g., below approximately 65 m min^{-1} (200 ft min^{-1})). Experiments have shown that in many cases the rate of wear is greatly increased if a jet of oxygen is blown at the tool during cutting. The use of this method is helpful in determining those areas where the atmosphere has access to the interface and those where complete seizure occurs. Water and liquids such as CCl$_4$ that contain chlorine also often greatly accelerate wear in the sliding region at relatively low cutting speeds (Trent, 1967b).

It is clear that in relation to rates of tool wear seizure is often a beneficial condition at the interface. If it were possible to prevent seizure and produce sliding conditions at the interface, in many cases tool wear rates would be accelerated to an unacceptable degree.

VI. Interfacial Layers

Very large differences in tool wear rate may be caused by the formation of intermediate layers at the interface between tool and work material. These layers have been observed at those parts of the interface in which there are conditions of seizure.

One type of layer is formed from certain minor phases in the work material which behave in a plastic manner in the flow zone adjacent to the tool surface. Certain sulfide and silicate inclusions in steel have been found to behave in this way. The addition of relatively large proportions (up to 0.3%) of sulfur to steel has long been known to improve its "machinability." The sulfur is effective when in the form of relatively large inclusions of MnS, which behave in a plastic manner during machining. At high cutting speed they are drawn out into long thin ribbons on the flow zone. When cutting steel at speeds higher than those required to form a built-up edge using steel-cutting grades of cemented carbide tool containing TiC or TaC, some of the MnS makes its way to the interface, where it is consolidated into a layer adhering to the tool surface in the area of seizure. The layers, which are of the order of 1–2-μm thick, appear to reduce the wear rate in a certain range of cutting speed. Although layers of MnS are frequently observed on parts of the tool surface adjacent to the contact area when cutting with steel or WC–Co tools, it is only on the steel-cutting grades of carbide that they adhere at the interface during cutting and act as a protective layer (Trent, 1967b).

The effect of certain silicates in acting as protective layers is still more marked. The type of nonmetallic inclusions present in steel is dependent upon the deoxidation procedure in the final stages of steel making. When the steel is finally deoxidized with certain mixtures of Ca, Al, and Si, the inclusions formed are calcium–aluminum silicates which behave in a very ductile manner in the flow zone when cutting steel. On the steel-cutting grades of carbide tools they form layers 1–4-μm thick at the interface under conditions of seizure in a certain speed range. These layers, of a

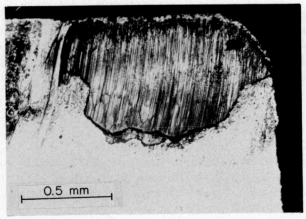

Fig. 37. Silicate layer on rake face of steel cutting grade of carbide tool after cutting steel containing calcium–aluminum silicate inclusions.

glassy character, can be observed on the tool surface after cutting (Fig. 37). The speed range in which these layers are effective is considerably higher than that obtaining for sulfides. They are very effective in raising the maximum cutting speed with steel-cutting grades of carbide tool and in reducing the wear rate at high speeds by a factor of 5 or even more (Opitz *et al.*, 1962; Trent, 1967b).

These silicates appear to function as do the glass lubricants in the hot extrusion of steel, but the only way of effectively introducing them in machining is as inclusions in the work material. Only inclusions that are deformed plastically in the flow zone will function in this way, and they will adhere only to certain tool materials under conditions of seizure. It seems most probable that the adhesion is caused by atomic bonding at the interface, but this has not been proved.

Patents have been taken out covering the deoxidation procedures to produce steels that contain the optimum silicate inclusions required for long cutting tool life (Pietekainen, 1970), and some of these "specially deoxidized steels" have been made commercially. However, other considerations often dictate the deoxidation practice, and as a consequence the rates of tool wear when cutting commercially available steel at high

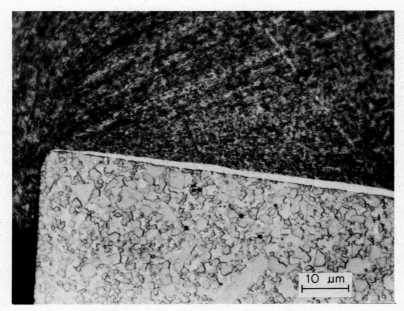

Fig. 38. Layer of metallic character on rake face of cemented carbide tool after cutting steel. These layers are insoluble in HCl and are probably carbides.

cutting speed vary widely, depending on the nonmetallic inclusions present, the amount and distribution of sulfide particles, and probably other factors that have not yet been studied.

Another type of interfacial layer has been observed when cutting steel. Layers of a metallic character, usually about 1–2-μm thick, are often formed at the interface, starting from the tool edge and extending back along the rake face for a distance of a few tenths of a millimeter (Fig. 38). They are formed on steel tools as well as on all types of cemented carbide tool when cutting at moderate speeds. If cratering occurs, they may be found between the crater and the cutting edge but are not observed in the crater. Etching characteristics and microprobe analysis suggest that these layers consist of cementite. They do not appear to have a pronounced effect on the rate of tool wear. Their occurrence, like that of the sulfide and silicate layers, is of interest in demonstrating the formation, under conditions of seizure, of interfacial layers related to phases present in the work material. In view of their potential influence on tool wear rates, an understanding should be sought of the laws governing the formation of such layers.

VII.   Other Tool Materials

A.   Alumina Tools

While the great majority of all metal-cutting applications is carried out with high-speed steel or cemented carbide tools, other tool materials are available and have proved their superior performance in certain applications. The most important examples of these tool materials are aluminum oxide (King and Wheildon, 1960) (generally known as "ceramic") and either diamond or synthetic cubic boron nitride (Hibbs and Wentorf, 1974). The wear of these tools has not been studied in the same detail as has that of the conventional materials, but they have been in use long enough to permit some general conclusions concerning the wear in particular applications.

Aluminum oxide tools are made by either sintering or hot pressing Al_2O_3 powder to more than 98% theoretical density. The alumina is either pure or with the addition of MgO or a small percentage of some other agent to maintain a fine grain size (of the order of 5 μm or finer). Unless the grain size is fine, the wear rate is high in many applications. With fine-grained alumina of high density it is possible to machine both cast iron and steel at speeds of over 700 m min^{-1} (2100 ft min^{-1}) for long periods with a very low rate of wear on the tool flank. This is a much higher speed than that possible with cemented carbide tools. There is little published evidence as to the occurrence of seizure at the interface with alumina tools, but there is some evidence that the rate of flow of the work material is slow very close to the interface (King, 1970). It seems probable, therefore, that a flow zone exists and that this is the heat source, as with carbide and steel tools. The ability of alumina tools to cut at speeds higher than those of conventional tool materials is probably caused by two properties of the alumina—maintenance of high strength at very high temperatures and high resistance to solution in, or reaction with, steel or cast iron. It is known that Al_2O_3 has extremely low solubility in iron up to or even above its melting point, which is why aluminum is an effective deoxidizing agent in steel making. It has been suggested that at those parts of the interface where oxygen has access wear can take place by interaction to form a spinel $FeO \cdot Al_2O_3$, which is then sheared away (King, 1970), but there is little evidence as to the effectiveness of this reaction in promoting wear.

In industrial practice alumina tools are used mainly to cut cast iron at speeds sometimes exceeding 700 m min^{-1} (2100 ft min^{-1}), and consistently long tool life is obtained. They are less commonly used for cutting steel because of the incidence of premature failure by fracture unless all precautions are taken to maintain rigidity and avoid accidental impact damage to the cutting edge. When tools are repeatedly engaged and disengaged in cutting, the repeated thermal and mechanical stresses may cause local cracking (Pekelharing, 1961). This occurs also with cemented carbide tools, but with alumina tools it is more likely to cause rapid wear, fine fracture, and failure.

Recently, "throw-away" tool tips have been developed, consisting of a cemented carbide substrate coated with TiC with a very thin layer of Al_2O_3 over this. These are claimed to give longer tool life than TiC-coated tools, probably by making use of the resistance of Al_2O_3 to diffusion wear, without the sensitivity to mechanical and thermal stresses which limits the use of alumina tools.

B. Diamond Tools

Natural diamond is used as a cutting tool material for certain difficult applications to secure exceptionally high surface finish or in cases for which a very low rate of wear is required to ensure small dimensional changes in a component being machined, or when machining very hard or abrasive materials. One example of its use, in spite of the high cost of diamond, is in the machining of the high-silicon–aluminum alloys previously discussed. The diamond can resist the abrasive action of the large silicon crystals and give a tool life many times longer than that of carbide tools at several times the rate of metal removal.

Synthetic diamonds in the form of a layer, usually about 0.5–0.75 mm thick, bonded to the surface of a cemented carbide tool tip, may also be used for this purpose. However, neither the natural diamond nor the tools with compacted surface layers give exceptionally low wear rates when cutting steel at high speeds. As with ceramic tools, studies of the wear mechanisms have not been reported in great detail, but it is likely that seizure conditions exist with a flow-zone heat source and that the relatively rapid wear when cutting steel is caused by reaction between the diamond (carbon) and the steel at the high temperatures of the interface. For the same reason, diamond tools are not recommended for cutting nickel-based alloys since rapid wear occurs, which may be attributed to the reactions at interface temperatures over 900°C (Hibbs and Wentorf, 1974).

C. Cubic Boron Nitride

Cubic boron nitride can be synthesized and bonded at ultrahigh pressure to the surface of cemented carbide for use in metal cutting. It is much more stable than diamond at temperatures over 800°C and reacts less readily with iron, nickel, and their alloys. For this reason it is much more wear resistant than diamond when cutting both steel of high hardness and nickel-based alloys. For the latter alloys particular advantages have been claimed (Hibbs and Wentorf, 1974). For the most creep-resistant nickel alloys, permissible cutting speeds with carbide tools are very low, e.g., 12 m min^{-1} (40 ft min^{-1}), and machining costs are extremely high for this reason. Using cubic boron nitride, good tool life when cutting the same alloys has been demonstrated at speeds up to 190 m min^{-1} (600 ft min^{-1}). The superior performance must be attributed to high strength maintained at high temperature combined with very low solubility in, or reaction rate with, the nickel alloys at the interface temperatures.

References

Amini, E. (1968). *J. Strain Anal.* **3**, 206.

ASM Metal Handbook (1948). 1st Supplement, p. 141.

Braiden, P. M., and Dugdale, D. S. (1970). I. S. I. Rep. No. 126, p. 30.

Eggleston, G. M., Herzog, R., and Thomson, E. G. (1959). *J. Eng. Ind.* **81**, 263.

Freeman, R. (1974). Thesis, Univ. of Birmingham, England.

Heginbotham, W., and Gogia, S. (1961). *Proc. Inst. Mech. Eng.* **175**, 892.

Hibbs, L. E., and Wentorf, R. H., Jr. (1974). *Plansee Semin., 8th* Paper No. 42.

Jonsson, H. (1975). *Conf. New Tool Mater. Cutting Tech., London* (Engineers Digest), Paper No. 3.

Kato, S., Yamaguchi, K., and Yamada, M. (1972). *J. Eng. Ind.* **94**, 683.

King, A. G. (1970). I.S.I. Rep. No. 2, p. 162.

King, A. G., and Wheildon, W. M. (1960). "Ceramics in Machining Processes." Academic Press, New York.

Lenz, E. (1971). *Int. Cemented Carbide Conf., 1st* S.M.E., Dearborn, Paper No. MR 71-905.

Loladze, T. N. (1958a). "Wear of Cutting Tools," pp. 105–119, Moscow (in Russian).

Loladze, T. N. (1958b). "Wear of Cutting Tools," pp. 118–130, Moscow (in Russian).

Mayer, J. E., Moskowitz, D., and Humenik, M. (1970). I. S. I. Rep. No. 126, p. 143.

Opitz, H., and König, W. (1967). I.S.I Rep. No. 94, p. 35.

Opitz, H., Gappisch, M., and König, W. (1962). *Arch. Eisenhüttenwes.* **33**, 841.

Pekelharing, A. I. (1961). C.I.R.P. Meeting, Prague.

Pesante, M. (1967). *Proc. Sem. Met. Cutting, OECD, Paris, 1966* p. 127.

Pietekainen, J. (1970). *Acta Polytech. Scand.* **91**.

Roth, R. N., and Oxley, P. L. B. (1972). *J. Mech. Eng. Sci.* **14**, 85.

Schintlmeister, W., and Pacher, O. (1975). *J. Vac. Sci. Technol.* **12**, 743.

Smart, E. F., and Trent, E. M. (1975). *Int. J. Prod. Res.* **13**, 265.

Taylor, F. W. (1907). *Trans. ASME* **28**, 31.

Trent, E. M. (1959). *J. Inst. Prod. Eng.* **38**, 105.

Trent, E. M. (1963). *J. Iron Steel Inst.* **201**, 1001.

Trent, E. M. (1967a). I. S. I. Special Rep. 94, p. 11.

Trent, E. M. (1967b). I.S.I. Special Rep. 94, p. 77.

Trent, E. M. (1968a). *Proc. Int. Conf. M.T.D.R., Manchester, 1967* p. 629.

Trent, E. M. (1968b). *Proc. Int. Conf. M.T.D.R., Manchester, 1967* p. 638.

Trent, E. M. (1969). *Powder Metall.* **12**, 566.

Trent, E. M. (1970). I.S.I. Rep. No. 126, p. 15.

Trent, E. M. (1977). "Metal Cutting." Butterworth, London.

Usui, E., and Takayema, H. (1960). *J. Eng. Ind.* **82**, 303.

Williams, J. E., and Rollason, E. C. (1970). *J. Inst. Metall.* **98**, 144.

Williams, J. E., Smart, E. F., and Milner, D. R. (1970). *Metallurgia* **81**, 3, 51, 89.

Wright, P. K., and Trent, E. M. (1973). *J. Iron Steel Inst.* **211**, 364.

Wright, P. K., and Trent, E. M. (1974). *Met. Technol.* **1**, 13.

Zlatin, N. (1971). *Int. Cemented Carbide Conf., 1st, Chicago* Paper 1Q71-918.

Zorev, N. N. (1963). "International Research in Production Engineering," p. 42. American Society of Mechanical Engineers, Pittsburgh.